RARE BOOK   £9.99

Engineered fills

Proceedings of the conference *Engineered fills '93*, held
on 15 -17 September 1993 in Newcastle upon Tyne

Edited by B. G. Clarke, C. J. F. P. Jones
and A. I. B. Moffat

Conference sponsored by the British Geotechnical Society, in association with the British Dam Society and the International Society of Soil Mechanics and Foundation Engineers

Organizing Committee: Dr B. G. Clarke, Professor C. J. F. P. Jones, A. I. B. Moffat, and V. Adams

Published by Thomas Telford Services Ltd, 1 Heron Quay, London E14 4JD

First published 1993

Distributors for Thomas Telford books are
USA: American Society of Civil Engineers, Publications Sales Department, 345 East 47th Street, New York, NY 10017-2398
Japan: Maruzen Co. Ltd, Book Department, 3-10 Nihonbashi 2-chome, Chuo-ku, Tokyo 103
Australia: DA Books and Journals, 648 Whitehorse Rd, Mitcham, 3132, Victoria

A catalogue record for this book is available from the British Library

Classification
Availability: Unrestricted
Content: Collected papers
Status: Invited authors' opinion
User: Geotechnical engineers, dam engineers, foundation engineers, highway engineers, materials engineers

ISBN: 0 7277 1967 X

Preface

INTRODUCTION

1.   Building on or with compacted or uncompacted fill forms
a major part of the construction industry's activities.
Further, it is an essential part of any surface mining
activity. The largest man made structures in the world, dams,
are built of fill; moving and compacting fill forms the major
part of any road project; excavating, moving, storing and
replacing fill is an essential part of any mining operation;
levelling building sites includes an element of filling. Thus,
the need to understand the behaviour of fill and methods of
placing and improving it, are an important part of the
continuing education of practising engineers. Development of
new materials, revised specifications, improved testing
methods, ground improvement techniques and performance
monitoring are continually taking place. Against this
background it was considered very timely to hold this
International Conference.

THE PAPERS

2.   There are 43 papers in the proceedings three of which
are invited theme papers on aspects of highway, dam and
specifications. Papers have been submitted from the UK,
Australia, Egypt, France, Iran, Japan, Poland, Norway and South
Africa thus giving a truly international flavour to these
proceedings.

3.   The first invited paper by Ekins et al (1) covers
aspects of highway construction including ground investigation,
analysis and design, ground treatment, construction and
performance. Examples are given of the cost of failures of
earthworks during and after construction emphasising the need
to specify and control earthworks correctly.

4.   The second invited paper by Ervin (2) is concerned with
specifications, their development and application highlighting
the problems that can occur which, in some instances, lead to
litigation.

i

5. The final invited paper, by Charles (3), covers types of fill, their uses, method of construction and control and performance through selected examples.

6. The remaining 40 papers have been grouped into the three categories; theory and testing, specifications and materials and case histories. An index of the authors and main topics covered is given at the end of the volume. The index refers to the paper numbers shown in the contents list.

THEORY AND TESTING

7. This, the most extensive group, covers the research taking place into the understanding of the behaviour of partially saturated compacted material.

8. The 16 papers deal with mathematical models (4, 5, 7, 8, 13), laboratory tests (4, 10, 12, 13, 14, 15, 16, 17, 18, 19) and model tests (9, 11, 17). They all deal with compacted fill which includes clay, sand and gravel, residual soils, volcanic soils and rock. Parameters covered include earth pressure (9, 10, 11, 14, 15), density (16), stiffness (5, 6, 12) and strength (16, 17, 18, 19).

SPECIFICATIONS AND MATERIALS

9. This group covers new materials and specifications for acceptability and compaction both for the new materials and existing materials; all deal with compacted fill.

10. The 13 papers in this group include several state of the art papers (20, 21, 22). Specification topics covered include acceptability criteria (20, 22, 23, 24, 30) and compaction criteria (21, 24, 28). Materials covered include clay (23), till (21, 24, 25), sand and gravel (25, 31, 32), volcanic soils (26, 27), pulverised fuel ash (29), chalk (20), rockfill (30), polystyrene (22) and refuse (31, 32).

CASE HISTORIES

11. This group includes examples of the performance of fills both compacted and uncompacted. A number of papers deal with the behaviour of opencast backfill showing the performance over a considerable length of time

12. Their are 11 papers in this group covering highways (39, 40, 42, 43), dams (37, 38) and foundations (39) as well as observations of the behaviour of general backfill (33, 34, 35, 36). One paper deals with loose fill (34) and the remainder with compacted fill which includes rockfill (33, 34, 35, 36, 37, 39), till (41), sand and gravel, (40, 42, 43), pfa (42, 43), refuse (39, 42) and slag (38). Most of the papers include performance data showing settlements resulting from load (34, 39, 41, 42, 43), creep (33, 34, 35, 36, 42) and inundation (33, 34, 36).

ACKNOWLEDGEMENTS

13. The quality of the proceedings reflect the effort and time given by the authors who not only produced technically outstanding papers but also produced them promptly and according to instructions. Without this, these proceedings would not have been such a success. Special mention must be made of the invited authors who produced the theme lectures.

14. All the papers were reviewed in great detail by a number of people including Professor G T Houlsby, Professor J H Atkinson, Dr S Wheeler, Dr J A Charles and staff at BRE, Mr K Cole, Dr J Temporal and staff at TRL, Dr N Taylor, Dr M Davis and Mr S Corbett

15. The theme for this Conference was first proposed in 1991 and was given the support of the British Geotechnical Society, the British Dam Society and the International Society for Soil Mechanics and Foundation Engineering. The encouragement of Mr S Thorburn, Dr T W Cousens, Dr K Rainbow, Dr D Greenwood, the late Mr B Cripwell, Mr K Cole, Dr J Temporal is appreciated.

16. Finally, these proceeding would not have been completed without the considerable support and effort of the organising committee and their assistants who included Professor C J F P Jones, Mr A I B Moffat, Mrs V Adams, Ms C Currie and Mrs S Dodd. They assisted with the reviewing and editing process as well as maintaining contact with the authors.

B G Clarke

University of Newcastle upon Tyne, UK

Contents

Theme papers

Theory and laboratory studies

Specification and materials

Case histories

1. Highway embankments—their design, construction and performance

J. D. K. EKINS, R. CATER and A. D. HOUNSHAM, Hampshire
County Council, UK

SYNOPSIS. This paper gives an overview of the design,
construction and maintenance of highway embankments from
the point of view of a Highway Authority. It draws widely
from examples in Hampshire: problems, failures and
remedial measures are covered, and references made to
published works. Throughout the paper, the needs of the
road user are considered.

INTRODUCTION
1. It may not be an overstatement to say that the
provision of an adequate highway network is amongst the
most important features of a civilised state. Indeed, the
Roman Empire is remembered for its legacy of highways which
remained in use well after the collapse of the civilisation
which had built them. Those roads were planned, designed
and constructed in a manner which bear more similarities
with today's practices, than with the random creation of
paths by the passage of feet which went before. Routes
were carefully chosen to permit reasonable horizontal and
vertical alignments and, where necessary, earthworks were
carried out to ensure that this was achieved. Pavement and
fill materials were selected and placed to ensure that the
required performance was achieved. Particular attention
was paid to drainage.

2. These standards of construction and the performance
consequently achieved permitted the road users of the past
to gain benefit from these works over a period of many
years. Similarly, it is the objective of the present day
Highway Engineer to provide a comparable service to the
road users of both the present and of the future.

3. Ground conditions and the behaviour of embankment fill
materials are complex subjects and, even after 2000 years
of experience, are still of primary concern to Highway
Engineers. Failures still happen, both during construction
and in service.

4. For example, on one 9km length of motorway constructed in Hampshire in the late 1970's (Reference 1), the direct cost of remedying earthworks failures during construction was about 5% of the tender sum. Consequential losses due to the delays in completion of the works were additional to this. Further subsequential sums have been expended since on repairing in-service failures.

5. A study of 10 major road construction contracts totalling 37km carried out in 1983 showed that the average additional construction costs were some 35% of the tender totals. A significant proportion of this was due to difficulties with earthworks, (Reference 2). A greater understanding of the behaviour of the ground and the available fill materials at the design stage could well have prevented many of these problems.

6. Although some years have elapsed since the examples quoted above, Highway Engineers are still experiencing problems with embankment construction and stability and with associated ground conditions. The cost of these problems ultimately falls onto the general public, our customer. The Institution of Civil Engineers set up the Site Investigation Steering Group, under the chairmanship of Professor Littlejohn, in 1991 to consider means of improving the procurement of ground investigations with a view to reducing consequential costs. Hopefully their recommendations will contribute to further reductions in the problems experienced when designing and constructing highway embankments, especially by the use of properly qualified and experienced personnel.

7. There are many interlocking stages in the design, construction and maintenance of highway embankments. This paper gives an overview of the more significant of these, together with references to case studies, primarily from the County of Hampshire, and other published works. It is of increasing importance to consider, and allow for, the impact of our work on the motoring public. Delay and inconvenience should be minimised, and benefits enhanced. More expensive solutions to engineering problems may be applicable where there is a consequential benefit to the customer.

GROUND INVESTIGATION
8. Any civil engineering project should have an adequate ground investigation carried out, and highway embankments are no exception. Twenty to thirty years ago it was common practice to sink a series of evenly spaced boreholes along the centreline of a proposed road regardless of

expected ground conditions. This approach led to many expensive problems arising from unforeseen ground conditions.

9. An example of this was the Sutton Coldfield Bypass in Warwickshire, where the 100m spaced boreholes could not adequately indicate the variability of the glacial outwash deposits at the site which in some places varied metre by metre. By the mid 1970s the shortcomings of this approach were generally recognised, and improvements made to the planning and execution of ground investigations. However, we have been left with a legacy of ageing earthworks where problems are being experienced.

10. Any ground investigation should commence with a properly structured planning process combined with a desk study. Much useful information is available to guide selection of type and location of exploratory holes, and highlight issues for investigation. Fieldwork should only commence once this process is complete. There are many instances of cases where problems could have been avoided by the most cursory of desk or walk-over studies.

11. For example, in the mid 1980's Hampshire County Council geotechnical staff were called in to investigate a length of recently constructed housing estate access road at Headley, Hampshire which had settled severely. A quick look at the surrounding area was sufficient to reveal that the road (and the housing) had been built across a 20 to 30 year old domestic refuse tip. Unfortunately, solving this problem was much more complex than identifying the cause, and the local residents still have the inconvenience of an undulating road.

12. The investigation should identify and explore the extent of any particular features and problem areas identified during the desk study, in addition to the routine investigation of underfill conditions. Consideration may also be given to investigating potential borrow-pit areas. Care should be taken throughout the site investigation operations to build and maintain good relations with landowners, other residents and the public generally. This is often the first time that the proposed scheme makes a physical impact on the public, and especial care devoted here will often repay dividends later in the project. Reinstatement of investigated areas and access routes is important, especially for sensitive sites.

GEOTECHNICAL ANALYSIS AND ENGINEERING DESIGN

13. The information obtained from the ground investigation should be carefully analysed so that all geotechnical issues are identified. Only then can appropriate design decisions be made. Consideration should be given to the possibility of shortcomings and errors in the investigation, and the consequential impact on the design. Additional investigations may be carried out into any anomalous or unexpected findings.

14. For example, boreholes into the fine sands of the Bracklesham Beds for the Bitterne Bypass in Southampton revealed unexpectedly low SPT values. The implication was that the sands were in a relatively loose state, and could be expected to settle under the planned major embankment. However, additional investigations by a different driller gave higher SPT values indicating medium dense conditions. It was considered that the original drillers were at fault, possibly by not maintaining a positive head of water in the casing during the SPT tests.

15. A further example also relates to the Bracklesham sands at Southampton, this time on the M27. Testing of samples recovered from boreholes indicated a clean but wet sand which was expected to possess a reasonable permeability. Stockpiling to allow drainage was thought to be enough to render it suitable as fill. Again, fortunately, the true nature of the deposit was revealed by limited additional work, when the sand was found to contain significant amounts of silt, which reduced the permeability markedly. The proposed drying technique was abandoned.

16. In designing an embankment, it is necessary to consider the nature and strength of both the fill material and the natural ground below the embankment. Except for very large projects, it may not be practicable to carry out enough soil strength testing and of large enough test specimen size to obtain a statistically significant set of parameters. Past experience, and results from other projects on similar geologies can give valuable supplementary information.

17. Traditionally, analysis of embankments has been carried out for the short term case only, as the gain of strength with consolidation leads to an increasing factor of safety with time. However, some fill materials can reduce in strength with age under certain conditions, and parameters may be selected to model this in the analyses. The Department of Transport has published a comprehensive advice note on the design and preparation of contract documents for highway earthworks (Reference 3).

18. The Transport Research Laboratory has carried out a programme of surveys of highway earthworks failures (Reference 4). These relate the probability of failure to factors such as slope angle, geology and age. Recommendations are given for embankment slopes formed of various soil types to limit the probability of failure to an acceptable level. This information provides a valuable supplement to the analytical results when making design decisions.

TREATMENT OF THE UNDERFILL ZONE

19. Sometimes the nature of the underfill is such that it is preferable to replace or modify it, rather than to design the embankment works in accordance with the existing characteristics of the underfill. For example, embankments on peat and other alluvial deposits are liable to substantial and irregular settlements which may continue for many years. Embankments may need to be reinforced or be of special fills to remain stable and resist the effects of these settlements. It can often be cheaper and more satisfactory simply to remove the soft underfill and replace it with something of better quality.

20. Alternatively, where soft or weak underfill has to be left in place, there are various techniques for improving its strength. Vibroflotation introduces a regular pattern of granular columns into the underfill, with the dual effect of both reinforcing and consolidating the ground. This technique was used for the Six Dials highway scheme in Southampton where a low embankment had to cross an area much disturbed with pockets of fill and other consequences of a long history of occupation. See Photograph One.

21. Dynamic Compaction may be used for loose underfills where magnitude of settlement is expected to be a problem. This technique was used at Marchwood, Hampshire, where a new road was constructed across a disused refuse tip.

22. Occasionally, underfills possess adequate strength but are both compressible and of such low permeability that settlements may continue for many years. This may be accelerated by use of vertical drains, which are available in various forms.

23. Before selecting any of these improvement techniques, consideration should be given to the effect such works may have on the public. Heavy plant movements and ground vibration can be unacceptable for certain site locations.

5

Photograph One – Ground Improvement by Vibroflotation at Six Dials, Southampton, Hampshire.

EMBANKMENT FILLS AND CONSTRUCTION METHODS
24. It is most often the case that the most appropriate source of fill material, on cost, convenience and environmental grounds, is that won from on-site excavations which are themselves necessary as part of the works. The ground investigation should have identified the nature and quantities of available fill materials, and the works should be designed so that use may be made of these. It will be necessary to set acceptability limits based on the required performance and available quantities of the fill materials.

25. Where on-site materials are inadequate or insufficient, or where special properties are required, it will be necessary to import fill materials. This may be done by simply specifying required parameters, and leaving

it to the Contractor to arrange a source, although increasing public awareness of environmental issues and shortages of good quality materials may dictate a different approach. In some circumstances it may be appropriate to specify a particular source of fill.

26. Sometimes, specific sites require specialised solutions, or combinations of solutions. Expert advice is often necessary before identifying an optimum solution. The following projects are examples of the sort of techniques that are available:

(a) The M275 motorway was built across tidal mud flats in Portsmouth Harbour, Hampshire by constructing a series of parallel chalk bunds to a level above the high water mark. The chalk was transported 4km from Portsdown Hill on a belt conveyor. The chalk tended to degenerate into a slurry whilst being conveyed, but quickly gained strength after placing (Reference 5). The embankments between the bunds were formed from sea dredged sand, which was hydraulically placed in its final required position.

(b) Lightweight P.F.A. fill was extensively used for the embankments of the Southampton Western Approach Road, Hampshire where the line of the road passed across soft ground. Surcharging, instrumentation and settlement periods were all used to ensure settlement was substantially complete before pavement construction commenced.

(c) The earthworks for the Great Yarmouth Bypass were reinforced with geogrids and geotextiles, with one area adjacent to a railway bridge being formed from expanded polystyrene blocks (References 6 and 7).

(d) The 70m high Scammonden Dam in Yorkshire carries the M62 on its crest, and is constructed from rockfill with an inclined clay core. See Photograph Two.

(e) Embankments for the A3(M) north of Portsmouth had to cross an area of swallow holes. Each was dug out until solid chalk was exposed around the perimeter of the hole, and then capped with a reinforced concrete slab. Fill was then placed to bring the level back to that required. See Photograph Three.

(f) Reinforced earth walls have been used extensively in highway embankments over the last 15 to 20 years. A trial wall using soft chalk was constructed at Portsmouth in the mid 1980's with generally satisfactory results (Reference 8).

Photograph Two – The Scammonden Dam, Yorkshire, carrying the M62 on its crest.

(g)   Poor fill material may be modified by addition of cement, lime or chopped fibres. At Fareham, Hampshire the Town Quay Road embankment fill slurrified, but was treated with lime, giving a satisfactory and compactable material (Reference 9). See Photograph Four.

8

Photograph Three – Capping of Swallow Holes for the A3(M).

EMBANKMENT CONSTRUCTION

27.   Requirements for the selection and placing of fills for highway embankments are given in the DoT Specification for Highway Works. The detailed requirements and nature of the control testing have varied with the introduction over the years of the various editions of this document. Many Engineers still retain as treasured possessions the various pink, orange, blue, brown etc. volumes! Under the terms of the latest edition of the DoT Specification (the "grey" 7th Edition), the Engineer will generally make the contractor responsible for the selection, testing and classification of fill materials to determine acceptability and compaction requirements. In addition, it is usual for the Engineer to carry out his own testing for record purposes.

Photograph Four — Lime Modification of Slurrified Fill, Fareham Town Quay Road, Hampshire.

28. The acceptability parameters are laid down in the contract by the Engineer, and may be in terms of grading, moisture content as a function of optimum, moisture condition value (MCV), saturation moisture content, etc. depending on the classification of the fill material, (Reference 10).

29. Compaction method requirements for each class of fill are laid down in the Specification, and as such there is no specified final density requirement. However, checks may be carried out on densities achieved by use of driven core cutters or the Nuclear Density Gauge.

30. Many otherwise satisfactory fill materials will be rendered unacceptable by the addition of excess water. Small quantities of water can dramatically alter the behaviour of silty fine sands, for example. Such water may be the result of excavation or placing of fill below water, or simply from inclement weather. Thin layers of fill which have become too wet may be left to dry naturally, or even

treated with lime or cement. High porosity chalk often contains enough water in its natural state to slurrify on crushing, but appropriate handing measures can reduce the risk of this happening (Reference 11).

31. Dewatering measures can have a dramatically beneficial effect on the nature of some potential fill materials. About 600,000 cubic metres of silty fine sand fill for the Blackwater Valley Road in Hampshire were obtained from up to 13m below groundwater level following wellpoint dewatering. The natural moisture content of around 23% was reduced to less than the optimum value of around 14%. If this source had not been used, imported fill would have been required, which would have had a serious effect on local traffic conditions for many months, as well as direct cost implications.

32. Sometimes it is necessary to specify surcharge loads and/or settlement periods for embankments, even where underfill conditions and fill types are not unduly poor. For example, on the A31 Alresford Bypass in Hampshire it was necessary to construct an 11m high, 1 in 5 sideslope, chalk embankment across a river valley. There were no stability problems, and anticipated settlements were not such as to affect the pavement construction. However, the embankment formed the approaches to a 3 span continuous bridge with piled piers, and bankseats resting in the chalk fill. Differential settlement constraints meant that the bridge deck could not be cast until any settlement was substantially complete. Monitoring was carried out by means of precise leveling of the embankment top and also be use of mercury settlement gauges in the underfill. These showed settlements to have ceased within three months of completing embankment construction, thus allowing bridge deck construction to proceed.

IN-SERVICE PERFORMANCE

33. In general, properly designed and constructed embankments give entirely satisfactory long term performance. However, certain combinations of characteristics can lead to localised or even major failures.

34. There have been a number of instances of embankment failure in areas of previously existing landslips (Reference 12). One such took place shortly after completion of embankment construction at the Hulbert Road Interchange on the A3(M) north of Portsmouth (Reference 1). A 13m high embankment with a sideslope of 1 in 2.5 had been constructed on a natural slope of firm becoming stiff fissured London Clay. After the failure, which affected a

100m length of the embankment, investigations revealed relic slip surfaces in the underfill. Redesign was carried out using residual strength parameters, and the embankment reconstructed with flatter slopes, additional drainage measures and a toe berm. See Photograph Five.

35. Shallower slips in the sideslopes of embankments constructed from stiff overconsolidated clay fills are not uncommon. The stress relief caused by excavation from depth and placing in the upper parts of embankments leads to swelling, take up of available water and softening. Side slope failures, typically to depths of about 1.0m to 1.5m, are often then experienced, (Reference 13). The same mechanism also leads to softening of the formation level and a consequential reduction in the life expectancy of the road pavement. These problems are difficult to avoid with such fill materials. The only solutions are either to design in full knowledge of this effect, or to avoid the use of such fills in the upper parts of an embankment.

36. Water entry into embankments can lead to various types of failure. Water will generally exit from the toe area, causing localised softening. This can be particularly marked with fine sandy fills, for example on the M27 west of Southampton where many hundreds of metres of embankment toe have failed.

37. Larger volumes of water flowing through an embankment can lead to much more extensive softening and even washout of fines. Such water may enter from pavement cracks and joints, defective drainage, central reserves, verges and/or sideslopes. A 20 year old clay and fine sand embankment on the A33 at Eastleigh, Hampshire suffered a deep-seated failure in the late 1980's due to excessive water entry. During the remedial excavation works, large pockets of water were released from within the embankment. A similar failure on the M27 north of Southampton was found to be due to leaking central reserve drainage. It is clearly important to take positive measures to limit water ingress. Hardening of central reserves and regular drainage inspections can play their part in this.

38. Naturally occurring chemicals within the fill can become involved in reactions with deleterious results. Sulphates will react with lime stabilised subgrades, leading to a loss of strength and reduced pavement life, and are often present in soils such as London Clay. Pyrites (Iron Sulphide) can be broken down by bacteria in the presence of oxygen, producing both acidic conditions and iron oxides which have been known to precipitate out

and obstruct drainage systems. Significant concentration of Pyrites have been found in parts of the Bracklesham Beds in Hampshire, (Reference 14).

Photograph Five - Embankment Failure at Hulbert Road Interchange, A3(M), Hampshire.

EMBANKMENT REMEDIAL MEASURES

39. Failures of in-service highway embankments constitute a maintenance problem, although they seldom form a threat to life. It is important to carry out regular inspections in order that failures are identified and dealt with at an early stage before larger slips can form. It is essential to identify the cause of any failure in order to implement effective and economic remedial works. Advice on inspection and simple remedial works has been produced by the DoT (Reference 15).

40. The most common reinstatement technique remains the excavation of the failed fill material and its replacement by coarse graded granular fill. It has the merit of being straightforward and relatively quick to carry out, although

larger failures may need to be dealt with in bays to avoid further failures. The cost effectiveness depends largely on the availability and price of off-site tipping and imported fills. However, the cost of delay to the motorist must also be considered. A recent embankment failure on the A31 in the New Forest, Hampshire necessitated a lane closure which caused traffic tailbacks of several miles until emergency repairs could be completed. A prudent Highway Authority will have procedures in place for the immediate supply of plant and materials for such repairs.

41. Less threatening failures which do not affect traffic flows can be dealt with in a more considered way. Investigations to determine the cause, extent and appropriate remedial measures should be carried out. It may be appropriate and cost effective to adopt a remedial solution which reuses the slipped material. Various methods have been experimented with, some of which have been reported by the TRL (Reference 16). This study showed that the relative costs depend on many site specific factors, but in general the methods which reused on-site materials were more cost effective and less disruptive to traffic.

42. If additional land is available, it is possible to reinstate embankment failures by importing fill to slacken side slopes or to construct a toe berm. It may be possible to avoid excavating the slipped material. Excavated slipped fill material may be replaced into the slipped area after improving its strength by lime or cement modification. Alternatively, the fill can be contained in layers by use of appropriate geofabrics. Both of these methods require the fill to be recompacted in layers with incorporation of suitable drainage measures, and can therefore be rather laborious. Failed slopes may be supported or reinstated, and slopes flattened, by use of some form of retaining wall at the toe. Crib walling units and gabion baskets are two ways of doing this, and some reinstatements have been carried out using a wall of old vehicle tyres.

43. Where sections of embankment slope are considered to be under threat of failure, possibly where failures have already taken place nearby, it may be appropriate to consider some form of preventative measure. This could take the form of slope drains, rock counterforts or geomesh secured to the slope with ground anchors. Alternatively, slope stability may be enhanced by selected use of vegetation (Reference 17).

14

EMBANKMENT WIDENING

44. The DoT's programme of motorway and trunk road widening means that many highway embankments are having to be extended. In many cases it is not possible or desirable to acquire more land, so selected fills and/or retaining structures are necessary to permit steeper slopes. Such embankment extensions need to be carefully interfaced into the existing fill to avoid problems with differential settlements. A primary factor in any such works is the need to ensure that the travelling public is not unduly delayed by the temporary traffic measures.

CONCLUSIONS

45. The Roman Empire's highway works have remained with us, in many cases, until the present day. If modern Engineers are to achieve anything like comparable performance from highway embankments, they must observe a number of basic geotechnical and engineering principles in design, construction and maintenance. Additionally, emphasis on environmental considerations, value for money, resistance to the effects of heavy traffic and public involvement are now relevant issues.

46. The ground investigation should be carefully planned and executed by competent personnel to reveal all relevant information. This should then be analysed, with full understanding of the geological processes involved, to arrive at a clear picture of the ground conditions and how they relate to the proposed embankment. Design work should be carried out in accordance with current best practice, making use of past experience and performance data on existing embankments. Treatment of the underfill and selected fill materials may be necessary, as may be innovative materials or construction techniques. Construction work should be carefully controlled.

47. Embankments in service should be inspected regularly to give early identification of incipient failures, thus allowing prompt repair work. Consideration may be given to innovative repair methods where appropriate. Slopes considered to be at risk may have preventative works carried out.

48. It is recognised that many embankments will have to be widened in the near future, often within existing land constraints. The difficult nature and highway public profile, together with the vital necessity to maintain traffic flows, demand particular care in the performance of these works.

49. In all this, the customer's needs remain paramount. It has never been more important to consider the impact of construction work on road users. The balance of priority between engineering convenience, environmental factors, cost, and the delay and inconvenience caused to the road user, needs to be continually assessed. These are, I suspect, issues that the Romans did not have to consider!

REFERENCES
1. CATER R. Stability problems with earthworks construction in Hampshire. Synopsis I.C.E. Southern Association Conference on Earthworks. 1979.

2. TYRRELL A.P., LAKE L.M. AND PARSONS A.W. An investigation of the extra costs arising on highway contracts. TRL Report SR814, 1983.

3. DEPARTMENT OF TRANSPORT ADVICE NOTE. HA 44/91 Earthworks Design and Preparation of Contract Documents 1991.

4. PERRY J.A. Survey of slope condition on motorway earthworks in England and Wales. TRL Report No RR199. 1989.

5. LEWIS W.A. AND PARSONS A.W. The application of belt conveyors in road earthworks. Proc.Inst.Civ.Eng. Part 1, 54, 425-450. 1973.

6. WILLIAMS D AND SANDERS R.L. Design of reinforced embankments for Great Yarmouth Bypass. Source unknown. 1984/85.

7. NEW CIVIL ENGINEER. Load bearing first for polystyrene. pp 26, 27, 14 November 1985.

8. BRADY K, BARRATT D.A. AND CATER R., Performance of an experimental earth retaining structure using soft chalk backfills. TRL Report RR 260, 1990.

9. CATER R, Fareham Creek, Hampshire - Relating Highway Design to Geology. Journal of the Institution of Highways and Transportation. February 1986.

10. BAIRD H.G. Earthworks control - assessment of "suitability", Ground Engineering, May 1988.

11. INGOLDBY H.C. and PARSON A.W. The classification of chalk for use as a fill material. TRL report No. LR806, 1973.

12. CHANDLER R.J., PACHAKIS M., MERCER J AND WRIGHTMAN J. Four long-term failures of embankments founded on areas of landslip. Q.J. Eng. Geol. Volume 6, pp 405-422, 1973.

13. GREENWOOD J.R., HOLT D.A. AND HERRICK G.W. Shallow slips in highway embankments constructed of overconsolidated clay. Failures in earthworks. Thomas Telford Ltd, London, 1985.

14. SANDOVER B.R. AND NORBURY D.R. On an occurance of abnormal acidity in granular soils. Quarterly Journal of Engineering Geology, 26 pp 149 to 153. 1993.

15. DEPARTMENT OF TRANSPORT ADVICE NOTE HA 26/83. Maintenance of highway earthworks. 1983.

16. JOHNSON P.E. Maintenance and repair of highway embankments : Studies of seven methods of treatment. TRL Research Report 30. 1985.

17. BARKER D.H. Enhancement of slope stability by vegetation. Ground Engineering, April 1986, pp 11 to 15.

2. Specification and control of earthworks

M. C. ERVIN, Golder Associates Pty Ltd, Australia

SYNOPSIS.   The requirements for specification and control of earthworks is often poorly understood and implemented, with poor performance of completed works and litigation often the result. The paper presents some background to compaction and control testing procedures, together with some discussion on specification requirements and the engineering properties of compacted fill.

INTRODUCTION

1.   Soil and rock are the most abundant construction materials available, and form an integral part of virtually every civil engineering project, whether as the foundation to a structure or as part of the structure, as in the case of roadworks and dams. However, the earthworks required for many projects are poorly specified and poorly controlled during construction. This is believed to be due to a lack of understanding by many designers, specifiers and construction engineers of the fundamentals of compaction, and of the influence on the engineering properties of compacted fill of such important factors as material type, placement moisture and achieved compaction.   As a result poor engineering performance may occur, and litigation sometimes results due to differences between the contractor's and the designer's expectations.

2.   The author believes much of this lack of understanding stems from the scant treatment of compaction given in undergraduate courses, its rare attention in the literature and the unfortunate trend that the design and specification of the earthworks for many projects is undertaken by civil or structural engineers who have little fundamental understanding of earthworks.   In this way past poor practices are perpetuated and often it is only when problems arise on site, or when litigation results, that the shortcomings of the design and specification for a project are realised.

3.   This paper presents some background to the theory and practice of compaction, the test procedures used in the control of compaction, and some of the author's experiences in relation to matters which have resulted in litigation associated with earthworks. Some discussion is also presented with respect to the engineering properties of compacted earth fill.

4.   For the purposes of this paper, the term earthworks is used to describe the placement and compaction of fill. Issues relating to excavation, cut batter stability, control of groundwater, and other similar matters which often need to be considered in the wider context of earthworks will not be discussed. Similarly, the construction of pavements, a particular case of compacted fill, will not be considered in other than very general terms.

Engineered fills. Thomas Telford, London, 1993

THE NEED FOR ADEQUATE CONTROL OF EARTHWORKS

5.   Rarely are building sites sufficiently flat that no site levelling is required. In many cases, particularly for large commercial or industrial developments, such site levelling will involve filling. Such fill may be imported, but where practicable on-site excavated materials will be utilised.

6.   As prime development land becomes harder to obtain, the economics of developing more steeply sloping or flood prone sites is becoming more attractive. Hence, the importance of properly specified and controlled earthworks is also increasing. Once the depth of fill exceeds about 0.5 metres, its implications on the design of any structures proposed for the site becomes significant. Similarly, placement of fill may affect both the surface and subsurface drainage of a site, and will be relevant to the design and construction of any pavements needed for the site.

7.   In the author's experience, there is a wide divergence of opinion amongst civil and structural engineers as to what engineering properties can be assigned to filling. Many will argue that under no circumstances should any structure be supported on filled ground, even if this filling has been placed and compacted under controlled conditions. At the other extreme are those who believe that by reducing the applied bearing pressure sufficiently, their proposed structure will perform adequately on uncontrolled fill. Such people also often have the view that because fill is "old" it will have mysteriously improved engineering properties, even though it is known to have been uncontrolled at the time of its placement.

8.   Where engineered fill is prohibited as a foundation, the result is that in some instances gross over-conservatism may have been employed, probably at considerable cost to the owner. However, in these instances the structure presumably performs adequately, and the owner believes he has received the product he sought, unaware that it cost him more than was necessary. On the other hand, where the designer is content to support his structure on uncontrolled fill, excessive total or differential settlement may occur, and the owner becomes unhappy. This may even be after the designer has advised the owner that there is some risk involved, or that he stands to save significant money by adopting this higher risk approach. Unfortunately, owners tend to forget such earlier advice when their asset is not what they expected, and litigation results.

9.   In the case of large civil engineering projects, the attention given to the specification and control of any earthworks is usually greater than that for, say, commercial development. However, in these types of project larger contracting groups are usually involved in performing the earthworks and they are selected usually after competitive tendering. If the materials encountered are perceived to be different from those expected, or if difficulties are experienced in achieving the specified compaction requirements, such contractors soon express their concerns. If these matters are not able to be resolved quickly, a claim often results. All too often these apparent difficulties arise due to the failure to adequately consider at the design stage the earthworks component of the project. This lack of understanding often continues into the assessment of the claim, with the frequent result that no satisfactory resolution can be reached and litigation follows.

10. It is during the subsequent litigation proceedings that there is close scrutiny of the specification, of the materials used and of the testing procedures applicable to the project, usually by experts who it is assumed have a sound understanding of the principles involved. This scrutiny very often reveals inadequacies in the specification for the project, and in the application of test procedures to assess compliance with the specification. Again, this is often as a result of a lack of understanding of the processes involved by those most commonly responsible for design and specification of earthworks. Regrettably the advice of the experts called in when a problem has already arisen is not sought often enough at the design and specification stage of the project, when the factors that have lead to the litigation may have been resolved.

11. It therefore is considered that our teaching institutions need generally to give more attention to both the fundamental principles and the practicalities of compaction and the performance of earthworks. Geotechnical reports should be required by civil and structural engineering clients to address the earthworks component of projects. This advice should then be incorporated into the design and specification of the works. Once work commences, site engineers should recognise the specialist nature of many aspects of earthworks control and not try to bluff their way through any confrontation with a contractor. Similarly, contractors should be familiar with testing procedures and have realistic expectations of their equipment or the materials involved and not try to blame the specification or testing as soon as they perceive they are not achieving their planned production rates.

12. If only some of these ideals can be achieved, it is considered there would be more competitive pricing of works, due to increased confidence, and there would be fewer claims. This would result in savings on projects and to the community generally. The increased confidence and understanding may also dispel some of the concerns regarding the suitability of filled ground as a foundation, and of the likely performance of structures supported on such fill.

13. It is interesting to note that civil engineers have long accepted the use of compacted earth fill for construction of dams. Strength and compressibility parameters are assigned to the compacted materials and analysis is performed to evaluate economical designs. A high degree of confidence is then assigned to these designs, even though the implications of failure may often be very serious. This confidence is only maintained due to close materials and compaction quality control during construction. However, these experiences do not appear to be translated to other civil projects, particularly where buildings are involved.

COMPACTION

14. Fundamental to drafting an appropriate specification for an earthworks project is an understanding of the process of compaction and the basis for the adopted compaction requirements and associated control testing procedures. A knowledge of the materials to be compacted and in particular any characteristics of these materials which may influence their performance is essential. Such factors as breakdown on compaction, breakdown with time after exposure to air or water, swelling and shrinkage characteristics and loss of strength on saturation all may be relevant.

15.   Hilf (ref 1) presents a useful discussion on the process and theory of compaction. In essence, compaction is the process whereby the density of a soil mass is increased by mechanical, and usually dynamic means. This typically involves tamping, rolling or vibration or a combination of these processes. This results in a relocation of the soil particles and in the expulsion of air from the soil mass, but usually without significantly altering the amount of water in the soil. Compaction should not be confused with consolidation, which is a different geotechnical process, and these terms should not be interchanged in specifications.

16.   The degree of compaction which can be achieved for a given compaction effort is dependent, *inter alia*, on the moisture content of the soil being compacted. This is discussed by Proctor (ref 2) in his landmark papers on the principles of soil compaction and its control. Although the moisture content of a soil mass usually is not significantly altered by the compaction process, it is common to add water or to dry out the soil mass, to achieve a suitable moisture content for achieving the compaction required.

17.   The degree of compaction is usually considered in terms of the achieved dry density of the compacted soil mass, for a given soil, with the peak dry density which can be achieved referred to as the maximum dry density, and the moisture content at which this occurs as the optimum moisture content. Beyond the optimum moisture content a reduction in achievable dry density occurs because the voids in the soil mass become occupied with a greater volume of water (that is the void volume increases) and the compacted soil approaches saturation. In practice, under the influence of mechanical compaction it is not possible to achieve saturated soil conditions, and typically a soil mass compacted at moisture contents higher than the optimum moisture content will contain about 2% air voids.

18.   Although the terms maximum dry density and optimum moisture content are often referred to, these are not unique properties of a given soil. They are dependent upon the energy input during compaction, and on the nature of this compactive effort. By the same reasoning, different soils will respond differently to the same compactive effort, even though they are similar in appearance.

COMPACTION CONTROL TESTING

Historical Approach
19.   The influence of moisture and applied forces in improving the strength and compressibility of mechanically transported soil has doubtless been recognised for years, and applied by such master builders as the Romans in achieving some of their remarkable feats. However, compaction appears to be rarely referred to until the introduction of railroads. The requirements of flat grades and accommodating the heavy loads imposed by locomotives resulted in compaction becoming a recognised requirement in construction of permanent ways. Prior to the development of steam powered rollers in the late 1850's, compaction was performed by light horse drawn rollers or by trafficking placed soil with animals (the original sheepsfoot roller!).

20.   As the use of large steam powered rollers became widespread, the need to place soil at the appropriate moisture content to avoid settlement and failure became apparent, but still without any fundamental understanding of the principles

involved. Presumably observations such as if the soil was too dry, settlement would occur during the following rainy period and if the soil was too wet the compaction equipment became "bogged down" and failures would occur resulted in the pragmatic establishment of a suitable moisture content.

Laboratory Compaction Test

21.   Probably the earliest reported experiments on the relationship between density and soil moisture under compactive effort were those reported by Kelso (ref 3) and carried out during the construction from 1927 to 1930 of Silvan Dam, near Melbourne, Australia. Kelso adopted a 3 inch cube mould and used 150 blows of a "standard cement - testing hammer - machine" to compact samples of soil prepared at a range of moisture contents. He recognised an optimum moisture content at which a maximum density was achieved.

22.   However, it is the work carried out by Proctor (ref 2) at about the same time as Kelso's studies which is better known. Proctor reports "The need of a more precise procedure for the design and construction of earthfill dams initiated an extensive investigation ..... for new methods of construction control ....." This investigation resulted in development of field and laboratory testing techniques that have remained virtually unchanged in principle since this work was carried out in the early 1930's. For this reason the laboratory dry density-moisture content relationship often is still referred to as the "Proctor curve".

23.   The laboratory compaction test devised by Proctor is now standardised in many countries and referred to as the Standard compaction test (eg. AS1289.5.1.1, BS 1377:Method 10, ASTM D698; refs 4, 5 & 6). Although minor variations exist in the requirements of the various procedures, the test method represents a highly standardised set of arbitrarily chosen procedures based on the initial work of Proctor. The Standard compaction test initially involved 25 blows of a 5.5 lb hammer falling through one foot on each of three equal layers so as to just fill a 1/30 cubic foot mould. With metrication, a one litre mould is typically adopted, but with the energy input per unit volume kept constant. Hence, a total of 75 blows of a 2.7 kg hammer falling through 300 mm and a 1 litre mould may be adopted, retaining the energy input to the soil constant at about 595 kJ/m^3.

24.   As engineering structures and compaction plant became larger, and in particular as the need to construct airfields and heavy duty pavements arose, it became apparent to engineers in the 1940's that specifying field compaction requirements in accordance with the Standard Proctor laboratory compaction test was unsatisfactory. Larger rollers operated more efficiently at moisture contents lower than the Standard optimum moisture content, and densities in pavement materials in excess of 100% of Standard compaction were required for satisfactory performance. This resulted in development of the "Modified" compaction test (ref 7) then commonly referred to as the Modified AASHO compaction test. In its metricated (Australian) form (AS1289.5.2.1 ; ref 4) this test requires the application of about 2703 kJ/m^3 of energy to the compacted soil mass, or about 4.5 times the energy input of Standard compaction.

25.   The above discussion has related to the compaction relationship applicable to cohesive or well graded granular soils, where a well defined moisture-density relationship can be established. However, in cohesionless soils

(clean sand), this is not practical, and it is usual to compare the achieved field density to a "loosest" and "densest" state into which the soil can be placed. This comparison results in a relative density or density index value. Again it should be appreciated the so called loosest and densest states of the soil are based on completion of arbitrary, but standardised, procedures (eg. AS1289.E5.1 ;ref 4) and do not represent actual extremes.

Field Density Testing

26.   To establish whether a satisfactory level of field compaction has been achieved, the most common practice is to compare the dry density of the compacted soil to a maximum dry density achieved in either the Standard or Modified laboratory compaction tests.

27.   To allow these comparisons to be made it is necessary to reliably determine the achieved field density and moisture content. The field test procedure adopted needs to be relatively simple to perform using robust equipment and be sufficiently repeatable to minimise potential argument as to the reliability of the results when specification requirements are not being met.

28.   The most commonly adopted means of measuring the field density include the sand replacement test and the nuclear density meter. Nuclear density meters have gained wide acceptance in recent years because of the rapidity with which the test can be performed, the assumed non destructive nature of the test (only a 15 mm diameter hole is required in the fill to allow the test to be performed) and general acceptance that the available calibration techniques are reliable.

29.   Because nuclear density meters obtain a measure of soil density (wet density) and moisture content by indirect methods, calibration of these meters is essential. In Australia the trend is to not rely on the manufacturer's calibration, but to establish a calibration for each meter using several (usually five) blocks of known density. The blocks need to be such that their range of densities are similar to those expected for the range of soils likely to be tested. Regular checks on the internal stability and drift of the nuclear meters are then routinely carried out, using a standard stone block.

30.   Similarly, the apparent moisture content relationship can be obtained using standard blocks of known moisture equivalence. These blocks typically contain varying amounts of polyethylene or polystyrene, the hydrogen ions in these materials substituting for the hydrogen of water. However, because hydrogen ions may be present in a soil mass in other than the pore water, calibration of the nuclear meter for each soil type being tested is required, to establish the offset from the ideal moisture calibration. Clearly, in organic soils the nuclear meter is unreliable for measurement of moisture content.

31.   This need to calibrate the meter for each new soil type (as defined by geological origin, geographic location, etc and not simply whether a clay or a sand) means that for many control testing assignments, it is still necessary to excavate a hole in the tested soil to obtain a sample for laboratory moisture determination. This is in contradiction to the often assumed non destructive nature of the test procedure. As will be discussed subsequently, in most naturally occurring soils a sample of the as-completed material should be recovered anyway, to permit a representative laboratory compaction to be performed.

Relative Compaction

32.   Having obtained a measure of the field dry density and moisture content, and subsequently determined a laboratory maximum dry density and optimum moisture content, the relationship between these values is then usually determined. The ratio of the field dry density (FDD) to the laboratory maximum dry density (MDD) is commonly referred to as relative compaction or the dry density ratio (DDR), usually expressed as a percentage. Hence,

$$DDR = \frac{FDD}{MDD} \times 100\% \tag{1}$$

33.   The moisture variation is usually calculated by subtracting the Optimum moisture content from the field moisture content. That is, a negative moisture variation indicates the compacted soil is drier than the optimum moisture content in the laboratory test that is being used for comparison purposes.

34.   In the case of clean sands, the relative compaction is expressed in terms of density index or relative density, and is calculated using the relevant maximum and minimum density values obtained in the laboratory test (ρ_{max} and ρ_{min})

$$DI = \frac{\rho_{max}(FDD - \rho_{min})}{FDD\,(\rho_{max} - \rho_{min})} \tag{2}$$

Control of Earthworks by Other Methods

35.   Other control testing procedures also exist, with some in widespread use. For example, because of difficulties experienced in performing the above test procedures in clean sands, the use of penetration resistance using a hand held dynamic cone penetrometer or similar as a measure of density (strength) is sometimes used. (eg. ref 8). Similarly, in dam construction, shear strength has sometimes been adopted as a measure of the suitability of the achieved compaction. (eg. ref 9).

36.   It is important to recognise that any in-situ field density test and its associated laboratory compaction only provides information on the condition of the fill at the test location. It therefore is critical that the uniformity of the compaction process is also monitored as the earthworks are carried out. Variables such as layer thickness, number of passes of the compaction equipment, uniformity of moisture conditioning and the consistency of the materials being compacted all need to be controlled. In addition to any field density testing and these visual observations of performance, test rolling to check for uniformity of fill placement is also commonly adopted. However, unless a full time inspection and testing service is provided, it is likely that no one can provide a statement on completion of the work with respect to its compliance with specification requirements. If less than full time inspection is provided, designers might reasonably argue that they do not have sufficient assurance that the fill will be suitable, for example, as a foundation for a building.

37.   Using simple field and laboratory testing procedures to assess the effectiveness of earthworks compaction is usually restricted to fine and medium grained soils. Although various attempts have been made to scale-up the test procedures described above so that coarse gravels and some rockfill could be tested, the effort involved and the time required to perform these tests virtually precludes their routine use. As a result it is believed that once the percentage of gravel sized particles which are coarser than about 40 mm exceeds 15% to 20%, it is generally accepted no suitable routine compaction control test procedure is available. In these circumstances, it is common to resort to closer control of roller routines, layer thickness and moisture application to obtain confidence that the placed fill will perform in accordance with its design requirements. Test rolling with a large roller to confirm that there is no visible deformation or springing under the action of the test roller may then be adopted as a final acceptance criteria.

APPLICATION OF STATISTICAL METHODS TO COMPACTION CONTROL

38.   The use of statistical evaluation of the results of control testing has gained increasing acceptance over the past ten years or so, in particular on larger earthworks projects such as for major highway construction. Although statisticians may argue the same statistics principles can be applied to small projects where there are only a few control tests performed, the adoption of statistical control procedures in these applications is believed to be rare.

39.   Davis (ref 10) was probably the first to suggest that statistical control testing procedures, widely used in the manufacturing industry, could be applied to earthworks control. Turnbull et al (ref 11) presented data on field density and moisture content from various earthworks projects, and concluded that substantial and consistent variations occur, with these variations considered to be larger than might have been expected by the designers. Their data also suggested that although earthworks may have been completed on the basis that all compliance tests indicated specified minimum values had been achieved, it was probable that areas of non compliance would have been present within the compacted fill. Notwithstanding this, Turnbull et al recognised that structures built without statistical control had performed satisfactorily. Although recommending that statistical variations in soil properties should be recognised in specifications, Turnbull et al acknowledged that the continued application of requirements for a minimum test value in specifications was expected to be satisfactory.

40.   A problem with minimum value specifications is the need for judgement in interpreting the test results when enforcing the specification requirements. Hence, a very experienced pragmatic engineer who has confidence in the compaction methods being used on a project might allow some low test values to be ignored. On the other hand, an inexperienced engineer might consider only the test results given to him by his laboratory and reject any work in which a low test value was obtained. The application of statistical control procedures provides a basis for eliminating these potential inconsistencies in administration of a contract.

41.   A psychological barrier often exists with both contractors and superintending engineers when statistic control testing procedures, or lot testing, are applied to a project. The historical approach of acceptance on the basis of "not

one test to fail" may be perceived to indicate that accepted work is all in accordance with the specification requirement. However, as indicated in the work of Turnbull et al (ref 11) and others since, the achieved relative compaction does vary such that there will very often be areas which, if tested, would have failed to meet the specified criteria.

42.   In most studies, such as those of Turnbull et al, there is an underlying assumption that the calculated value of relative compaction for a given test site is a "true" value.   That is, that the test procedures used to obtain this value are accurate and repeatable to within close tolerances.   This has been variously challenged (eg. ref 12).   Recent work by Servais and York (ref 13) has shown that the precision of the laboratory compaction test is poor, and that there is an appreciable likelihood that the specification compliance of work will be judged wrongly when the true quality of the work lies within a substantially wide band centred on the specified limit.   Servais and York conclude that in practical terms the reproductibility of the maximum dry density is about 3% for the Modified compaction test and 4% for the Standard test.   Added to this uncertainty is the precision with which the field density test is performed (ref 14).

43.   Because of this lack of precision in both the field and laboratory tests, Servais and York argue that it is inappropriate to directly compare the results of parallel testing for achieved relative compaction, such as might be agreed to between a Contractor and a principal in the event of a dispute.

44.   The use of statistical evaluation of test results, or lot testing, aims to reduce the influence of test procedure variability as well as eliminate inconsistencies in contract administration.   Through lot testing, the contractor and principal share the risk of test results indicating compliance or non-compliance when the "true" situation may be different.

45.   Statistical acceptance schemes in common use in Australia typically define a lot, or area submitted for testing, as being a portion of work which is essentially homogeneous with respect to material type, general appearance, test rolling response, moisture condition during compaction, compaction technique and the state of the underlying materials.   If these criteria are met, the size of a test lot generally is not a criterion, although restricting a lot to a maximum of one day's production is common.

46.   Having nominated a lot for testing, the test sites need to be selected essentially at random.   Where the testing is being performed on natural soils, it is then essential that a laboratory compaction is performed for each field density test, thus eliminating the influence of inherent material variability in the evaluation of the lot. For manufactured products from hard rock sources (ie crushed rock), it may be possible to demonstrate that any variability in the laboratory measured maximum dry density is due to testing procedures and not material variability.   In this situation, one laboratory compaction test for one field density may not be required, and a single value of the maximum dry density may be assigned to the material placed and compacted in the lot (eg AS1289.5.4.2 ,ref 4).

47.   It is common for six test sites to be adopted for each lot, although this will vary. Having obtained the relative compaction value at each test site, the characteristic relative compaction CRC, for the lot is then calculated as

$$CRC = X - fS \qquad\qquad (3)$$

where X is the mean value of the individual test results and S is the sample standard deviation of the test results. The factor f is selected based on the number of test sites in the lot. The value of f varies between various schemes, but is chosen so that on the basis of the test results following an essentially normal distribution, a certain percentage of values will be greater than the characteristic value. For six test sites on a lot, a value of $f = 0.92$ suggests 80% of the lot will be at a density above the characteristic value.

48.   A typical lot testing acceptance criteria may be that if the calculated characteristic value of relative compaction (CRC) exceeds 99% the lot is accepted unconditionally.   If CRC is in the range of 98% to 99% the work may be conditionally accepted, subject to a re-coil, but if less than 98% the lot under consideration is required to be re-worked and re-submitted for testing.   The method of adopted for calculating the CRC determines the degree of use accepted by the contactor and the principal. That is, for different values of f, different risks are assigned to the two parties.

49.   This sharing of risk may be considered in the light of a low test value being obtained and included in the statistical evaluation of the work. In this case a lower CRC would be obtained than if it had been a high test value.   The probability of the low and high values occurring could be the same. However, in the low value case, the specification requirement may not have been met, and the contractor is asked to re-work the area.   In the high value case, the work is accepted and the contractor proceeds with the next layer. However, if a larger sample of tests had been performed, the influence of the out-lying value would have diminished and the "true" value of the characteristic compaction obtained. This may or may not be consistent with the specification requirement. Hence, in the case of the high value, the owner accepts some risk that his completed product will not be consistent with his expectations.   This may result in increased maintenance costs or unsatisfactory performance.   For the low values, the contractor accepts the risk, and faces additional costs because of it (ie. the reworking). It therefore is important that the statistical acceptance scheme that is devised is fair to both parties without compromising the required end product or resulting in excessive costs due to excessive conservatism.

50.   Hilf (ref 1) discusses some alternative approaches to statistical evaluation of compaction control testing. There are doubtless many other in-house schemes devised by various authorities. Each will have its disadvantages and advantages. However, the author considers that when applying statistical acceptance criteria to earthworks control, the need for care in performing the individual tests is increased , with errors or omissions in test procedures having the potential to significantly influence the evaluation of performance.

PITFALLS IN CONTROL TESTING PROCEDURES

51.   Inevitably, as with any repetitive task, time or labour saving modifications to the procedures specified for control testing of earthworks may be adopted by testing personnel. A number of commonly adopted short cuts have the potential to introduce errors into the evaluation of the achieved compaction. This may adversely effect either the owner or the contractor, depending on the nature of the error. A number of these issues have been central to the arguments put forward in litigation associated with earthworks, and with which the author has been involved. Based on these experiences, and a consideration of the principles involved in compaction control testing, it is considered that as a minimum the following guidelines should be adopted:-

52.   The sample for the laboratory reference density test should be taken of the as compacted material and should comprise the material recovered from the field density determination, or from that volume considered in the field density determination in the case of a nuclear density meter. Only if it can be established that the influences of field compaction or material variability do not affect the value of the laboratory reference density is it acceptable to sample from stockpile or from the test area before compaction.

53.   If the material being tested breaks down significantly during the laboratory compaction process, and this affects the compactability of the material and therefore the achieved maximum dry density, the relative compaction criteria may need to be changed. Sampling for the laboratory compaction prior to field compaction is not an acceptable alternative.

54.   Where the mass of soil excavated from the field density test site is insufficient for the laboratory test, additional material should be recovered from immediately around the zone in which the field density determination was made, over the full depth considered in the test but no deeper. Recovery of material from elsewhere on the project is unacceptable.

55.   A laboratory reference density should be established for each field density determination, unless the material being tested is sufficiently uniform in its properties and compaction characteristics that any variability in the laboratory reference density is due solely to test procedure variability. This will usually only be applicable to manufactured products from hard rock quarry sources.

56.   Although test procedures often provide a basis for "correcting" the measured value of field density for coarse material eliminated in the laboratory reference density test, where this percentage of oversize material exceeds about 15%, the calculated value of relative compaction may lack relevance. That is, the correction applied allows for mathematical removal of the oversize material, but without recognising the effect the oversize may have on the compaction characteristics of the material. The percentage of oversize for which a correction may be made is subject to argument, with some arguing no more than 10% should be permitted. However, there appears to be general consensus that 20% is the maximum, that should be permitted, even though some would argue it is possible to allow for up to 40% if an additional correction factor is applied. Houston and Walsh (ref 15) consider various correction methods, but conclude that significant differences exist in the calculated relative compaction for the various methods. They also recommend that the method to be adopted is spelt out in the specification for the works.

57.   Any use of "one-point" laboratory compactions to obtain a laboratory reference density should only be permitted where an accurate result is not essential.   Where the result may affect contractual matters or be potentially deleterious to the performance of the completed project, this approach should not be used.

58.   For routine earthworks compaction control, procedures for moisture curing of laboratory reference density samples may be relaxed from the strict requirement of air drying, crumbling of aggregations of soil to sand size and then allowing several days for uniform moisture conditioning. However, where clay samples are at moisture contents considerably higher or lower than the desired value for compaction, longer curing times may be required. Clay soils should be broken up to finer than about 10 mm before laboratory compaction.

59.   Under no circumstances should samples of material taken from two or more field density test sites be combined and used for the laboratory compaction test. The value of maximum dry density obtained will lack relevance.

60.   Mechanical excavating tools for field density tests should only be used with caution due to the possible compactive effect of this equipment.

61.   A number of these guidelines are presented in AS3798(ref 16). This Standard was developed in response to an increasing amount of litigation associated with earthworks projects. This litigation often resulted from neglect of some or all of the above principles. Even where the difficulties experienced on site may have been due to the contractor not performing the work satisfactorily, it was not possible to adequately identify if this was the case because of these types of shortfalls in the control testing procedures. Legal argument then inevitably followed and the only true winners were the legal profession!

PROPERTIES OF COMPACTED FILL

General

62.   Without the designer having some knowledge of the influence of placement moisture or achieved relative compaction on the performance of the completed project, the specification prepared for the project may be inappropriate, and possibly result in inadequate performance. All too often the specification used for the last project is adopted, but without any review for the particular conditions and needs of the project under consideration.

63.   It is unrealistic to expect that in a paper such as this, it would be practical to present any comprehensive discussion on the properties of compacted fill. However, some of the factors which should be considered in the design and specification of earthworks will be discussed below.

Strength

64.   In general terms, the strength of compacted soil will increase with increased dry density. For cohesionless soils this tends to follow logically because many of the relationships used in routine geotechnical engineering relate, for example, relative density of sands to angle of friction. It follows that a well compacted sand or gravel will be stronger than one which is only loosely dumped.

65.   The effective strength of cohesive materials also follows this same general rule of increased strength with increased density, or relative compaction.

Figure 1 presents the results of some consolidated undrained triaxial tests with pore pressure measurement performed on compacted samples of weathered granite (Ipoh Expressway, Malaysia). Although only a few data points are available the trend for increased strength with increased relative compaction is apparent. To allow comparison of the various data on this figure, strength has been calculated as $S = c' + \sigma \tan \phi'$, with the applied stress, σ, taken as 50 kPa. The classification test data for the four samples are presented in Table 1.

Table 1.   Properties of Weathered Granite.

Sample	Unified Classification	% Fines	$W_L(\%)$	$I_P(\%)$	MDD(t/m³)	OMC (%)
			Atterberg Limits		Compaction	
1	GC	23	70	39	1.72	14.5
2	SM	36	45	16	1.66	15.5
3	SC	32	73	44	1.65	16.5
4	SC	30	96	59	1.66	16.0

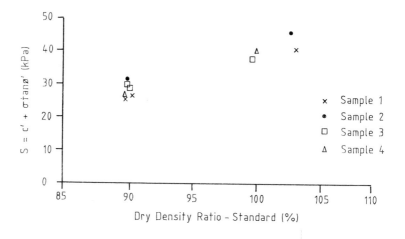

Fig. 1. - Effective Strength vs Relative Compaction (data courtesy Maunsell & Partners Pty Ltd)

66.   For cohesive materials, the undrained shear strength is a function of both achieved compaction and of moisture content. For example, a clay compacted when quite dry may be of low density but be of high apparent, or undrained, strength. However, with time it is probable the moisture content of this fill will increase through natural equilibrium processes, a leaking service, ponded water, etc, and significant loss of strength will occur. This moisture dependence of compacted clays may result in a misleading assessment of the performance of a contractor if such techniques as test rolling are used as the sole indicator that the required compaction has been achieved.

30

67.   Figure 2, adapted from ref. 17, shows the influence of relative compaction and moulding moisture content on the undrained strength. Whilst this is for a particular experiment in which samples of silty clay (CL) were saturated under constant volume, this type of behaviour is exhibited in other test programmes. Probably the most commonly reported strength vs moisture content and density relationship is for the California Bearing Ratio, CBR of compacted soils. Figure 3 illustrates this relationship for an extremely weathered basalt (GC). (WL) = 42%, Ip = 20%, 13% fines).

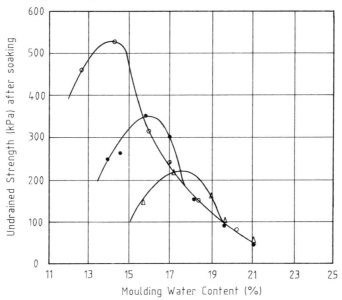

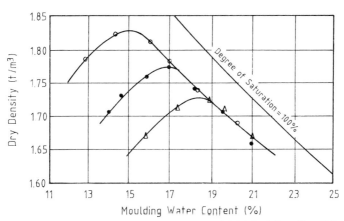

Fig. 2. Undrained Strength after soaking vs Initial Moulding Conditions

68.   The influence of moulding moisture content on the as compacted undrained shear strength is illustrated in Figure 4. These data were obtained as

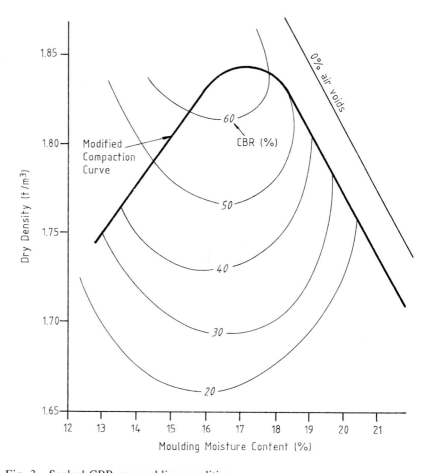

Fig. 3. Soaked CBR vs moulding conditions.

part of a study into the compactability of some clayey soils being placed in a dam. A dispute arose as to the control testing procedures being used by the owner, in particular in his interpretation of the optimum moisture content. The contractor claimed he was being forced to place the material at moisture contents in excess of Standard optimum moisture content, when the Specification permitted placement within the range ±2% of Standard Optimum. A careful laboratory testing program showed that the moisture content determined by the field laboratory was in error, and that the contractor's claim was valid. The error had arisen because the laboratory compactions had been performed after breaking down the clay to finer than 19 mm, and then drying back using rapid methods. This resulted in crusting on the clay lumps, and they then behaved more like a "soft gravel" than a clay when compacted. By requiring the contractor to place the clay several percent wet of Standard optimum meant the minimum relative compaction requirement of 95% Standard was very difficult to achieve, and the compaction equipment was experiencing difficulty simply trafficking the material to be compacted due to its low strength. Figure 5 shows the results of a laboratory study of the energy required to achieve the specification requirement for different moulding moisture contents. The result is what would be expected from the general principles of compaction. However, the owner for this project took considerable convincing that his test procedures were wrong, and that it should not be necessary for compaction plant to be ploughing their way through the fill material to achieve the specification requirements. Of course, a claim was made by the contractor!

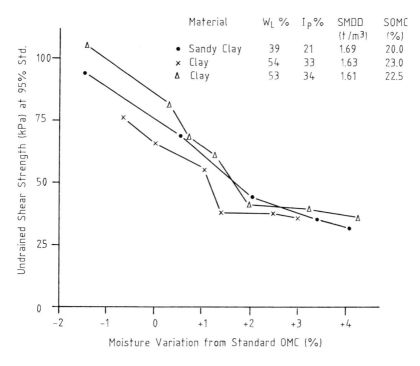

Fig. 4. Undrained Shear Strength vs Moulding Moisture Content

Compressibility

69.   The settlement of compacted earthfill on a rigid foundation may comprise:-

-immediate settlement, as a result of elastic deformatios but without change in moisture content. Well compacted high strength fill will be less compressible than loosely dumped fill.

-consolidation settlement, occurring as a volume reduction associated with expulsion of pore air and pore water under the influence of an applied load. The magnitude of this form of settlement is a function of the initial compaction of the soil and its placement moisture content.

-collapse settlement, occurring in an initially partially saturated soil as a result of saturation. For sample, soil lumps in a dry but loosely placed fill may have adequate strength to support an applied load with little settlement. However, on

33

saturation softening of the clays will occur and the available bearing contact between the lumps may be insufficient to support the load. Collapse into voids of the soil fabric will then occur with an associated volume reduction, or settlement.

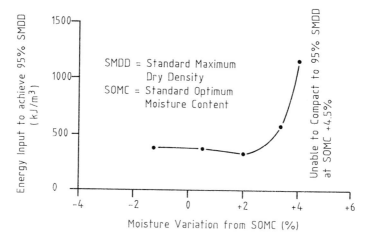

Fig 5.   Compaction Energy required to achieve 95% Standard at various Moulding Moisture Contents

70.   The immediate and consolidation settlement will occur as the fill is constructed and loads are applied. The collapse type settlement may occur at any time after construction, due to such external influences as a leaking service, saturation due to flooding, prolonged wet weather, etc, and is often more dramatic with respect to structures built on the fill due to this time lapse.

71.   These forms of settlement are illustrated on Figure 6. This data was obtained from a laboratory study (ref. 18) of the possible long term settlement performance of a proposed 17 metre deep fill to be placed under construction traffic only, albeit well disciplined with respect to coverage of the fill area. Oedometer tests were performed on three expected material types (two sandy clays of low plasticity, and a clayey sand), moulded at various densities and moisture contents. Figure 6 presents the data for one of the soil types.

72.   The data from the three material types has been averaged and presented as Figure 7. This data was used to predict the expected performance of the completed fill, allowing for a uniform mix of each material type and at placement moisture contents within the range of 0% to -3% of Standard optimum.

73.   This data in Figures 6 and 7 indicates that provided a minimum relative compaction of 95% Standard has been achieved, under modest applied stresses the resulting settlements will be small. At lesser relative compaction, the amount of settlement increases substantially. This is further illustrated in Figure 8, which presents the results of oedometer tests performed on samples of a clay soil of medium to high plasticity.

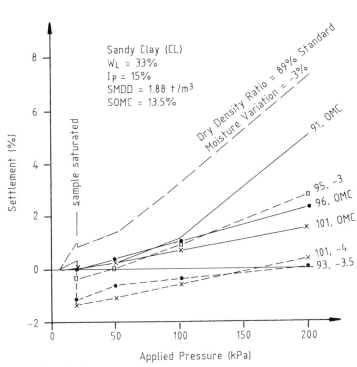

Fig. 6. Settlement vs Applied Pressure

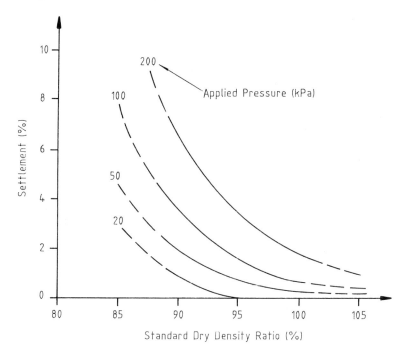

Fig. 7. Settlement vs Relative Compaction

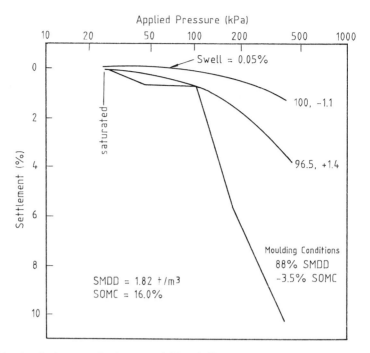

Fig. 8. Oedometer Settlement of Clay Soils

Reactivity

74.   In many applications the reactivity of a compacted soil (ie. its susceptibility to undergo volume change with variations in moisture content) is just as important as the settlement characteristics of poorly compacted fill. Even clay of low plasticity when placed at commonly specified relative compaction levels but dry of Standard optimum may swell if saturated at low confining stresses. For domestic structures founding at shallow depth this situation may arise and therefore needs to be considered in footing design. For clays of high plasticity, the amount of swell will be even greater. Figure 9 presents the results of some laboratory swell tests performed on a clay of high plasticity, moulded at Standard maximum dry density but at varying moisture contents. The samples were placed in an oedemeter cell under an applied pressure of 109 kPa. They were then inundated and the swell observed. After swelling ceased, the applied pressure was reduced to 55 kPa and the swell again observed. This was then repeated under 6 kPa. The differences in swell between placement at Standard optimum and four percent drier is significant. Also shown on Figure 9 is the corresponding drying shrinkage of samples at the various moulding moisture contents.

75.   At lower placement densities the swell is less for the same moisture content. Hence, where the reactivity of placed fill may be of importance to the performance of a subsequent development, careful control of both placement moisture and relative compaction may be necessary. In particular, it may be necessary to specify a maximum relative compaction as well as a minimum.

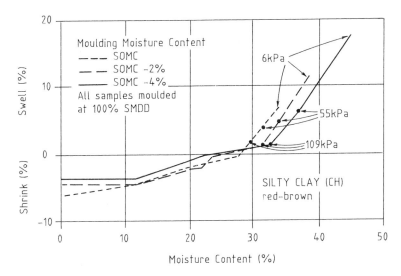

Fig. 9. Swell/Shrink vs Placement Moisture Content

76.   Although no directly comparable data has been collected, the author also believes that for residual soils, such as basaltic clays for which the data in Figure 9 was obtained, the reactivity of the remoulded soil is higher than for intact soils. The study from which Figure 9 was taken also included a measure of the reactivity of undisturbed soils from the same area. In general, the reactivity of the undisturbed soil was lower than that measured for the remoulded soil. It therefore is important when designing footings for reactive clay sites, that account is taken of this potential for higher reactivity if any fill is proposed.

Modified vs Standard Compaction

77.   The author has been concerned for some years that there is a trend amongst designers and specifiers of earthworks to adopt the Modified compaction test as the basis for control testing. The reason for this is not known, other than a possible belief by specifiers that because the Modified test applies approximately 4.5 times as much energy to the soil specimen as the Standard compaction test, or that it is a later version of the original Proctor test, that it must be better. There is also an unfortunate belief that there is a direct equivalence between the Modified and Standard maximum dry densities, and that a relative compaction of 95% Modified is about equivalent to 100% Standard.

78.   Figure 10 presents comparative data collected by the author, and indicates this simple equivalence does not exist. It follows from Figure 10 that for the Modified test, the optimum moisture content for a crushed rock (say MMDD = 2.3 t/m³) may be 2% or 3% less than the corresponding Standard optimum. However, for high plasticity clays (MMDD = 1.6 t/m³) the difference may be up to about 8%.

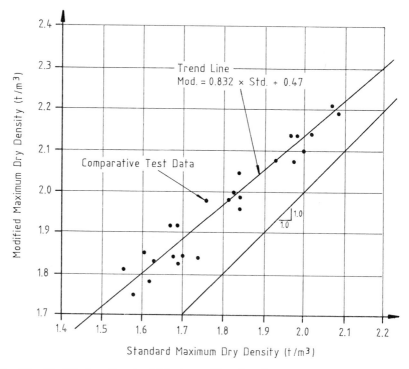

Fig. 10. Modified vs Standard Maximum Dry Density

79. The trend line shown on Figure 10 has been manipulated to produce Figure 11. This shows that the assumed equivalence of 95% Modified being equal to 100% Standard is not valid. Hence, naively specifying earthfills or road subgrades to be compacted to, say, 95% Modified in the belief that it is equivalent to 100% Standard (as may have been previously used) is wrong, and may incur both financial penalties and potential engineering problems.

80. The financial penalties will be associated with the requirement to compact the soil to higher densities than might reasonably be required for satisfactory performance. For example, for a high plasticity clay, the specified level of compaction could correspond to about 110% Standard, well in excess of that required for satisfactory performance with respect to strength or compressibility.

81. Furthermore, to achieve these high densities, it will be necessary to compact the soil at a moisture content which is several percent dry of Standard optimum. The potential for subsequent swell of the placed fill is then high. In many temperate climates, the equilibrium moisture content of soil, below any zone of desiccation, is near to Standard optimum. Clay fills placed at close to Modified optimum are therefore likely to swell. This may result in uneven pavement surfaces, poor performance of building footings if founded in the clay fill and possibly lower long term strengths than might have occurred if the soil was initially compacted at a moisture content closer to the expected equilibrium condition.

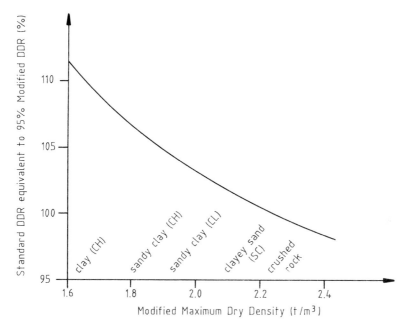

Fig. 11. Equivalence of 95% Modified and Standard Compaction

SPECIFICATIONS

82.   It is not considered appropriate to present typical specification clauses in a paper such as this. In the preceding discussion an attempt has been made to identify those factors which need to be considered when preparing a specification for an earthworks project. What is important is that the specification for a previous project is not simply adopted without consideration of any different factors which may be relevant to the particular project under consideration.

83.   Factors which need to be considered, even if obvious, will include:-
- Stripping requirements:- In addition to removal of the vegetation and root affected topsoil is it necessary to remove additional material? For example, is the subsoil silty and unsuitable for use as fill, or as a foundation?
- Removal of wet or unsuitable material:- Is there an expectation of weak or unstable material being exposed after compaction of the prepared subgrade?
- Nature of materials:- Are the materials to be excavated suitable for use as fill? Do they need special consideration with respect to moisture conditioning?
- Compaction control testing:- Can the proposed fill materials be tested using routine compaction control testing procedures? Is there anything unusual with respect to the behaviour of these materials? eg:- breakdown on compaction; highly porous stone which might affect moisture conditioning; soils are very dry or very moist in-situ making uniform moisture conditioning difficult.
- Is fill to be placed on sloping ground?
- Are there any drainage problems which need to be addressed by installation of surface or subsurface drains.

- What compaction standards are to be specified? Is reactivity an issue? Is settlement or strength of the compacted fill a criteria which needs to be considered?
- Are the earthworks to be carried out at a time when prolonged wet weather may be of concern?
- What level of site supervision is proposed? Will there be full time engineering supervision of the fill placement or periodic visits and occasional control tests?
- What type of contractor is likely to bid the project? Will there be a high or low fill placement rate?
- Will the project be carried out under a Quality Assurance programme?

84.   The last consideration is becoming more prevalent, with the onus for satisfactory performance being placed with the contractor rather than the owner or designer. The success or otherwise of these Quality Assurance schemes for earthworks is yet to be fully assessed. Whilst in principle such schemes ensure that the same end product is achieved, unless the contractor is dedicated to the achievement of the same goal as the designer, the author believes the completed product will typically be to a lesser standard than if it were controlled by the owner. This is not to suggest that unscrupulous behaviour is a necessary result of a Quality Assurance scheme, but rather than a contractor will place a different emphasis to the owner on completion of the project, with human nature inevitably playing a role in acceptance criteria.

CONCLUSIONS

85.   An embankment, or filled area, constructed with due regard to the nature of the material used as fill, and of the proposed end use of the area constructed, can have engineering properties at least as good as those of natural soils. However, it is important to recognise in advance the importance of the earthworks on the overall project, and to provide adequate attention to preparation of specifications, choice of fill material, selection of contractor, supervision and testing. If this is done, footings may be constructed within the fill at acceptable bearing pressures, and construction may proceed with little further need to consider that a site has been filled. Similarly a road embankment or earth dam can be constructed with an expectation of satisfactory performance.

86.   However, if development is allowed on inadequately compacted or controlled earthworks, the potential will exist for total and differential settlement, heave, and softening of the fill. The consequent risk that the long term performance of the fill, and of any structures supported on it, will be unsatisfactory is then significant. This then opens the doors for the legal profession, and litigation will often result. That is, placing a sign at the front of a site indicating "Good Clean Fill Wanted" and coming back when the site is filled, may be an unwise and ultimately expensive way of obtaining a level site!

REFERENCES
1.   HILF, J.W. Compacted fill. Foundation Engineering Handbook, ed. Hsui-Yang Fang, 2nd Edition, Van Nostrand Reinhold, 1990, p 249ff.
2.   PROCTOR, R.R. The design and construction of rolled earth dams. Engineering News Record, 1933, Vol. III, August 31, September 7, 21 and 28.

3. KELSO, A.E. The construction of Silvan Dam, Melbourne Water Supply, Proc. ICE, Paper 4824, 1934, Vol. 239, 403-446.

4. STANDARDS AUSTRALIA - AS1289. Methods of testing soils for engineering purposes. 1977 to 1993.

5. BRITISH STANDARDS INSTITUTION - BS1377:1990. Methods of testing soils for civil engineering purposes.

6. AMERICAN SOCIETY OF TESTING AND MATERIAL. Part 14, Concrete and mineral aggregates.

7. US WAR DEPARTMENT. Office of the Chief of Engineers. Compaction tests and control. A. Methods of compaction tests. Engineering Manual, Chapter XX. 1943.

8. GLICK, G.L. and CLEGG, B. Use of a penetrometer for site investigation and compaction control at Perth, W.A. Proc. Eng. Conf. of Institution of Engineers, Australia, 1965, Perth. pp 114-118.

9. DENNEHY, J.O. The remoulded undrained shear strength of cohesive soils and its influence of the suitability of embankment fill. Clay Fills, ICE, 1978. pp 87-94.

10. DAVID, F.J. Quality control of earth embankments. Proc III ICSMFE, Zurich. 1953.

11. TURNBULL, W.J., COMPTON, J.R. and AHLVIN, R.G., Quality control of compacted earthwork, ASCE J. SMFE, SM1, 1966. pp 93-103.

12. METCALF, J.B. Methods of specifying and controlling compaction. Jour. of Australian Road Research Board, Vol.4, No.2, 1969. pp 4-17.

13. SERVAIS, S.E.C. and YORK, K.I. The precision of two laboratory compaction tests and a particle density test. Paper submitted to Australian Road Research Board, 1993.

14. SERVAIS, S.E.C. and YORK, I.K. Comparison of the precision of two methods of determining field density of earthworks. Australian Road Research, 20(2), June 1990. p 23-37.

15. HOUSTON, S.L. and WALSH, K.D. Comparison of rock correction methods for compaction of clayey soils. J. Geotechnical Engineering, ASCE, Vol.119, No.4, 1993, pp 763-778.

16. STANDARDS AUSTRALIA - AS3798. Guidelines on earthworks for commercial and residential developments. 1990.

17. SEED, H.B., MITCHELL, J.K. and CHAN, C.K. The strength of compacted cohesive soils. Proc. ASCE. Research conference on shear strength of cohesive soils. Boulder, 1960.

18. COFFEY & PARTNERS PTY LTD. Assessment of fill settlement. Unpublished report, 1985.

3. Engineered fills: space, time and water

J. A. CHARLES, Building Research Establishment, UK

SYNOPSIS. Some fundamental questions are addressed concerning the nature of engineered fills, the purposes that they serve, the behaviour that is required of them and the means adopted to obtain the required performance. In providing answers to these questions, some key elements of fill behaviour are highlighted. It is emphasised that the ingress and seepage of water often have a dominant effect on performance. This theme lecture focuses principally on two applications of engineered fills; embankment dams and foundation fills. Fill performance is illustrated with descriptions of the behaviour of some major engineered fills in England.

INTRODUCTION
    1. Many of the papers presented at this conference provide detailed information on the behaviour of specific types of engineered fills under particular conditions. Such field records are a vital element in the advancement of knowledge and understanding of fill performance. However it can be helpful to examine the subject of engineered fills in a wider context.
    2. The well known definition of civil engineering provided by Thomas Tredgold (Chrimes, 1991), for the first charter of the Institution of Civil Engineers, which was granted in 1828, was given in the following terms:

".... being the art of directing the great sources of power in Nature for the use and convenience of man...."

The charter thus sees civil engineering works as contributing to the wellbeing of mankind.
    3. Engineered fills form a sub-division of civil engineering activity in which the earth's surface is modified by earthmoving for the use and convenience of man. Such activity does not meet with universal approval as indicated by the following extract from a letter, which was published in New Civil Engineer on 25th February 1993 and which refers to road construction at Twyford Down:

"True, large quantities of excavated material will be landscaped. However, destroyed habitats cannot be recreated."

4. The construction of large dams involves very substantial modifications to the environment as large areas of land are submerged. The dam thus has an effect over a much greater area than the sites of the excavation and placement of the engineered fill. Large dams have acquired a particularly strong worldwide network of opposition groups and literature (Goldsmith and Hildyard, 1984; Pearce, 1992). Environmental objections need to be addressed and it cannot be assumed that they never have any merit. Dr Pircher, President of the International Commission on Large Dams, made a valuable contribution to the debate when he gave the Geoffrey Binnie Lecture at the 7th British Dam Society Conference on the theme "36 000 Large Dams - and still more needed" (Pircher, 1992).

5. There are two types of opposition. There are those who consider that in the past the impact of engineering works was not adequately assessed. They assert that a more thorough analysis of all relevant factors would show that many engineering works do not provide a net benefit. In terms of Tredgold's definition, they challenge the view that certain civil engineering works are of use and convenience to man. It may be that in the future major schemes will need not only economic and environmental evaluations, but also analyses of energy and entropy will be required (Hudson, 1992).

6. The second form of opposition is based on an outright denial of Tredgold's assumption that man is justified in adapting the physical world for his use and convenience. Those who follow this philosophy will not be convinced by any number of analyses.

7. It is important to have a clear perspective at a time when a substantial and growing body of opinion appears to oppose much, if not most, civil engineering activity on supposedly environmental grounds. It may seem ironic that in England in the eighteenth century, Capability Brown carried out major earthmoving operations with the sole object of improving the environment for his wealthy clients (Hinde, 1986) and most, although not all, his efforts have been widely appreciated for some 200 years. All engineered fills effect a change to the environment not only where they are placed, but also at the location where they are excavated; borrow pits and spoil heaps are inevitable.

8. To develop a balanced view of the subject, engineered fills are now examined in a framework provided by a number of basic questions.

(a) What is an engineered fill?
(b) What purposes do engineered fills serve?
(c) What behaviour is required of engineered fills?
(d) How is the required behaviour to be achieved?
(e) How do engineered fills behave in practice?

It has been necessary to be selective in providing answers to these questions. The major emphasis has been given to two applications of engineered fills; embankment dams and foundation fills. The paper concludes with brief descriptions of the behaviour of some major engineered fills in England.

DEFINITION
9. What is an engineered fill? At the outset it is well to make sure that there is a rational and helpful definition of the subject. This is approached in two steps; firstly the differences between fills and natural soils are identified and then secondly the distinction between engineered fills and non-engineered fills is discussed.

Fills and natural soils
10. The following propositions are implicit in the decision to hold a conference on the subject of engineered fills which thus excludes consideration of natural soils.

(a) Fills and natural soils can be defined and distinguished.
(b) Fills and natural soils have significantly different modes of behaviour which give the distinction between them some engineering importance.
(c) Fills are being widely utilised and their behaviour is therefore of major interest to engineers.

The third proposition will be examined in the reply to the question relating to the purposes which engineered fills serve. The first two propositions are examined at this stage.
11. The term "fill" is used to describe ground that has been formed by material deposited by man. Thus fill, or "made ground" as it is sometimes called, results from human activity, in contrast to "natural soil" which has its origin in geological processes. There is thus a clear distinction in principle between fills and natural soils. In practice it will be easy to distinguish between them in many cases but not necessarily in all cases; a rockfill may be readily identifiable as fill whereas a sandfill might have little to distinguish it from a natural sand deposit. In some situations there will be historical evidence of fill placement. Many aspects of the engineering behaviour of fills are similar to natural soils. For example, the principle of effective stress is as relevant to fills as it is to natural soils.
12. The behaviour of fills can be compared and contrasted with the behaviour of natural soils, with the following factors kept in mind.
13. Nature of the material. The fill may be composed of the same type of material as a natural soil (eg clay, sand or rock) but it could be composed of something quite different (eg industrial, chemical, building, commercial or domestic wastes).
14. Method of deposition. The fill may have been deposited in a manner that is quite similar to the deposition of some natural soils (eg slow sedimentation under water) but it may

have been deposited in a quite different manner (eg compaction in thin layers with a heavy vibrating roller). No study of natural soil behaviour would be complete without some consideration being given to the origin of the soil deposits because the mode of formation affects subsequent engineering behaviour. However the geological processes involved in the formation of a natural soil deposit have to be inferred from a study of the deposit in its present condition. In contrast the formation of a fill deposit may have been witnessed and there may be recorded information on its origin. The study of the formation of fills is therefore different from the study of the formation of natural deposits in that the evidence is of a different nature and the depositional processes are often of a different form.

15. Location and extent. The fill may be placed over a wide area in a similar fashion to many natural deposits. Alternatively the fill deposit may be very localised with vertical or near vertical boundaries (eg where docks or clay pits have been backfilled).

16. Heterogeneity. With a high degree of control exercised during placement, the fill may be no more variable than many natural deposits. Without adequate control fills may be extremely variable.

17. Age. Fills have been placed during the historical period, often very recently. Consequently they are likely to be significantly younger than natural soils. The behaviour of natural soils and rocks may be strongly influenced by structure arising from cementation or bonding together of particles so that strength is increased and the stress domain over which the soil exhibits stiff behaviour is enlarged (Leroueil and Vaughan, 1990). In fills formed from natural soils, such structure will have been largely destroyed during excavation and placement of the fill. Often construction takes place relatively soon after fill placement with little time for structure to develop in the fill. Consequently fill behaviour may be inferior to that of an apparently similar natural soil.

Engineered and non-engineered fills

18. The nature of the difference between engineered and non-engineered fills has to be clarified before the utility and importance of the distinction can be assessed (Charles, 1993a). Engineered fill is fill that has been selected, placed and compacted to an appropriate specification in order to achieve some required engineering performance. Thus the fill has been designed and built with a specific use in mind. Through an appropriate specification and enforcement of that specification during placement, a fill has been produced which has engineering properties which are known and which are considered adequate for the purpose for which the fill has been placed.

19. Non-engineered fills generally arise as the by-product of human activities associated with the disposal of waste materials. The fill has not been placed with a subsequent engineering application in view. Little control may have been

exercised and consequently there is the possibility, and in some cases the probability, of extreme variability. This makes it very difficult to characterise the engineering properties and predict behaviour. Where non-engineered fills are used as foundation materials problems may be experienced and considerable caution is essential. Some form of ground improvement may be required prior to construction.

20. As defined here engineered fill is a relative term and is applied to a fill in relation to a particular application for the fill. The definition is thus related to purpose. The closely specified and controlled fills used, for example, in a modern embankment dam are clearly engineered fills. Many waste disposal operations have created non-engineered fills. Some cases may not be so easy to classify and two examples are presented to illustrate this.

(a) A refuse fill may be engineered for waste disposal purposes with respect to gas and leachate control but may not be engineered with a subsequent use for the site, such as building, in view. Waste Management Paper No 27 (Department of the Environment, 1991) defines "landfill" as follows;

"The engineered deposit of waste into or onto land in such a way that pollution or harm to the environment is minimised or prevented and, through restoration, to provide land which may be used for another purpose."

(b) Old British embankment dams had a central clay core of puddle clay and shoulders of general embankment fill. Although neither type of fill was placed with modern earthmoving machinery and compaction plant, both fills can be considered to be engineered. The puddle clay was of a requisite very low permeability to provide the watertight element and the shoulders were built to support the core.

FUNCTION
21. What purposes do engineered fills serve? The definition of an engineered fill that has been adopted is related to purpose. Although there are a great many applications of engineered fills, each of which is the subject of an extensive literature, a large number of these applications can be grouped under two main headings.

Engineered fills which control the flow of liquids
22. Fills are used to form embankment dams which retain water reservoirs. The objective of the dam is to impound water which may be required for hydropower, water supply to towns, irrigation, flood control etc. Similar embankments can be used to retain canals. An embankment dam will usually be composed of several different types of fill; low permeability fill to form the watertight element, stronger fill to support the

watertight element, and fills to act as filters, drains and transition materials. Possible hazards to the dam include internal erosion, slope instability and overtopping during floods. Breaching of the embankment could lead to an uncontrolled release of the impounded water and could be a hazard to public safety.

23. Fills also are used to form embankment dams which retain waste lagoons. The objective of these dams is to retain the sedimented waste material from some industrial or mining activity and to allow the water to drain through the embankment. The composition of the embankment could be similar to a water retaining embankment dam but may be quite different in cases where the embankment is built in stages as waste disposal progresses.

24. Fills are used to form impermeable barriers around landfills. The objective is to protect the environment surrounding landfills by preventing or controlling the ingress and egress of liquids and gases. Compacted clay linings and compacted clay cappings are frequently used at landfill sites as seepage of leachate and the migration of noxious gases can pollute the ground and the groundwater at a considerable distance from the site. A very high level of control is required in placing fills to form such barriers.

25. Fills are used to form breakwaters. Rubble mound breakwaters can provide shelter for harbours, boat moorings, shorelines etc. High quality quarried rock may be required and placement to a required profile presents obvious difficulties.

Engineered fills which support roads and buildings

26. Fills are used to form embankments which support highways. The objective is usually to reduce gradients on roads and many engineered fills are placed during cut and fill operations for highway construction. There is limited scope for material selection as material excavated in a cutting needs to be placed in the adjacent embankment unless clearly unsuitable.

27. Fills are used as foundation materials for buildings. The objective is to safely support buildings without the occurrence of damaging settlement. This situation usually occurs where old excavations are infilled but in some circumstances it could involve embankments being built above the level of the surrounding ground. Where settlement sensitive structures are to be built on the fill, a high quality fill which is not vulnerable to large post-construction movement will be required.

28. Fills are used as backfills to retaining walls and bridge abutments. The pressure exerted by the backfill on the wall is a key factor in the design and subsequent safety of the wall and compaction pressures created during fill placement may be significant. Where there is filling behind a bridge abutment, it is important to avoid undesirable differential settlement in the road surface at the approach to the bridge.

29. Not all fills can be classified under these two

headings. For example fill may be used to infill voids to prevent collapse. Reinforced fills can be used either as part of a liquid retention structure or to support roads. Reinforced fill comprises tensile reinforcing elements embedded in a compacted mass of fill and includes the connections and facings necessary for stability. A reinforced fill structure has a vertical or near vertical face. Many retaining structures now use this concept of fill reinforcement and the complex behaviour of the composite structure has been the subject of much research.

PERFORMANCE REQUIREMENTS

30. What behaviour is required of engineered fills? The required behaviour is closely linked to the purpose of the engineered fill and the processes and hazards to which it may be exposed. There may be performance requirements related to a wide range of geotechnical properties such as shear strength, stiffness, compressibility, permeability. Two applications of engineered fills are briefly examined; embankment dams and fills which support buildings. The effects of the ingress and the flow of water through such fills form a theme which links certain aspects of the performance of these engineered fills.

Embankment dams

31. Safety and the environment have become dominant themes in the world today. Embankment dams have major consequences for both the environment and for public safety.

(a) The formation of large reservoirs has a uniquely massive effect on the environment and therefore large dam construction is in the forefront of environmental conflict.

(b) The storage of large quantities of water upstream of populated areas poses a major potential hazard to public safety at a time when public awareness of these matters is acute. Design, construction and operational safety of embankment dams are controlled by strict regulations in many countries.

32. An embankment dam must be sufficiently watertight to store water with leakage restricted to an acceptably small quantity. However in considering performance requirements, public safety must have priority. It must be ensured that the dam does not breach catastrophically with an uncontrolled release of the water stored in the reservoir. Embankment dams have been built from the earliest periods of the history of mankind and many of these structures have exhibited remarkable longevity. Nevertheless some notable disasters have occurred.

33. Embankment dams are relatively complex structures which can fail due to several different causes such as slope instability associated with high pore pressures or inadequate shear strength, internal erosion associated with seepage and leakage through the embankment and external erosion associated

48

with overtopping of the embankment. The different fill types incorporated in the structure will have different performance requirements concerned with, respectively, low permeability to limit seepage losses, appropriate filter properties to prevent or halt internal erosion, sufficient shear strength to maintain slope stability, and compatible volume change and deformation properties.

34.   Internal erosion has been one of the most common causes of failure. Unlike slope instability which often occurs during construction or with a low reservoir level, an embankment dam is most vulnerable to internal erosion with the reservoir at top water level. Failure due to internal erosion is therefore likely to be catastrophic. The safety of modern embankment dams depends to a large extent on the proper design and construction of filters and much research has been carried out into the properties of granular filters required to prevent or halt erosion. The assessment of the safety of older dams built without such filters is a different and, in some ways, a more difficult matter. The significance of hydraulic fracture in clay cores is now widely appreciated and investigations of the stress conditions in typical UK puddle clay core dams have been carried out (Charles and Watts, 1987).

35.   Where internal erosion does occur it is not easy to identify and analyse the causes for a number of reasons.

(a) It is a hidden internal phenomenon and there may not be any external evidence until a late stage.
(b) It is a weak link phenomenon that may be very localised within the embankment.
(c) It occurs at locations where there is low effective stress and this type of soil behaviour is difficult to investigate.

36.   Because of these factors, there is often considerable debate about the precise cause and the development of internal erosion failures. This is true, for example, not only of the failure in the nineteenth century of Dale Dyke Dam in England but also of the more recent failure of Teton Dam in the USA in 1976.

37.   It may seem much more satisfactory to analyse slope stability using well developed techniques than to attempt to analyse the hazard posed by internal erosion. Nevertheless for embankment dams in service internal erosion is the major threat. Safety evaluation should be related to the relative risks posed by different hazards and not the ability to carry out intellectually satisfying analyses of those hazards.

38.   Penman and Milligan (1993) have critically reviewed the performance of embankment dams and have concluded that the longevity of any embankment dam cannot be assumed over the long term, unless some degree of surveillance and maintenance exists. Charles (1993b) has recommended the adoption of the observational method, as described by Peck (1969) in the ninth Rankine lecture, for the evaluation of the safety of embankment

dams against the hazards that occur under normal operating conditions as the dam ages in service. The problems posed by the hazard of internal erosion make this the most feasible way forward for old embankment dams.

Foundation fills

39. In contrast to embankment dams the impact of foundation fills on public safety and the environment is likely to be minimal. In many cases they are used as part of reclamation schemes in areas affected by industrial dereliction. The performance requirements for foundation fills are in some ways less onerous than those for embankment dams. They are not required to sustain high hydraulic heads and internal erosion should not usually be a hazard. However as they support structures, there are often severe restrictions on acceptable movements. As it is differential settlement rather than total settlement which will damage the building, uniformity of properties will be important as well as the absolute values of those properties.

40. Where fills support structures, volume change within the fill is a major hazard. There are many potential causes of damaging ground movements. Volume changes within fills occur due to a variety of conditions and processes including the following:

(a) change in effective stress due to placement of additional fill,
(b) change in effective stress due to change in ground water level,
(c) change in effective stress due to weight of the structure built on the fill,
(d) increase in moisture content of partially saturated fill causing collapse compression,
(e) change in moisture content of a shrinkable clay fill,
(f) dynamic loading,
(g) decomposition of biodegradable fill,
(h) chemical reactions.

41. Bearing capacity and settlement due to structural loads are not usually the main concern when building on fill and the major hazard is generally associated with settlement due to effects other than structural loading. In many cases the greatest hazard to buildings on fill is associated with collapse compression on inundation. Most types of partially saturated fill are susceptible to collapse compression under a wide range of applied stress when first inundated if they have been placed in a sufficiently loose and/or dry condition. This reduction in volume can occur without any change in applied total stress. The causes of collapse compression in fills on inundation fall into 3 basic categories:

(a) weakening of interparticle bonds,
(b) weakening of particles in coarse grained fills,

(c) weakening or softening of aggregations of particles in fine grained fills.

42. When construction is about to take place on fill, susceptibility to collapse compression may be the most significant hazard as collapse compression after construction has taken place may seriously damage buildings. Inundation can result from a rising ground water level or from downward infiltration of surface water (Burford and Charles, 1991; Charles et al, 1993). An important objective of the specification and control procedures adopted for a foundation fill will be to eliminate or at least minimise volume change within the fill during and, especially, subsequent to building on the fill.

43. If settlement due to structural loads was the principal hazard it would rarely be necessary to do more than ensure that the upper 2 m to 3 m of fill had adequate load bearing characteristics. This is illustrated in Figure 1 which shows the distribution of vertical stress (σ_v) with depth, as predicted by elastic theory, on the centre-line of strip footings of different widths each with a net loading of 50 kN/m run. Only one metre below foundation level there is little difference in vertical stress despite large differences in footing width. The structural load ceases to apply a significant increase to the overburden pressure a little below 2 m depth. However damaging settlement is generally caused by effects such as collapse compression which can occur at much greater depths and shallow treatment is not likely to be adequate.

FORMATION

44. How is the required behaviour to be achieved? This question has to be supplemented by a further question. What assurance can be provided that the required properties have been achieved? These two questions reflect two basic elements of quality assurance:

(a) the provision of a product with the required quality,
(b) evidence that the product has the required quality.

Quality assurance involves an appropriate specification and rigorous control procedures. It is not possible to prevent some variability as there will be a degree of heterogeneity in the source material and some segregation during placement. It is necessary to determine how the required fill properties can be achieved with an acceptable degree of uniformity.

45. Some of the wide variety of uses of engineered fills have been reviewed. Clearly fill formation will differ substantially in some of these applications. Where there is a requirement for very low permeability or for specific filter properties, very strict specification and control procedures will be required. With fills placed to act as filters, segregation during placement may be a major hazard.

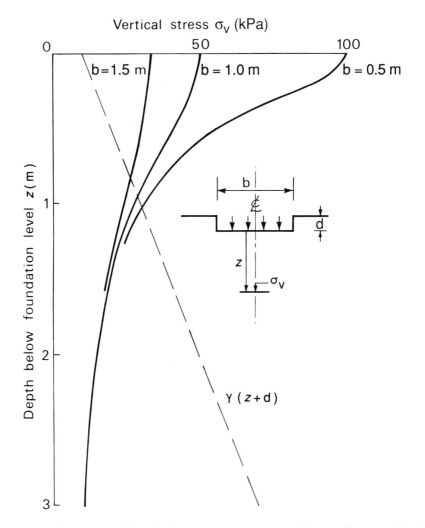

Figure 1. Vertical stress on centre-line of strip footing
b = width of footing (m)
d = depth of foundation (m)
z = depth below foundation level (m)
γ = bulk unit weight of fill (kN/m^3)
σ_v = vertical stress (kPa)

Placement and compaction

46. Placement at an appropriate moisture content in thin layers with heavy compaction is the method usually adopted to obtain the required performance from an engineered fill used in dam, highway or foundation applications. There are three basic approaches to the specification of such fills:

(a) a method specification,

(b) an end product specification,
(c) a performance specification.

47. In a method specification, the precise procedure to be used is described. The type and mass of the compactor, the number of passes and the layer thickness are all specified. These will be related to fill type. Reliance is then placed on close inspection to ensure compliance with the specification.

48. An end product specification is based on a required value for some property or properties of the fill as it is placed. The basic measurements of the in situ state of compaction are density and moisture content. However these measurements on their own are not adequate indicators and the density needs to be interpreted in terms of the density at a specified moisture content under some standard type of compaction. Alternatively, for a clay fill, the specification could be in terms of % air voids or undrained shear strength. Compliance is tested as filling progresses.

49. In a performance specification some facet of post construction behaviour of the fill is specified such as a permissible post construction settlement of the fill or a load test result. With this approach the specification is directly related to one or more aspects of the performance requirements.

50. The relative merits of the different approaches have been widely debated. The performance specification may appear to be a superior approach as it provides a direct link with performance requirements. However, although compliance with a post construction performance requirement may be easy to check, it may be difficult at such a late stage to obtain any adequate redress where non-compliance with the specification is established. Where an as placed property of the fill is specified, it can be difficult to effectively check and enforce the specification during a major earthmoving operation in which large quantities of fill are being placed very quickly. If a method specification is used, it is essential that adequate data is available on which to base the specified method so that there is confidence that a fill placed to the specification will have the required behaviour.

51. The successive editions of the Department of Transport (DOT) earthworks specification have been widely used for foundation fills. However the DOT specification was prepared for engineered fills for highway embankments and therefore may not always be appropriate for foundation fills. If a fill is to be built on, elimination of collapse potential may be the main requirement. The compaction and moisture content at which this is achieved may not be obvious and it may require careful testing to establish the critical conditions.

Quality control and testing
52. Quality control procedures should be implemented to ensure that the specification is complied with. The nature of these procedures will depend on the type of specification that has been adopted. The method specification will require close

inspection throughout fill placement. The end product approach will require an appropriate type and quantity of testing of the fill during placement. Verifiable records including adequate documentation and photographs are required for quality assurance.

Preloading

53. A non-engineered fill may be effectively converted into an engineered fill by in situ ground treatment subsequent to the completion of fill placement, although the limitations of such post-construction treatment should be recognised. Unsuitable material contained in the fill is still present after treatment. Preloading with a surcharge of fill can often form a suitable treatment method. The properties of over-consolidated natural soils are generally superior to those of comparable normally consolidated soils; preloading during their geological history has made them stiffer under subsequent applied loads than the same soil in a normally consolidated condition. Similar benefits may be obtained for fills. The beneficial effect of preloading is illustrated in Figure 2 which shows the behaviour of three samples of a sandstone rockfill, denoted (a), (b) and (c), compacted into a large oedometer to different initial densities. After an initial loading and unloading, the loosest sample (a) was reloaded and this is denoted (ar). The reloading (ar) has been replotted in Figure 2 as though it was a new sample (ie from zero strain). It is seen that the constrained modulus is much increased due to preloading and is greater for (ar) than for the first loading of the initially densest sample (c).

54. Over-consolidation can be achieved by temporarily preloading with a surcharge of fill. For uncompacted fills with large air voids, compression is largely immediate and it is not necessary to leave the surcharge in position for an extended period. The surcharge fill can be moved around the site in a continuous earthmoving operation and this can greatly reduce the amount of fill which is required and make efficient use of the earthmoving plant. The method has a wide field of application but there are some restricting factors. A relatively large area is needed for preloading with a surcharge of fill to be practical. The cost depends on the haul distance for the surcharge fill and consequently a local supply is usually required for the method to be economic.

BEHAVIOUR

55. How do engineered fills behave in practice? A few examples are presented to illustrate aspects of fill behaviour. It seemed appropriate to select English examples for an international conference held in England. Of the six engineered fills which are described, two are foundation fills, three form embankment dams, three merited inclusion in the 1969 edition of the Guinness Book of Records (McWhirter and McWhirter, 1969). Outline cross-sections of the earthworks are presented in Figure 3 to facilitate a comparison of their size and shape.

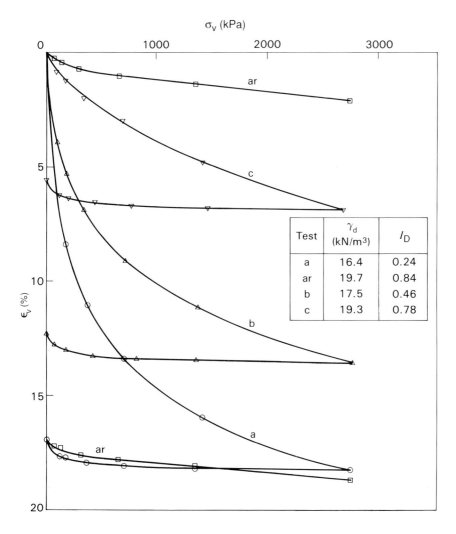

Figure 2. Compressibility of sandstone rockfill during loading, unloading and reloading
γ_d = dry unit weight of fill (kN/m^3)
I_D = relative density (density index)
ε_v = vertical strain (%)
σ_v = vertical stress (kPa)

Silbury Hill
56. The mound is part of the complex of Avebury Monuments which are located in Wiltshire. Archaeological evidence suggests that it is more than 4000 years old. A Roman road swerves to avoid it.
57. Function. This is unknown. Major excavations between 1776 and 1970 have found out much about the construction of the mound, but nothing about its purpose. It has been assumed that

55

it is a burial mound, but no graves have been found. A suggestion that it has astronomical significance has not been substantiated.

58. <u>Performance requirements</u>. With an unknown purpose, it is difficult to identify performance requirements. Durability was probably a main objective.

59. <u>Formation</u>. The mound is 40 m high, the diameter of its base is 155 m and it has a fill volume of 250×10^3 m^3. It appears to have been built in three stages but as a single project. It was carefully engineered in a series of six stepped horizontal layers created by concentric rings of chalk block walls. There are radial as well as circumferential walls with the compartments between them infilled with chalk rubble. The steps in the outer slope were infilled in the final stage of construction to give a smooth slope.

60. <u>Observed performance</u>. It appears that the mound has retained its original shape and size for over 4000 years.

61. Silbury Hill is the largest prehistoric man made mound in Europe (McWhirter and McWhirter, 1969) and has been described as one of the most remarkable early civil engineering achievements in Europe (Vatcher and Vatcher, 1976). It demonstrates the durability of well built fills.

Clifford's Tower Mound

62. The mound was built in York in 1069 by William I during his campaign to subdue the north of England. It was built on low lying ground adjacent to the river.

63. <u>Function</u>. The mound provided a suitable elevation for the construction of a stronghold. The original tower was made of timber. Clifford's Tower, as it stands today, dates from the middle of the thirteenth century.

64. <u>Performance requirements</u>. A stable foundation was required for the fortification; this was initially a timber structure but later a stone tower was built.

65. <u>Formation</u>. The mound is about 15 m high with a base diameter of 70 m and it contains some 40×10^3 m^3 of fill. It was built in horizontal layers of a fill which has been described as comprising stones, gravel and clay. As construction was part of a military campaign, speed of construction would have been a vital consideration.

66. <u>Observed performance</u>. The present irregular shape of the mound is largely due to erosion which it suffered during floods in the middle ages. Soon after the erection of Clifford's Tower, severe floods in 1315-16 softened the fill in the mound (English Heritage, 1987). In 1358 the tower was described as "cracked from top to bottom in two places". These cracks, which were repaired at great expense before 1370, are still visible. Although little was done to remedy the damage to its foundations until it was underpinned in 1903, the tower remained standing.

67. It is significant that the damaging settlement was not linked to inadequate bearing capacity of the fill or slope instability, but rather the softening of the fill due to

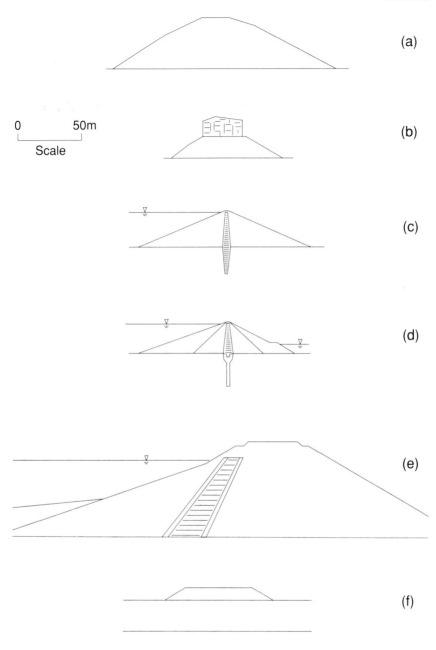

Figure 3. Six engineered fills
(a) Silbury Hill
(b) Clifford's Tower Mound
(c) Dale Dyke Dam
(d) Ramsden Dam
(e) Scammonden Dam
(f) Preloaded Opencast Backfill at Snatchill

inundation during flooding. Thus the history of the mound provides an early example of collapse compression in a fill causing serious building damage.

68. It is interesting to compare this case with the recent damage to the Round Tower at Windsor Castle which is also founded on a 15 m high mound built by William I. The fill at Windsor comprises loose to medium dense gravel sized angular chalk fragments in a matrix of remoulded chalk. Following heavy rainfall, serious foundation movements occurred in January 1988 causing cracking of walls. Ford and Chartres (1992) concluded that the most likely cause of movement was a circular slip within the mound, although it was also suggested that a limited bearing capacity failure could have occurred beneath the tower walls.

Dale Dyke Dam

69. This large embankment dam was built near the village of Bradfield, close to Sheffield in Yorkshire. It followed the traditional British form of embankment dam construction with a narrow central core of puddle clay as the watertight element.

70. Function. The embankment was built to impound a reservoir to supply water to Sheffield. The reservoir capacity was 3.2×10^6 m^3 (Binnie, 1981).

71. Performance requirements. The water retaining structure required an adequate degree of watertightness and the embankment needed to be stable and resistant to internal erosion.

72. Formation. The embankment had a maximum height of 29 m, a crest length of 382 m and a fill volume that was probably in the region of 300×10^3 m^3. This was a large embankment for its time. The upstream and downstream slopes were 1 in 2.5. The narrow vertical puddle clay core had a top width of 1.2 m, batters on both faces of 1:16, and a maximum width at ground level of 4.9 m. The fill on either side of the core was open and permeable. Embankment construction commenced in 1859. The cut-off, a puddle clay filled trench with a maximum depth of 47 m below the crest, was finished in 1861 and the embankment was completed in 1863.

73. Observed performance. By 10th March 1864 the water level was 0.7 m below the crest of the weir. In the late afternoon of the 11th March 1864 a crack was observed along the downstream slope near the crest of the dam. At 23.30 a collapse occurred and the dam was breached. The failure was recently reassessed by Binnie (1978, 1981). It is probable that failure was caused by internal erosion following hydraulic fracture of the thin clay core and cut-off.

74. The failure of the Dale Dyke Dam was the worst disaster in the history of British dam construction (McWhirter and McWhirter, 1969). In his "Complete history of the Great Flood at Sheffield on March 11 and 12 1864", Samuel Harrison described the catastrophe as follows.

"An overwhelming Flood swept down from an enormous reservoir at

Bradfield, carrying away houses, mills, bridges and
manufactories, destroying property estimated at half a million
sterling in value, and causing the loss of about two hundred
and forty human lives." (Harrison, 1864)

Ramsden Dam
    75.   The majority of the traditional British embankment dams,
which have central vertical cores of puddle clay, have a good
safety record and their long term performance has generally
been very satisfactory.   Ramsden Dam provides an interesting
example of this category of structure.
    76.   Function.   The embankment was built to impound a
reservoir with a capacity of 400×10^3 m^3 to supply water to
Batley.
    77.   Performance requirements.   The water retaining structure
required an adequate degree of watertightness and the
embankment needed to be stable and resistant to internal
erosion.
    78.   Formation.   The dam was constructed between 1879 and
1883.   The dam is about 25 m high with a crest length of 320 m.
The upstream slope is 1 in 3 and the downstream slope 1 in 2.
Selected fill was placed on either side of the central puddle
clay core.   The reservoir did not come into service until 1892
because the concrete filled cut-off trench had to be extended
to prevent water escaping around the eastern end of the trench.
    79.   Observed performance.   There are visible signs that
considerable settlement has occurred.   The Building Research
Establishment (BRE), in collaboration with the dam owner,
Yorkshire Water, has carried out extensive monitoring.   Of
particular interest are the movements which occur during
reservoir drawdown (Tedd et al, 1990).   The maximum settlement
on a complete drawdown of the reservoir in 1988 was 58 mm which
is a relatively large movement for an old dam.
    80.   In assessing the safety of old embankment dams, crest
settlement may be used as an indicator of performance.   It is
therefore important to ascertain whether or not the relatively
large movements which may occur on reservoir drawdown indicate
some malfunctioning of the dam and a potential problem or
whether they can be regarded as normal behaviour.
    81.   Based on a simple one-dimensional compression model,
Figure 4 contrasts the behaviour of the upstream fill during
reservoir drawdown for dams which have a central watertight
element with dams which have an upstream watertight element
(Tedd et al, 1994).   Where there is a central watertight
element, reservoir drawdown increases the effective stresses in
the upstream fill and produces settlement.   This is a maximum
at the crest.   Where there is a watertight element on the
upstream slope, reservoir drawdown reduces the effective
stresses in the upstream fill and produces heave, but this is
zero at the crest.   Thus a major reservoir drawdown can be
expected to cause significant crest settlement if the dam has a
central watertight element, but not if it has a watertight
element on the upstream slope.

(a)

Settlement units $\dfrac{(\gamma-\gamma')H^2}{D}$

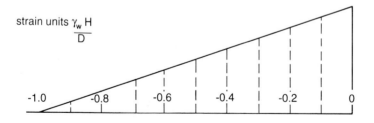

strain units $\dfrac{(\gamma-\gamma')H}{D}$

(b)

Heave units $\dfrac{\gamma_w H^2}{D}$

strain units $\dfrac{\gamma_w H}{D}$

Figure 4. Deformation of upstream fill of embankment dams
during complete drawdown of reservoir
(a) embankment with central watertight element
(b) embankment with watertight element on upstream slope
γ = bulk unit weight of fill
γ' = submerged unit weight of fill
γ_w = unit weight of water
H = height of embankment
D = constrained modulus

Scammonden Dam

82. This large modern embankment dam is located close to Huddersfield in Yorkshire. It was built of sandstone and mudstone rockfill with a rolled clay core.

83. Function. The embankment serves a dual purpose; it impounds a reservoir with a capacity of 7.9×10^6 m^3 and the M62 Pennine motorway passes over its crest. It has an unusually wide crest to accommodate the motorway.

84. Performance requirements. As a water retaining structure there are performance requirements in terms of watertightness; as a motorway embankment there are performance requirements in terms of permissible deformations.

85. Formation. The embankment has a maximum height of 76 m, a crest length of 624 m and the fill volume is 4.3×10^6 m^3. It was completed in 1969. The shoulders were built of compacted sandstone and mudstone rockfill. It has an upstream sloping rolled clay core in the upstream part of the embankment.

86. Observed performance. The embankment deformations were monitored during construction (Penman et al, 1971) and subsequently for over 20 years with the reservoir in service. The settlement of the crest is shown in Figure 5 (Charles, 1990).

87. When built this was the highest embankment dam in the United Kingdom (McWhirter and McWhirter, 1969). Although this is now no longer the case, it remains the highest embankment dam in England. The crest settlement measurements indicate a very consistent pattern of movements. The crest rests on compacted sandstone and mudstone rockfill and the large crest width and upstream sloping clay core should ensure that reservoir level fluctuations have only a minor effect on crest settlement. The measurements of crest settlement thus give an interesting illustration of the long term creep settlement of a rockfill embankment where the settlement is little affected by changes in reservoir level (Charles, 1990).

Preloaded Opencast Backfill at Snatchill

88. A large area of land in the vicinity of Corby has been used for opencast ironstone mining. Ironstone was extracted for over 100 years until 1980 when it became uneconomic. Some of this land is now required for building developments.

89. Function. At the Snatchill site the 24 m deep opencast backfill was placed without compaction by dragline excavator ie it was a non-engineered fill. Now large areas of the site are being used for low rise housing.

90. Performance requirements. Only small differential movements are acceptable at foundation level.

91. Formation. It was considered advisable to adopt some form of ground treatment prior to building on the fill. A trial area of the backfill, 50 m x 50 m in plan, was preloaded in May 1975 with a 9 m high surcharge of fill. The surcharge was placed by towed scrapers over a 3 week period, left in position for a month and then removed. Houses were built on the preloaded fill later in 1975.

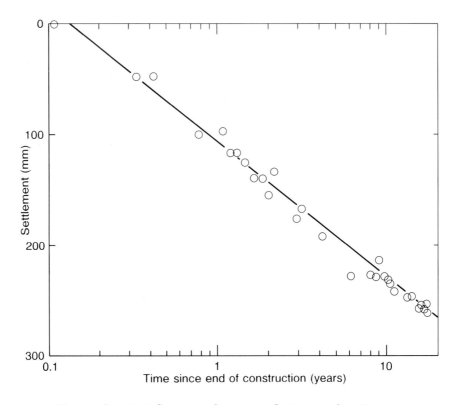

Figure 5.   Settlement of crest of Scammonden Dam

92. <u>Observed performance</u>. Substantial settlement occurred
as the surcharge was being placed. The movements measured
while the surcharge was left in position were small. A small
amount of heave occurred as the surcharge was removed. An
average surface settlement of 0.41 m was produced by preloading
and the settlement versus depth profile suggested that the
surcharge was effective down to a depth of 10 m. The stresses
produced by preloading were much greater than those
subsequently applied by foundation loads. The settlement of
the houses has been small over a 15 year period; the maximum
settlement has been 22 mm and the mean settlement has been
10 mm (Burford and Charles, 1991).
93.   This example shows the beneficial use of fills in land
deleteriously affected by human activity. Preloading can form
a very effective ground treatment technique. A large housing
development has recently taken place at Corby on 10 ha of
opencast backfill which had been preloaded (Burford, 1991).

CONCLUSIONS
94.   Large scale earthmoving has been undertaken throughout
the history of mankind and has provided many durable reminders
of previous ages. Today this type of activity, visually
intrusive and involving major modifications to the environment

particularly where large dams are built, meets increasing opposition. The effects of major earthmoving projects need to be fully assessed, but it is important that the benefits conferred by many types of engineered fills are not overlooked.

95. Where engineered fills form embankment dams which retain large quantities of water, a breach of the embankment resulting in an uncontrolled release of the water may endanger public safety. Internal erosion is one of the principle hazards for embankment dams.

96. Where engineered fills are built on, concern is usually focused on settlement of the fill damaging the buildings. Collapse compression of the fill when the fill is first inundated is a major hazard for buildings on fill.

97. In both these applications, embankment dams and foundation fills, the effects of the ingress of water into the fill are of crucial significance. In the past there was sometimes an over concentration on slope stability for embankment dams and bearing capacity for foundation fills and a consequent failure to give the above two phenomena the attention that they deserve.

98. Quality assurance is likely to become an increasingly important factor in the design and construction of engineered fills. It is important that quality control is related to those properties that have a controlling effect on fill behaviour.

99. Selected examples have been used to demonstrate hazards to the satisfactory performance of engineered fills. It is necessary to stress the durability of engineered fills in many situations and the effectiveness of preloading where fills support buildings.

REFERENCES
1. BINNIE G M (1978). The collapse of the Dale Dyke Dam in retrospect. Quarterly Journal of Engineering Geology, vol 11, no 4, pp 305-324.
2. BINNIE G M (1981). Early Victorian Water Engineers. Thomas Telford, London, 310pp.
3. BURFORD D (1991). Surcharging a deep opencast backfill for housing development. Ground Engineering, September, pp 36-39.
4. BURFORD D and CHARLES J A (1991). Long term performance of houses built on opencast mining backfill at Corby, 1975-1990. Ground Movements and Structures. Proceedings of 4th International Conference, Cardiff, July 1991, pp 54-67. Pentech Press, London.
5. CHARLES J A (1990). Laboratory compression tests and the deformation of rockfill structures. Advances in Rockfill Structures (ed E Maranha das Neves). Proceedings of NATO Advanced Science Institute, Lisbon, June 1990, pp 73-95. Kluwer, Dordrecht, 1991.
6. CHARLES J A (1993a). Building on fill; geotechnical aspects. Building Research Establishment report BR 230. BRE, Garston, Herts, 163pp.

7. CHARLES J A (1993b). Embankment dams and their foundations: safety evaluation for static loading. Dam Safety Evaluation. Proceedings of International Workshop held in Grindelwald, April 1993, vol 4, pp 47-75. Dam Engineering, Sutton, Surrey.

8. CHARLES J A and WATTS K S (1987). The measurement and significance of horizontal earth pressures in the puddle clay cores of old earth dams. Proceedings of Institution of Civil Engineers, Part 1, vol 82, February, pp 123-152.

9. CHARLES J A, BURFORD D and HUGHES D B (1993). Settlement of opencast coal mining backfill at Horsley 1973-1992. This conference.

10. CHRIMES M (1991). Civil Engineering 1839-1889: a photographic history. Thomas Telford, London, 181pp.

11. DEPARTMENT of the ENVIRONMENT (1991). Waste Management Paper No 27: Landfill Gas (2nd edition). HMSO, London, 82pp.

12. ENGLISH HERITAGE (1987). Clifford's Tower. English Heritage, London, 24pp.

13. FORD C J and CHARTRES R (1992). Monitoring bored cast in-place piles installation beneath an historic castle. Piling: European practice and worldwide trends. Proceedings of conference organized by Institution of Civil Engineers, London, April 1992, pp 262-268. Thomas Telford, London.

14. GOLDSMITH E and HILDYARD N (1984). The social and environmental effects of large dams. Volume 1: Overview. Wadebridge Ecological Centre, Camelford, Cornwall, 346pp.

15. HARRISON S (1864). A complete history of the great flood at Sheffield. Republished in 1974, Evan and Longley Associates, Dewsbury, 160pp.

16. HINDE T (1986). Capability Brown: the story of a master gardener. Hutchinson, London, 224pp.

17. HUDSON J A (1992). Rock engineering systems: theory and practice. Ellis Horwood, London, 185pp.

18. LEROUEIL S and VAUGHAN P R (1990). The general and congruent effects of structure in natural soils and weak rocks. Geotechnique, vol 40, no 3, pp 467-488.

19. McWHIRTER N and McWHIRTER M (1969). The Guinness Book of Records (16th edition). Guinness Superlatives Limited, London.

20. PEARCE F (1992). The Dammed; rivers, dams and the coming world water crisis. The Bodley Head, London, 376pp.

21. PECK R B (1969). Ninth Rankine Lecture: Advantages and limitations of the observational method in applied soil mechanics. Geotechnique, vol 19, no 2, pp 171-187.

22. PENMAN A D M, BURLAND J B and CHARLES J A (1971). Observed and predicted deformations in a large embankment dam during construction. Proceedings of Institution of Civil Engineers, vol 49, May, pp 1-21.

23. PENMAN A D M and MILLIGAN V (1993). Longevity of embankment dams - a critical review. Dam Safety Evaluation. Proceedings of International Workshop held in Grindelwald, April 1993, vol 1, pp 33-51. Dam Engineering, Sutton, Surrey.

24. PIRCHER W (1992). Geoffrey Binnie Lecture: 36 000 Large dams - and still more needed. Dams and Reservoirs, Journal of British Dam Society, vol 2, no 3, October, pp 5-16.
25. TEDD P, CHARLES J A and CLAYDON J R (1990). Deformation of Ramsden Dam during reservoir drawdown and refilling. The Embankment Dam. Proceedings of 6th Conference of British Dam Society, Nottingham, September 1990, pp 171-176. Thomas Telford, London, 1991.
26. TEDD P, CHARLES J A, HOLTON I R and ROBERTSHAW A C (1994). Deformation of embankment dams due to changes in reservoir level. Proceedings of 13th International Conference on Soil Mechanics and Foundation Engineering, New Delhi.
27. VATCHER F de M and VATCHER L (1976). The Avebury Monuments. English Heritage, London, 47pp.

4. A state boundary for unsaturated soil

S. J. WHEELER and V. SIVAKUMAR, University of Oxford, UK

SYNOPSIS

A critical state framework has been developed for unsaturated compacted clays. Included within the framework is a state boundary, defined by a single equation relating four state variables. A possible form for the state boundary equation is proposed, based on an assumption of elliptical constant suction yield curves. Test paths predicted from this proposed state boundary relationship showed excellent agreement with those observed in controlled suction triaxial tests on compacted kaolin.

INTRODUCTION

1. The ability to understand and predict the stress-strain behaviour of unsaturated soil is vital for the safe and cost-effective design of any geotechnical construction involving compacted fill. It is now generally accepted that the mechanical behaviour of unsaturated soil cannot be related to a single "effective stress" and that total stress σ, pore air pressure u_a and pore water pressure u_w must be combined in two independent stress parameters (Bishop and Blight [1], Fredlund and Morgenstern [2]). The two parameters normally selected are the "net stress" $\sigma - u_a$ and matrix suction $u_a - u_w$ (referred to hereafter simply as "suction").

2. Until very recently volume change and shear strength were treated independently for unsaturated soil. In the last few years however progress has been achieved in the development of a generalised constitutive model for unsaturated soil based on a critical state framework (Alonso, Gens and Josa [3], Wheeler and Sivakumar [4]). The intention is that such a model should be capable of predicting both volumetric and shear strains for any stress path and any drainage conditions.

3. An important feature of the proposed critical state framework for unsaturated soil is the existence of a section of state boundary linking a

Engineered fills. Thomas Telford, London, 1993

normal compression "hyper-line" to a critical state "hyper-line" (the term "hyper-line" is used here to describe a locus of states defined in a four-dimensional mathematical space by two independent equations). Data presented in an earlier paper [4] demonstrated the existence of the normal compression and critical state hyper-lines. This current paper focuses on the questions of whether there is a section of state boundary linking the normal compression and critical state hyper-lines, and if so what form this state boundary takes.

CRITICAL STATE FRAMEWORK

4. Alonso, Gens and Josa [3] proposed a critical state framework for unsaturated soil involving four state variables : mean net stress p, deviator stress q, suction s and specific volume v, where q, and v were defined in the normal way and p and s were given by

$$p = \frac{(\sigma_1 + \sigma_2 + \sigma_3)}{3} - u_a \qquad (1)$$

$$s = u_a - u_w \qquad (2)$$

Wheeler and Sivakumar [4][5] suggested that water content w should be included as a fifth state variable, but more recent evidence [6] suggests that w is not a state variable.

5. Included within the proposed critical state framework is an isotropic normal compression hyper-line, representing soil states when isotropically loaded to virgin conditions. The isotropic normal compression hyper-line would be defined by two equations relating the four state variables. Also postulated within the proposed critical state framework is a critical state hyper-line (representing soil states when sheared to ultimate or critical conditions), defined by a second pair of equations relating the four state variables.

6. Linking the isotropic normal compression and critical state hyper-lines, a section of state boundary "hyper-surface" can be postulated. This state boundary hyper-surface would be defined by a single equation relating the four state variables

$$v = f(p, q, s) \qquad (3)$$

An elasto-plastic critical state model would suggest purely elastic behaviour if the soil state were inside the state boundary defined by Equation 3 and plastic behaviour if the soil state were on the state boundary.

EXPERIMENTAL PROCEDURE

7.   A series of controlled suction triaxial tests was performed on samples of unsaturated compacted speswhite kaolin in order to examine some of the unsaturated critical state concepts proposed above, including the existence of a unique state boundary hyper-surface of the form shown in Equation 3. The test programme is described in detail by Sivakumar [7].

8.   Triaxial samples, 50 mm in diameter and 100 mm high, were prepared by compaction in a mould at a water content of 25% (4% less than optimum). All samples were compacted in 9 layers, with each layer "statically" compacted in a compression frame at a fixed displacement rate of 1.5 mm/min to a vertical total stress of 400 kPa. All samples were compacted in an identical fashion, in order to produce the same initial soil fabric in every test.

9.   The tests were conducted in two double-walled triaxial cells designed for measuring the volume change of unsaturated soil samples (Wheeler [8]). Pore water pressure u_w was applied or measured at the base of the sample via a porous filter with an air entry value of 500 kPa. Values of pore water pressure were maintained above atmospheric, using the "axis translation" principle first proposed by Hilf [9]. Pore air pressure u_a was applied at the top of the sample via a filter with a low air entry value. The cell pressure σ_3, pore water pressure u_w and pore air pressure u_a were each controlled independently by stepper motors operating regulators on a compressed air supply. The stepper motors were operated by a computerised control and logging system which enabled any required stress path to be followed, while simultaneously logging not only cell pressure, pore water pressure and pore air pressure, but also deviator load, axial displacement, sample volume change and flow of water from the sample.

10.   After setting up in the triaxial cell each sample was brought to equilibrium at a mean net stress p of 50 kPa and a suction of zero, 100, 200 or 300 kPa. The samples prepared at zero suction became saturated during this stage, but still retained a heterogeneous soil fabric produced by the initial compaction process. Each sample was then isotropically consolidated to a pre-selected value of p, while maintaining suction constant, by increasing the cell pressure at a constant rate of 0.6 kPa/hour. The data recorded during this stage indicated that each sample was at a virgin state (on the isotropic normal compression hyper-line) after ramped consolidation, having been taken to a stress level that exceeded the yield stress produced by the compaction process.

11.   Following consolidation each of the 24 samples was sheared, with the cell pressure σ_3 held constant, in one of three different ways:

(A)   Constant v/constant s shearing. In these tests the computerised control system was used to increase u_a and u_w by equal amounts (maintaining constant suction) in such a way as to keep the sample volume constant during shearing.

(B)   Constant p/constant s shearing. In these tests the control system was used to increase u_a and u_w by equal amounts (maintaining constant suction) in such a way as to keep the mean net stress constant during shearing.

(C)   Fully drained/constant s shearing. In these tests u_a and u_w were both held constant (maintaining constant suction) during shearing.

12. Shearing was conducted at a constant displacement rate that gave a time to failure of approximately 15 days, resulting in almost complete equalization of pore water pressure throughout the sample at failure.   The use of a constant displacement rate did however result in significant non-equalization of pore water pressure in the early stages of shearing, particularly in the constant volume, constant suction tests (type A) in which the deviator stress reached almost 50% of the critical state value within the first 24 hours.

NORMAL COMPRESSION AND CRITICAL STATE HYPER-LINES

13.   It was shown in an earlier paper [4] that the soil states at the start of shearing fell on an isotropic normal compression hyper-line defined by

$$q=0 \tag{4}$$

$$v = N(s)-\lambda(s)ln\left(\frac{p}{p_{at}}\right) \tag{5}$$

Atmospheric pressure p_{at} (taken as 100 kPa) was introduced in Equation 5 for dimensional consistency and to minimise any errors in the evaluation of the intercept $N(s)$.   Both the slope $\lambda(s)$ and intercept $N(s)$ of the normal compression hyper-line were functions of suction, with best fit values (derived In [6]) given in Table 1 for the four experimental values of suction.

s (kPa)	$\lambda(s)$	$N(s)$	$M(s)$	$\mu(s)$ (kPa)	$\psi(s)$	$\Gamma(s)$
0	0.128	2.052	0.813	0.0	0.110	2.011
100	0.182	2.122	0.933	54.2	0.108	1.984
200	0.196	2.196	0.959	83.5	0.181	2.042
300	0.176	2.212	0.910	122.0	0.223	2.105

Table 1. Values of soil parameters for compacted kaolin

14.   The soil states at the end of shearing fell on a critical state hyper-line defined by

$$q = M(s)p + \mu(s) \tag{6}$$

$$v = \Gamma(s) - \psi(s)ln\left(\frac{p}{p_{at}}\right) \tag{7}$$

The parameters $M(s)$, $\mu(s)$, $\Gamma(s)$ and $\psi(s)$ were all functions of suction [6], and best-fit values are given in Table 1.

EVIDENCE OF UNIQUE STATE BOUNDARY

15.   An objective of the research was to investigate whether all test paths, such as those shown in Fig.1 for eight tests conducted at a suction of 200 kPa, followed a unique state boundary.

16.   The plots of q versus p from the four constant v/constant s shear tests (type A) conducted at a suction of 200 kPa are re-plotted as solid lines in Fig. 2. Also shown in Fig. 2 (as dashed lines) are constant v contours constructed by joining appropriate data points from the two constant p/constant s shear tests (type B) and the two fully drained/constant s shear tests (type C) conducted at a suction of 200 kPa. Inspection of the individual data points in Fig.2 shows a high level of consistency between the values of v from the constant p tests and the fully drained tests (types B and C). Comparison of the constant v contours (dashed lines) with the test paths from the constant v tests shows slightly less consistency, but still a reasonable level of agreement. The differences between the constant v contours and the test paths from the constant v tests could represent a slight lack of uniqueness in the proposed state boundary. The differences could however have been caused simply by

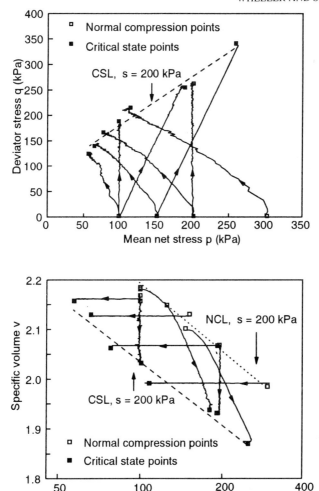

Fig.1 Test paths for shear tests at $s = 200$ kPa

the significant non-equalization of pore water pressure during the constant v tests (because of the rapid increase of deviator stress during the early part of shearing).

17. Data from shear tests conducted at suctions of zero, 100 and 300 kPa showed a similar pattern, with reasonable agreement between test paths from constant v tests and constant v contours constructed from constant p and fully drained tests (see Sivakumar [7]). This confirmed that, to a first approximation at least, all test paths traversed a unique state boundary hyper-surface defined by a relationship of the form given in Equation 3.

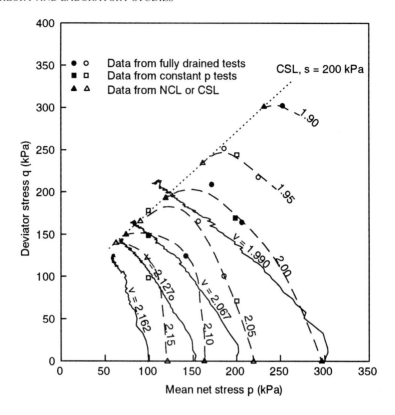

Fig.2 Constant v contours at s = 200 kPa

PROPOSED FORM FOR STATE BOUNDARY RELATIONSHIP

18. If the soil behaviour inside the state boundary hyper-surface is assumed to be elastic, any point on the state boundary is also on a yield surface in q, p, s space. Fig. 3a shows a constant suction cross-section of such a yield surface (a yield curve in the q, p plane). The yield curve forms the top of a constant suction elastic wall extending upwards from a constant suction swelling line in the v, p plane (Fig. 3b). The yield curve must therefore pass through points A and B, corresponding to the intersection of the elastic wall with the normal compression line for the relevant value of suction and the critical state line for the relevant value of suction.

19. If the elastic swelling index κ is assumed to be independent of suction, the value p_o of the mean net stress at the intersection of the current yield

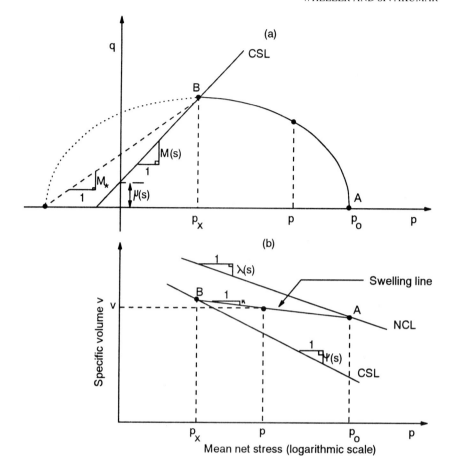

Fig.3 Constant suction yield curve

curve with the isotropic normal compression line (point A) can be related to the current values of p, v and s

$$\frac{p_o}{p} = \exp\left[\frac{N(s)-\lambda(s)ln(p/p_{at})-v}{\lambda(s)-\kappa}\right] \qquad (8)$$

Similarly, the value p_x of the mean net stress at the intersection of the yield curve with the critical state line is given by

$$\frac{p_x}{p} = \exp\left[\frac{\Gamma(s)-\psi(s)ln(p/p_{at})-v}{\psi(s)-\kappa}\right] \qquad (9)$$

73

20. A shape must now be selected for the yield curve passing through point A (with coordinates $p=p_o$, $q=0$) and point B (with coordinates $p=p_x$, $q=M(s)p_x + \mu(s)$). The most obvious choice is an elliptical yield curve with the apex at B, equivalent to the Modified Cam clay model for saturated soil (Roscoe and Burland [10]). If the ellipse is extended beyond B it will not pass through the origin in the q, p plane (see Fig. 3a), and the aspect ratio M_* of the ellipse (which is a function of s and the size of the yield curve) is given by

$$M_* = \left[\frac{M(s)p_x + \mu(s)}{p_o - p_x} \right] \tag{10}$$

21. The elliptical yield curve passing through points A and B is defined by the following expression

$$q^2 = M_*^2 \left(p_o - p \right)\left(p + p_o - 2p_x \right) \tag{11}$$

With p_o, p_x and M_* defined by Equations 8, 9 and 10 respectively, Equation 11 forms a first state boundary relationship of the type presented in Equation 3.

PREDICTION OF TEST PATHS

22. Using the state boundary relationship defined by Equations 8 to 11, together with the test conditions and the initial state of the sample, it was possible to predict complete test paths (in q, p, v space) for selected shear tests. Full details of the incremental procedure employed are given by Sivakumar [7].

23. The values of the relevant soil parameters used in predicting the various test paths are listed in Table 1. A constant value of 0.02 was used for the elastic swelling index κ, although a sensitivity analysis showed that the predicted test paths were relatively insensitive to the value of κ.

24. To predict a test path it was necessary to calculate values for p_o and p_x for all points in a shear test. In order to calculate p_o and p_x it was necessary, for substantial sections of many tests, to extend the normal compression line or the critical state line to values of p well outside the range of experimental data. Such extrapolation was clearly dangerous, and it was therefore decided to predict only those test paths for which the values of p_o

and p_x could be calculated throughout the entire test with relatively little extrapolation of the normal compression line or the critical state line.

25. Fig. 4a shows the predicted tests paths and experimental tests paths for two constant volume tests (2A and 4A) conducted at a suction of 200 kPa. The discrepancies between predicted and experimental stress paths may have been due, at least in part, to non-equalization of pore water pressure in these tests in which the deviator stress increased very rapidly during the early part of shearing. Predicted and experimental test paths for a constant mean net stress test (6B) conducted at a suction of 200 kPa are presented in Fig. 4b. The test path is plotted in q, v space, because p remained constant during shearing. Fig. 4c shows predicted and experimental test paths for three fully drained tests (26C, 9C and 17C) conducted at suctions of zero, 200 and 300 kPa respectively. Inspection of Figs. 4b and 4c indicates that good agreement between predicted and experimental test paths was achieved for type B and C tests.

26. In general therefore inspection of Fig. 4 indicates a pleasing level of success in the prediction of test paths for constant suction shear tests (bearing in mind that the experimental tests were designed to provide high quality critical state data but in some cases were conducted too fast to provide high quality test paths).

CONCLUSIONS

27. Analysis of experimental data from triaxial tests on compacted kaolin demonstrated the existence of a state boundary relationship, relating specific volume v to p, q and s. A possible form for the state boundary relationship was derived by assuming that constant suction yield curves were elliptical in shape in a plot of q versus p. This proposal for the form of the first state boundary relationship was used to predict test paths in q, p, v space for a variety of constant suction shear tests. The predicted test paths showed excellent agreement with those observed in the experimental programme.

28. The proposed critical state framework should ultimately provide a completely generalized stress-strain model for unsaturated compacted clays, capable of predicting both volumetric and shear strains for any stress path [6]. It could be used as simply a qualitative framework to aid understanding of unsaturated soil behaviour or as a formalized mathematical constitutive model within a finite element formulation [4]. The proposed model would therefore have considerable potential for application in the design and analysis of embankments, retaining structures and other constructions employing compacted clay fill.

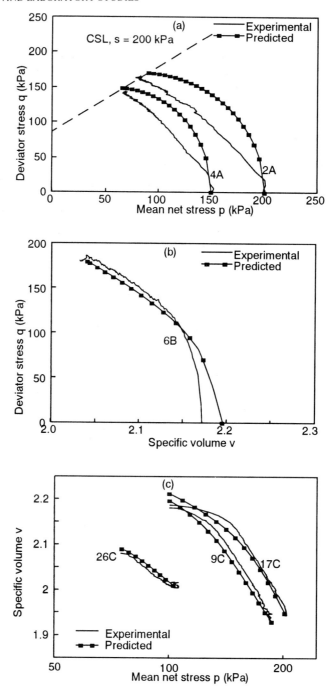

Fig.4 Predicted and measured test paths for (a) constant v tests, (b) constant p tests and (c) fully drained tests

ACKNOWLEDGEMENTS

29. The experimental research described in the paper was performed in the Department of Civil and Structural Engineering of the University of Sheffield. Financial support was provided by the Science and Engineering Research Council.

REFERENCES

1. Bishop, A.W. and Blight, G.E. (1963). Some aspects of effective stress in saturated and partly saturated soils. Géotechnique, 13(3), 177-197.
2. Fredlund, D.G. and Morgenstern, N.R. (1977). Stress state variables for unsaturated soils. J. Geotech. Eng., ASCE, 103 (GT5), 447-466.
3. Alonso, E.E., Gens, A. and Josa, A. (1990). A constitutive model for partially saturated soils. Géotechnique, 40(3), 405-430.
4. Wheeler, S.J. and Sivakumar, V. (1993). Development and application of a critical state model for unsaturated soil. Proc. Wroth Memorial Symposium on Predictive Soil Mechanics, Oxford.
5. Wheeler, S.J. and Sivakumar, V. (1992). Critical state concepts for unsaturated soil. Proc. 7th Int. Conf. on Expansive Soils, Dallas, 1, 167-172.
6. Wheeler, S.J. and Sivakumar, V. (in press). A critical state framework for unsaturated soil. Submitted to Géotechnique.
7. Sivakumar, V. (1993). A critical state framework for unsaturated soil. Phd thesis, University of Sheffield.
8. Wheeler, S.J. (1988). The undrained shear strength of soils containing large gas bubbles. Géotechnique, 38 (3), 399-413.
9. Hilf, J.W. (1956). An investigation of pore water pressure in compacted cohesive soils. US Bureau of Reclamation, Tech. Memo. 654, Denver.
10. Roscoe, K.H. and Burland, J.B. (1968). On the generalized stress-strain behaviour of "wet" clay. In Engineering Plasticity (Eds. J. Heyman and F.A. Leckie), Cambridge University Press, 535-609.

5. A new model for predicting the behaviour of partially saturated fills

G. HABIBAGAHI, Shiraz University, Iran, and M. SOULIE, Ecole Polytechnique, Montréal, Canada

SYNOPSIS. In this paper, a new model describing the stress-strain relationship of unsaturated soils is presented. The model being in incremental form with an integral term taking into account the loading history of the soil provides a unique relationship to deal with various mechanical aspects in partially saturated soils. The model is capable of predicting deformations resulted from a change in stress as well as those caused by a change in the degree of saturation. At this stage, a finite element program incorporating the model has been developed. With the aid of this program, behaviour of partially saturated soils under different loading conditions was studied. The predicted behaviour is promising and matches well with the available experimental results.

INTRODUCTION

1. Unsaturated soils may undergo excessive deformation at constant applied load if free access to water is provided. The amount of this deformation depends on different factors, such as placement condition, amount of change in degree of saturation and stress level at which free access to water is provided.

2. The behaviour of partially saturated fills is of particular importance in many geotechnical projects, examples of which are road embankments, earthfill and earth-rock dams. The importance of understanding volume change behaviour of compacted fills both during construction and afterwards is undoubted. During construction, with placement of each new layer on the top of the previously compacted layer, void ratio, and thus the degree of saturation and pore pressures change. After completion of the embankment, volume change may occur upon access to water . In the following sections, a new model is proposed which is capable of predicting volume change behaviour both as a result of change in applied stress as well as change in degree of saturation.

Engineered fills. Thomas Telford, London, 1993

BRIEF SUMMARY OF PREVIOUS WORK

3. Excessive deformation and volume change in embankments, fills, and earth dams upon an increase in degree of saturation are reported by many investigators, including Penman(1970), Jardin et al (1984), Hight and Farrar (1978) and Cox (1978). This additional deformation, also called collapse deformation, may severely influence the dam performance which may result even in piping and failure as explained by Mesri and Ali (1988). Different types of relationships for volume change prediction of partially saturated soils have been proposed by different investigators such as Radhakrishna (1967), Blight (1967), Barden et al (1969), Gili and Alonso (1988), Fredlund (1979), Nagaraj and Murthy (1985), and Toll (1990). Limitations and applicability of each of these models are well documented in literature.

PROPOSED MODEL

4. Habibagahi et al (1990) proposed the following relationship in order to predict volume change behaviour in unsaturated compacted soils:

$$d\sigma' = \{K_1(\varepsilon) + K_2(S,\varepsilon)\}d\varepsilon + \int_o^e \frac{\partial K_2(S,\varepsilon)}{\partial S} d\varepsilon dS \qquad (1)$$

where:

σ': effective stress, defined by Bishop (1959) as:

$$\sigma' = \sigma - U_a + \chi(U_a - U_w) \qquad (2)$$

S : degree of saturation

ε : strain

$K_1(\varepsilon)$: rigidity of the soil skeleton in fully saturated state.

$K_2(S,\varepsilon)$: contribution to the rigidity of soil skeleton from partial saturation.

U_a, U_w: pore air pore water pressure respectively

χ : a parameter which depends on degree of saturation, soil type, and hysteresis effect.

The integral term in equation (1) represents the effect of change in the rigidity of soil skeleton at any strain level. The mathematical form of equation (1) is evidently in disagreement with the principle of effective stress which states that any change in mechanical properties is solely related to a change in effective stress. On the other hand, in partially saturated soils an increase in the degree of saturation would result in a change in the rigidity of the soil. Hence, the soil structure before the

79

change in the degree of saturation and afterwards is not the same and the effective stress principle is not applicable for such a transition between two different soil structures. Limitation of effective stress principle in partially saturated soil has been previously reported by Jennings and Burland (1962). For detailed derivation of equation (1) and its application to 1-D problems the reader is referred to Habibagahi et al (1990). In the following paragraphs, extension of equation (1) to two dimensional problems is discussed.

5. Two dimensional plane strain in incremental form as suggested by Clough and Woodward (1967) was used :

$$\begin{Bmatrix} d\sigma'_x \\ d\sigma'_y \\ d\sigma_{xy} \end{Bmatrix} = \begin{bmatrix} M_B+M_D & M_B-M_D & O \\ M_B-M_D & M_B+M_D & O \\ O & O & 2M_D \end{bmatrix} \begin{Bmatrix} d\epsilon_x \\ d\epsilon_y \\ d\epsilon_{xy} \end{Bmatrix} \qquad (3)$$

where:

$$M_B = \frac{d\sigma'_b}{d(\epsilon_1+\epsilon_3)} \quad , \quad M_D = \frac{d\sigma'_d}{d(\epsilon_1-\epsilon_3)} \qquad (4)$$

$$d\sigma'_b = \frac{d(\sigma'_1+\sigma'_3)}{2} \quad , \quad d\sigma'_d = \frac{d(\sigma'_1-\sigma'_3)}{2} \qquad (5)$$

σ'_1 and σ'_3 are major and minor principal effective stresses. The two moduli M_B and M_D can be expressed in terms of Young's and bulk moduli as follow:

$$M_B = \frac{9K^2}{9K-E} \qquad M_D = \frac{3KE}{9K-E} \qquad (6)$$

6. Next, a hyperbolic formulation was used to find the tangent parameters, namely bulk and Young's moduli. The choice of hyperbolic formulation of the model was based upon several considerations, such as its suitability for volume change prediction and its widespread use for prediction of movements in different types of soil structures, particularly earth embankments and earth dams (Seed et al 1975). The hyperbolic bulk modulus was derived as explained by Selig (1988). Two separate sets of parameters must be used, one set corresponding to the soil in fully saturated condition which is equivalent to the rigidity $K_1(\epsilon)$ in equation (1), and another equivalent to the contribution of partial saturation to the rigidity, that is equivalent to $K_2(S,\epsilon)$ in the same equation. The second set of parameters must be based on tests on partially saturated specimens. Contribution of partial saturation is determined as the difference between results

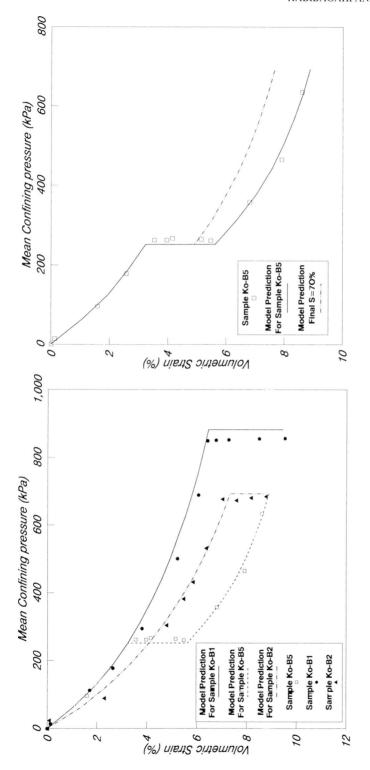

Fig. 2 - Effect of Partial Increase In Degree of Saturation On
The Amount of Collapse Deformation

Fig. 1 - Volume Change Prediction of The Model Versus Experimental
Results For Ko - Condition

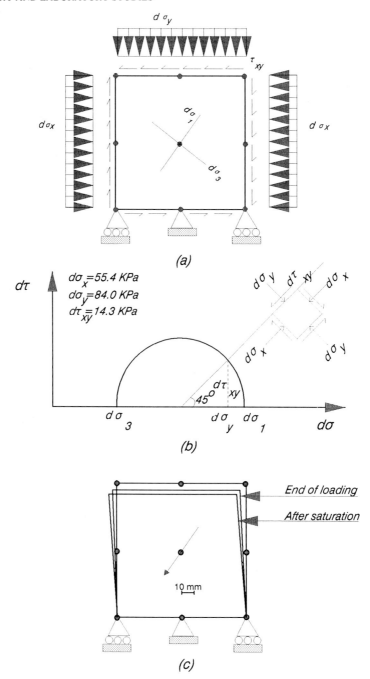

(a)

$d\sigma_x = 55.4\ KPa$
$d\sigma_y = 84.0\ KPa$
$d\tau_{xy} = 14.3\ KPa$

(b)

(c)

*Fig. 3 - Load Increments, Mohr Circle, and Distortion of The Element
For The General Loading Condition*

from similar tests but with different initial degrees of saturation.

7.   The model, as defined by equation (1) and using the hyperbolic formulation, was incorporated in a finite element computer program. In order to study the predictivity of the model, experimental data from K_o triaxial tests as reported by Radhakrishna (1967) were used. Since there were no shear strength data some of the hyperbolic parameters such as friction angle, cohesion, and failure ratio were estimated from the available soil properties and using well known empirical relationships in soil mechanics.

RESULTS AND DISCUSSION

8.   Radhakrishna (1967) performed a thorough experimental work on partially saturated compacted soil samples. The soil used was a mixture of 20% kaolin and 80% flint powder. His results from a series of K_o triaxial tests were used to calibrate the proposed model. Hence, the available data from sample K_o-B3, which was initially saturated, and sample K_o-B1 were used for the calibration purposes. Then the calibrated model was used to predict the behaviour of two other samples, namely K_o-B2 and K_o-B5. The model prediction versus experimental behaviour is shown in Fig. 1. The close agreement between the experimental data and the model prediction verifies the suitability of the model and its quantitative accuracy, bearing in mind the assumptions inherent in the model and input data.

9.   Since in practice full saturation is rarely encountered, expected behaviour for partial increase in degree of saturation was also examined. Fig. 2 shows the collapse prediction with the loading continued afterward for sample K-B5, when subjected to partial increase in degree of saturation to $S = 70\%$. As expected, the soil showed a smaller amount of collapse and a lower compressibility upon reloading compared to the case where the sample was fully saturated under the same loading.

10.   Furthermore, a soil element was tested under general state of plane loading. The applied load increment and state of stress are shown in Fig. 3a and Fig. 3b. Distortion of the soil element at the end of loading (14 load increments) and then after saturation are shown in Fig. 3c. The complete set of results for this case is shown in Fig. 4.

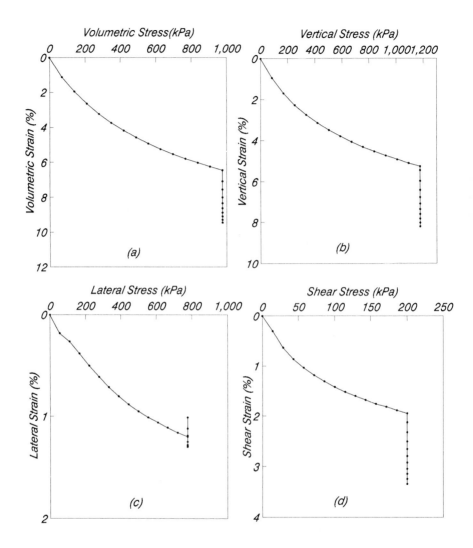

Fig. 4 - Collapse Deformations For The General Loading Condition

CONCLUSION

11. A model for volume change behaviour of unsaturated soils was presented. Suitability of the model for quantitative prediction of deformations resulted from change in degree of saturation was verified by comparison with experimental results. The finite element program developed is thought to be capable of dealing with subsidence problems often encountered in heavy structures resting on compacted fills and embankments, following wetting of the foundation. Research in this direction is currently being pursued by the authors.

REFERENCES

1. Barden, L., Madedor, A.O., and Sides, G.R. (1969), Volume Change Characteristics of Unsaturated Clay, ASCE, 95, J. of Soil Mech. and Found. Div., SM1, pp 33-51.

2. Bishop, A.W. (1959), The Principle of Effective Stress, Teknik Ukeblad, 39, pp 859-863.

3. Blight, G.W., (1967), Effective Stress Evaluation for Unsaturated Soils, ASCE, J. of Soil Mech. and Found. Eng., Vol. 93, SM2, pp 125-148.

4. Clough, R.W., and Woodward, R.J. (1967), Analysis of Embankment Stresses and Deformations, ASCE, J. of Soil Mech. and Found. Eng., Vol. 93, SM4, pp 529-549.

5. Cox, D.W., (1978), Volume Change of Compacted Clay Fill, Proc. of the Conf. held at the Inst. of Civil Eng., London, pp. 79-86.

6. Fredlund, D.G., (1979), Appropriate Concepts and Technology for Unsaturated Soils, Can. Geot. J., Vol. 16, No. 1, pp. 121-139.

7. Gili, J.A., and Alonso, E.E. (1988), Discontinuous Numerical Model for Partially Saturated Soils at Low Saturation, Numerical Meth. in Geomech., Rotterdom, pp 366-372.

8. Habibagahi, G., Soulie, M., Silvestri, V. (1990)., Proc. of Int. Iranian Cong. in Civil Eng. Shiraz University, IRAN.

9. Hight, D.W., and Farrar, D.M., (1978), Discussion: Eng. Properties and Performance of Clay Fills, Proc. of Conf. held at the Inst. of Civil Eng., London, pp 219-241.

10. Jardin, J., Baldit, R., and Delfaut, A. (1984), Constation Sur La Digue de Roussieres, Bult. Liaison Lab. Ponts et ch., 131, pp 29-44.

11. Jennings, J.E.B., and Burland, J.B., (1962), Limitations to the use of effective stress in partly Saturated Soils, Geotechnique 12, No. 2, pp 125-144.

12. Mesri, G., and Ali, S. (1988), Discussion, J. of Geot. Eng., ASCE, Vol. 114, No. 6, pp. 472-476.

13. Nagaraj, T.S. and Murthy, B.R.S., (1985), Compressibility of Partially Saturated Soils, ASCE, J. of Geot., Eng., Vol.111, No. 7, pp. 937-942.

14. Penman, A.D.M., (1970), Cracking of Clay Cores of Dams, Brit. Geot. Society, pp 115-117.

15. Radhakrishna, H.S. (1967), Compressibility of Partiaily Saturated Soils, Ph.D., Thesis, University of Waterloo, Canada.

16. Seed, H.B., Duncan, J.M., and Idriss, IM. (1975), Criteria and Methods for Static and Dynamic Analysis of Earth Dams, Proc. of an Int. Symp. held at Swansea, on Criteria and assumptions for numerical analysis of dams, pp. 564-589.

17. Selig, E.T. (1988), Soil Parameters for Design of Buried Pipelines, Proc. Pipeline Infrastructure Conf., ASCE, pp 99-116.

18. Toll, D.G. (1990), A Framework For Unsaturated Soil Behaviour, Geotechnique, 40, No. 1, PP 31-44.

6. Hydro-mechanical behaviour of partially saturated low porosity clays

M. AL MUKHTAR, J.-C. ROBINET, C.-W. LIU, University of Orléans, France, and F. PLAS, ANDRA, France

SYNOPSIS
Construction and utilisation of underground repository of radioactive waste in clay formation at high depth require the study of the behaviour of unsaturated zone around the facilities and the buffer materials placed in host medium. Partially saturated and highly compacted soils are characterized and controlled by the energy state of the interstitial water. Specialized oedometer and triaxial apparatus have been developed for controlling suction during mechanical loading and unloading. Analysis of experiment results led to the development of an elasto-plastic model with two yield surfaces, one of compressive and the other of expansive associate with hydraulic state of materials.

INTRODUCTION

The high level radioactive waste will be disposed in a designed underground repository constructed deep (may be more than 500 m in depth) in a selected geological environment, Pusch [1]; Sasaki et al [2]. Buffer materials placed between an overpack and the surrounding host medium will be initially unsaturated. Schneebeli et al [3] demonstrated the existence of an extended unsaturated zone in tunnel system built in a water-saturated granite formation due to ventilation at a relative air humidity below saturation. Unloading due to excavation in clay formation and the ventilation of these facilities induces also the desaturation, Robinet et al [4]. Ventilation of the facilities during operation contributes, to speeding up the mass transfer between the geologic formation and the surrounding air. The consequences of desaturation can be significant on the mechanical level, particularly by increasing the rigidity of the material, and on the hydraulic level, by altering the flow patterns. Buffer material proposed is a high density compacted clays, ($\gamma d > 1.6$ KN/m^3) because of its mechanical, chemical and hydraulic functions. Clay formations at a depth of more than 500 m have also a high density. It is recognized that the mechanical and hydraulic properties of clays are mainly controlled by their texture (microstructure), their mineral composition and their physico-chemical properties (interaction between interstitial fluid and solid matrix). Partially saturated and highly compacted soils are characterized by three distinct phases: solid matrix (mineral particles), liquid water and vapour.

In unsaturated soils, mainly in clay materials, if we consider all the possible forces exerted, the effective stress concept can be stated as (Lambe 1960):

$$\sigma = \sigma' + U_a + U_w + R - A$$

where:

σ: external applied forces on unit area

σ': mineral to mineral contact stress (effective contact pressure)

U_a: air-mineral or air-air contact stress

U_w: water-mineral or water-water contact stress

R & A: total interparticle repulsive and attractive forces acting across the interparticle area

The repulsive forces exerted by interacting double-layers of neighbouring stacks are counteracted by attraction caused by Van der Waals forces and also hydrogen bonds established between stacks at very small distances, as well as by external forces. There is no reliable method in the literature for a quantitative evaluation of all the parameters (particularly R and A forces) especially at close particle spacing. Israelachivili [5] demonstrates the dependence of attractive and repulsive forces on the distance between two mica surfaces in aqueous medium, figure 1.

The elementary solid particles in low density clays are highly anisotropic and, thus they tend to organize in stacks or aggregates more or less oriented. A bimodal distribution of porosity is then observed: macroporosity between the aggregates (pore size more than one micron) and microporosity inside the aggregates (pore size less than one micron). The application of mechanical stress on such materials induces the contraction of the solid matrix and subsequent decrease of the pore size due to rearrangement of particles closer to each other. Thus, in natural clays or in remoulded (compacted) clays of high density, the distribution of porosity becomes essentially unimodal, figure 2.

Bolt [6] noted that upon applying a load on water-clay system a certain amount of pore water is pressed out until the difference between the osmotic pressure of the system and the free liquid phase pressed out equals the loading pressure. When clay plates come very close, the osmotic pressure of the system might reach a value of 5-10 MPa. Hydraulic suction in unsaturated clays has an effect similar to that of isotropic stress; suction increase induces compaction on skeleton and suction decrease induces swelling. In fact, these phenomena's are irreversible and depend on the activity of clays and the state of soils (normally or overconsolidated). Hydraulic and mechanical stress induce in clays different effects due to nature (different dielectric properties) of pore fluids because of the existence of negative charges and the electrical attractive and repulsive forces on particle surfaces of clays, Sridharan et al [7].

The macroscopic behaviour of low porosity clays taking into account the influence of all these forces is analyzed in this paper. Experimental methods and specific devices developed for triaxial and oedometric paths are presented. Some aspects of numerical modelling will be introduced for highly compacted clays.

DESCRIPTION OF MICROFABRIC AND TYPES OF INTERSTITIAL WATER IN CLAYS

Based on scanning electron microscope (S.E.M) and porosimetry data, observation of natural soils, Mc Grown and Collins [8], Collins [9], indicate that two principal types of microfabric dominate in clays; matrix microfabric integrated by elementary particle arrangement of clay platelets and aggregation microfabric integrated by arrays of elementary particle configuration. Touret et al [10], figure 3, point three types of pore space:
- intra-particle pore space (inter-platelet pore)
- inter-particle pore space (intra-aggregate pore
- inter-aggregate pore

Water in clay mineral pores is bonded to the solid mineral surface by electro-chemical field, Rosenquist [11]. Hence, interstitial water shows particular physico-chemical properties. These properties are translated into different thermodynamic potentials permitting the definition of three major types of water:

i) bulk water: identified at 60-100 A° from skeleton surfaces

ii) bounded water: it 's characterized by strong links with solid matrix

    a- surface adsorption: identified at 10 A° from skeleton surfaces
- mono-molecular adsorption which corresponds to the fixation of the first water layer
-poly-molecular adsorption by piling up of several water layers with decreasing distance from skeleton
- capillary condensation obtained by the joining of the polymolecular layers.

    b- interlamellar water: identified between interlamellar space; 10 to 25 A from skeleton surfaces

iii) capillary water between bulk and bounded water

Push and Carlson [12] found that for montmorillonite of density 1,8 to 2,2 g/cm^3, the pore space is between 20 to 50 A and most of the pore water is not in "free state". Rosenquist [11] have also noted that in low porosity clays a large amount of the water is in the form of adsorbed water. Baldi and al [13] observed for Boom clays of specific weight of 17,4 KN/m^3 that interlamellar water may represent 24 to 49 % of total water volume. Push and Hökmark [14] indicate that for a saturated Na-montmorillonite of density 1,8 g/cm^2, more than 60% of interstitial water is a bonded water. Based on these results, one can conclude that: "As the porosity of clay decreased, the distance between clay platelets become smaller and an unimodal pore size distribution

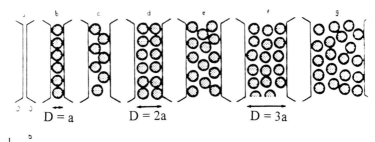

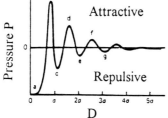

Fig. 1: Separation dependence of force between two mica surfaces in an aqueous medium (after Israelachivili 1978)

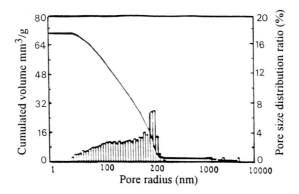

Fig. 2: Unimodal distribution of pore radius for "Bassin Parisien" clay of density 2.1 g/cm^3

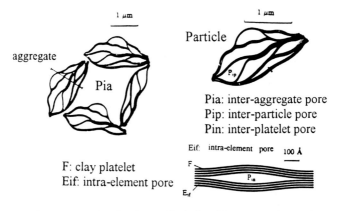

Fig. 3 Pore size in clay soils (after Touret et al 1990)

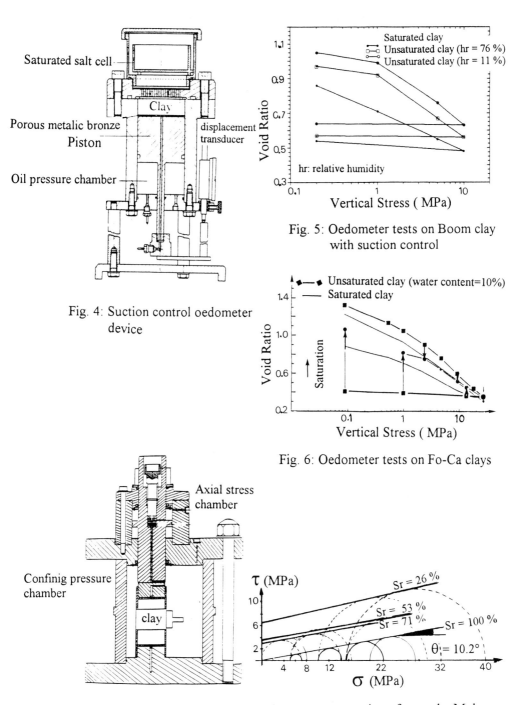

Fig. 4: Suction control oedometer device

Fig. 5: Oedometer tests on Boom clay with suction control

Fig. 6: Oedometer tests on Fo-Ca clays

Fig. 7: Triaxial apparatus for pressures up to 60 MPa. (R. Lahlou 1991)

Fig. 8: Representation of stress by Mohr circle for triaxial tests on Fo-Ca clays at different degrees of saturation

is observed. Matrix microfabric integrated by elementary particle arrangement of clay platelets dominate and a large amount of water is in the form of adsorbed in these clays. This water is held to the clay surface and is not considered part of the bulk water. The behaviour of adsorbed water and consequently that of low porosity clays is significantly altered by electro-chemical or electrical microstructural interaction".

PROGRAMME TESTS

Tests under controlled suction have been developed to describe the relationship between the void ratio, suction and stress paths of the clay materials and to develop a rheological model for the simulation of clay behaviour under various stress and hydraulic paths.

1 Suction Control in Low Porosity Clays

Different methods for controlling suction of more than -1 atm. (-0.1 MPa) in geomaterials have been developed and adapted for odometer, direct shear and triaxial apparatus. Among these methods are axe translation using a high air entry value porous stone, Escario [15] and Biarez et al [16], osmotic solution with semi-permeable membrane in contact with soils, Kassif et Shalom [17] and Delage et al [18]. Because of practical difficulties the suction maximum which can be applied with all these methods is about -2 MPa. In low porosity clays suction developed is very high (up to -100 MPa). It is known that the addition of a solute to water causes a lowering of the saturation vapour pressure over the solution. Because the extent of this decrease in vapour pressure depends on the solute and its concentration in solution, saturated solution of different salts are used to cover the required humidities (suction) variation. The great advantage of the saturated solution method is that the relative humidities so produced are fairly insensitive to temperature variation.

It is usually assumed that inside the Representative Elementary Volume (R.E.V), the phases are at equilibrium, which means that the thermodynamic potentials of the different constituents are equal. As a result, Kelvin's law can be applied to local macroscopic variables. From that moment this equilibrium is described by the isotherms of sorption/desorption which permit to link the water potential to the relative humidity or to the volumetric water content. Then with Kelvin's law it is possible to express the capillary negative pressure (suction) as a function of relative humidities.

$$h_c = -\frac{R.T}{g.M_v} Ln(h_r)$$

h_c	capillary pressure (suction)	h_r	relative humidity
M_v	molecular weight of water	R	ideal gas constant
T	temperature	g	acceleration of gravity

92

Table 1 gives relative humidity's and negative equivalent suction developed for some saturated salts and suction equivalentes.

Table 1: Saturated salt solutions for relative humidity control

Saturated solution of	Percentage relative humidity at 20° C	Capillary pressure or suction (MPa)
Potassium sulphate K2SO4	98	2.7
Potassium chloride KCl	85	23
Sodium Chloride NaCl	76	33
Calicum Nitrate Ca(NO3)$_2$.4 H2O	56	79
Calicum Chloride CaCl2.6H2O	32	158
Lithium Chloride LiCl	11	286
Phosphorus Pentoxide P2O5	$< 1 \times 10^{-4}$	>1000

2 Description of Oedometer with controlled suction

Oedometer cell, figure 4, is designed for applying pressure up to 40 MPa. Two metallic porous disc are placed at the top and bottom of clay specimen. Axial load is applied by a piston pushed by an oil pump (Gilson) . Diameter of samples tested in oedometer is of 65 mm and initial height of about 60 mm. Samples are directly prepared in the oedometer on beginning with a clay powders. The vertical displacement of the soil specimen are measured using linear voltage displacement transducer (LVDT). Oil pressure applied is measured by pressure transducer. This oedometer is adapted for the application of a controlled relative humidity on clay samples during external loading and unloading up to 40 MPa. Holes of 2 mm diameters have been realized in the cap of the oedometer to allow the transport of air at controlled humidities from the saturated salt solutions cell placed on the cap.

3 Oedometer Tests

Oedometer tests are conducted at room temperature. The experimental procedure consists of the preparation of the powder of reconstituted Boom clay (Table 2) by drying and crushing of the extracted material. For the study of fully saturated material clays is saturated with distilled water. These procedure are assumed to eliminate the natural clay stress history.

Two tests with relative humidities of 11% et 76% are presented in figure 5 and compared with an oedometer test on saturated materials. The principal results of these tests plotted in the form of void ratio versus log effective stress are:

- loading curve is composed of two parts of different slopes. The junction point of these two is termed "hydraulic consolidated stress". Method of determining this stress is the same used for the maximum past consolidated stress in the saturated non remoulded soils.

- swelling index increases with saturation. During unloading no swelling occurs in unsaturated soils.

- at constant axial pressure, void ratio decreases as suction increases.

Saturation of samples is realized at normally consolidated and over consolidated state at different and constant axial pressure as presented in figure 6 on Fo-Ca-smectit soils (table 2). It is observed that:

- saturation induces shrinking (settlement) in normally consolidated state and decreases with axial stress

- saturation induces swelling at overconsolidated state. This swelling increases as overconsolidated ratio increases

Table 2: Properties of clays

Clay	WL %	IP %	w %	S.S.* m²/g
Boom clay: Illite 25%, Kaolinite 29%, Smectite 22 % and quartz 30%	66	25	10	160
Fo-ca "Bassin de Paris": Undifferentiated mixed layer Kaolinite/smectite 80%, kaolinite 4%, non pyllosilicate mineral 16%	100	67	12	280

WL = Liquid Limit, IP = Plasticity Index, w = water content at saturation
*S.S. Spesific Surface

4 Triaxial Tests

Triaxial tests are conducted in a high pressure cell, Lahlou [19]. The vertical and confining stress were imposed by means of two separate oil pumps, figure 7. Main results of tests conducted on Ca-smectite clay at constant water content are:

- decrease in water content (suction increase) induces increase in deviator stress at failure.

- volumetric strain increase with suction and for high water content (very low suction) close to saturation, dilatancy occurs.

- representation of stress by the Mohr circle, figure 8, shows that slopes at failure are parallel and independent from suction

- at constant water content, effective stress increases as confining stress increases.

ELASTO-PLASTIC MODEL WITH HYDRAULIC HARDENING

Based on the analyses of experimental results a rheological model has been developed, Liu [20]. Alonso et al [21] have established an elastoplastic model for high porosity unsaturated clays. The Alonso model can simulate that the irreversible settlement occurs in normally consolidated state but not thus the plastic swelling occurs (mainly in active clays) in overconsolidated state. In our model the irreversible swelling observed in low porosity clays because of the existence of negative charges and the electrical attractive and repulsive forces on particle surfaces of clays is introduced by a yield surface called swelling unloading "SU". The second plastic mechanisms trait the settlement and called Loading collapse "LC". In figure 9 we present the schematisation of these two yield surfaces LC and SU. Variation of hydraulic consolidated stress (Pc*) has been approximated by a linear relation, figure 10:

$$Pc^* = k\ Pc$$

were K is a constant and Pc is the reference preconsolidation stress when the suction is zero.

The simulation of saturation at constant axial pressure in normally consolidated and overconsolidated states is done in figure 11.

CONCLUSION

Various types of microstructural investigation demonstrate that matrix microfabric integrated by elementary particle arrangement of clay platelets dominate and a large amount of water is in the form of adsorbed in low porosity clays. The behaviour of adsorbed water and as consequently that of low porosity clays is significantly altered by electro-chemical or electrical microstructural interaction.

Use of saturated salt solutions for controlling relative humidities covers the important variation of suction in low porosity clays.

An modified oedometer has been developed for controlling suction during loading and unloading. The analyses of oedometer results and triaxial tests realized at constant water content shows among other that as suction increase volumetric strain increases and deviator stress at failure, saturation at constant stress induces settlement at normally consolidated state settlement, and may cause failure. Saturation induces swelling at overconsolidated state mainly in active clay.

The elastoplastic model developed with two yield surfaces (Loading collapse and swelling unloading), simulates correctly coupling phenomena of hydraulic and mechanic stress in low porosity clays.

Acknoledgements

The authors are grateful to the "Agence Nationale pour la gestion des déchets radioactifs (ANDRA) - France for their interest and financial support for research on low porosity clays. Mr. Lecomte P. has given valuable assistance in making the new apparatus.

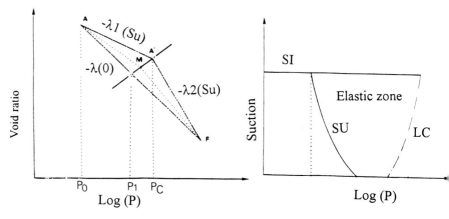

Fig. 9: Variation of hydraulic
consolidated stress (Pc) with suction

Fig. 10: Schematic representation
of two yield surfaces LC and SU

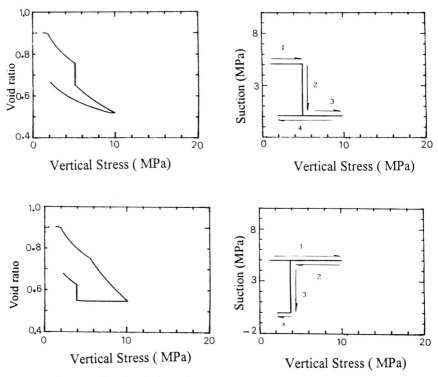

Fig. 11: The simulation of saturation at constant axial pressure
in normally consolidated and overconsolidated states

REFERENCES

[1] Puch, R. (1979), Highly cpmpacted sodium bentonite for isolating rock-deposit radioactive waste products. Nucl. Tech. 45: 153-157.

[2] Sasaki, N., Ishikawa, H., miyahara, K., Yamada, K. (1990), Development of the engineered barriers for the deep geological disposal of high-level radioactive waste. Int. High-level Radioactive Waste Management Conf. Las Vegas, Nevada, U.S.A.

[3] Scheneebeli M., Bear T., Wydler H. and Flühler H. (1991), "In situ measurement of water potential and water content in unsaturated granitic rock", Proc. of NEA-OECD Workshop on gas generation and release from radioactive waste repositories, Aix en provence pp 1-9

[4] Robinet J.C., AL-Mukhtar M., Plas, Rhattas M., Plas F. et Lebon P.(1993) Modèle de transfert de masse dans les argiles à faible porosité: application à la ventilation dans les galeries, Revue Française de Géotechnique janvier 1993.

[5] Israelachvili, J.N. and Adams G.E. (1987). Measurement of forces between two mica surfaces in aqueous electrolyte solution. Jour. Chem. Soc. 1, 74: pp 975- 1001.

[6] Bolt, G. H., 1956. Physico-chemical analysis of the compressibility of pure clays. Geotechnique, vol 6, pp 86-93.

[7] Sridharan, A. and Rao, V. 1973. Mechanical controlling volume change of saturated clays and the role of the effective

[8] McGrown, A. and Collins, K. (1975). The microfabrics of some expansive and collapsing soils. Proc. 5th. Pan. Am. Conf. SMFE1 : 323-332 Buenos Aires.

[9] Collins, K (1984) Caracterisation of expansive soil microfabric Proc. 5th. Int. Conf. Expansive Soils Adelaide 37-43

[10] Touret, O. , Pons, C.H, D. Tessier et Taroly. Y. (1990) " Etude de mla repartition de l'eau dans les argiles saturées Mg^{+2} aux fortes teneurs en eau," Clay Minerals: 25, 217-233

[11] Rosenquist , T. (1959) "Physico-chemical properties of soils in soil water system". ASCE Journal of the Soil Mechanics and Foundations Division, 85(SM2): 31-53

[12] Push R. and Carlsson, T. (1985), " The physical state of pore water of Na-smectite used as barrier component" Engineering Geology (Amsterdam) , 21 : 257-265.

[13] Baldi G., Hueckel T., Peano A. and Pellegrini (1990), "Developments in modelling of thermo-hydro- geomechanical behavior of Boom clay and clay based buffer materials", Final report for CEC, ISMES, Bergamo, Italy.

[14] Push R. and Hökmark H. (1990), "Basic model of water and gas flow through smectite clay buffers", Eng. geology n°28, pp 379-389.

[15] Escario, V.(1980), Suction controlled penetration and shear tests. Proc. 4th. Int. Conf. on expansive soils, Denver Vol. II.

[16] Biarez J., Fleureau J.M., Zerhouni M.I., et Soepandji B. (1988). Variation de volume des sols argileux lors de cycles de drainage-humidification, Revue Française de Geotechnique 41, pp 63-71.

[17] Kassif G. and shalom A. (1971). Experimental relation between swellpressure and suction. Geotechnique vol 21, no.3, pp 245-255.

[18] Delage P., Desilva G.P.R.S. and Delaure E. (1987), A new triaxial apparatus for non saturated soils" 9th Europ. Conf. Soil Mech. Dublin vol.1, pp25-28

[19] Lahlou R. (1991) Etude du comportement d'une argile gonflante et developpement d'une cellule triaxiale autonome: application au stockage des déchets radioactifs. Thèse de Doctorat - Université d'Orléans.

[20] Liu C.W. (1992) Elaboration d'un modèle elasto-plastique pour les argiles partiellement saturées avec écrouissage hydro-mécanique". Thèse de Doctorat - Université d'Orléans.

[21] Alonso E.E., Gens A. and Josa A. (1990). A constitutive model for partialy saturated soils. Geotechnique vol 40, no.3, pp 405-430.

7. Wet fills: evaporative dewatering techniqes applied to earthworks construction

B. R. THOMAS, Babtie Geotechnical, UK

SYNOPSIS. The presence of wet fills in earthworks construction may result in considerable problems during the works, often with cost implications.Wet fills cannot be compacted to the required dry density or maximum air voids content, but if moisture contents could be reduced then the materials may become acceptable. This paper considers the treatment of wet fills for use in earthworks contracts and, in particular, outlines the application of evaporative dewatering techniques which may allow the increased use of wet fills in the works,potentially resulting in substantial cost savings on large earthworks contracts.

INTRODUCTION

1. Earthworks construction problems commonly occur in the United Kingdom due to adverse climatic conditions and where wet fills comprise a major proportion of the earthworks, with interference by wet weather considered to be the principal cause of loss of output on earthmoving sites, Norman (ref.1). These problems are not new and may be particularly severe on higher ground in areas in the north and west of the country experiencing above average rainfall and low potential evaporation. Broadly speaking, rainfall amounts are strongly correlated with relief with average annual rainfall totals ranging from 500mm/pa in the south east to nearly 5000 mm/pa on some high ground in the north and west of the country.

2. For the purposes of this paper, wet fills may be defined as earthworks materials having an excess moisture content such that the material cannot be compacted to the required dry density or maximum air-voids content, but with a reduction in moisture content can become acceptable fill material. In other words the material is unsuitable only on account of its excess water content. The definition generally applies to cohesive fills and excludes materials which are unsuitable for forming load bearing fills irrespective of moisture content, principally highly organic clays and silts and peat.

3. Wet fills may occur due to the 'wetting-up' of originally suitable material in earthworks operations, where large horizontal surfaces are exposed to the elements during cutting operations. Wet fills may also occur in-situ in cut and borrow areas where the soils field capacity (equilibrium moisture content) exceeds a moisture content/plastic limit ratio of about 1.0 - 1.3, an upper moisture limit historically specified for cohesive fills.

4. It is not surprising that the estimation of suitable/unsuitable fill quantities on earthworks contracts can be problematical due to wet fill problems and it is not uncommon for large volumes of 'unsuitable' fill to be sent to tip or used for low grade fill e.g. in landscaping areas.

5. Problems associated with wet fills may include plant trafficability and a requirement to import replacement fill from new borrow areas resulting in large cost increases and claims for adverse ground conditions. Based upon the examination of a number of tenders, Farrar (ref.2) considers that on average, an allowance of about 20% of the total cost of earthworks is made for the replacement of unsuitable fill materials excluding peat.

6. This paper considers the treatment of wet fills for use in earthworks contracts and, in particular, outlines the application of evaporative dewatering techniques developed by the author (ref.3) which may allow the increased use of wet fills in the works.

TREATMENT OF WET FILLS

7. A number of methods are available for reducing the moisture content of wet fill materials:-

a) Consolidation. Consolidation may be accelerated by various well-established methods including surcharging and/or the incorporation of drainage layers within the fill (refs. 4-5). This method has been used in many dams, highways and land reclamation fills to reduce excess pore-pressures and moisture contents and to increase the undrained shear strength of the fill.

b) Mixing. Wet fills could be used after mixing with, or by placement in between, fill materials with low moisture contents.

c) Quicklime addition (stabilisation or modification). Certain wet cohesive soils can be rapidly dried with the addition of anhydrous granular quicklime which combines with up to one third of its weight of water to produce hydrated lime. The base exchange reaction results in the replacement of sodium and hydrogen ions by calcium ions leading to the release of loosely bound water, heat generation and improved soil properties. The principles and some applications of lime stabilisation are described by Bell (refs. 6-7) and the method is incorporated into Part 2 of the Department of Transport Specification for Highway Works (ref.8).

d) By evaporative drying. Grace and Green (ref. 5) consider that aeration techniques are not generally applicable for treating wet fills in the U.K. due to unfavourable climatic conditions. Work conducted by the author in Scotland (ref. 3) has demonstrated that wet soils can be successfully dewatered by evaporative dewatering techniques even in unfavourable climate conditions. Rodin (ref. 9) considers that evaporative drying of clayey or silty soils by aeration with disking and harrowing is usually impracticable in the U.K. due to the lack of prolonged dry periods, but for some low plasticity fills some drying may be practical on occasions.

8. The successful application of evaporative dewatering techniques for overseas earthworks has been described by various authors. Lieszkowszky (ref.10) describes moisture content losses from about 40% to 30% - 34% in one day for a highly plastic clay which was broken up by disk harrow to depths of 150-200mm during the construction of Fort Creek dam. Kuno et al (ref. 11) describe reductions in moisture content of about 20% for harrowed spread Kanto loam - a high plasticity clay soil used for motorway embankment construction. Grace and Green refer to the successful use of drying techniques at the Sasuma dam in Kenya.

9. Reeve (ref. 12) describes the moisture losses necessary to dry a 200mm thick layer of soil to moisture contents suitable for earthmoving on land restoration projects in the UK. The method uses a volumetric transformation between the plastic limit and field capacity to assess the time required to dry the soil to the plastic limit. This approach, whilst initially attractive in its simplicity, assumes constant rate drying and does not take into account the effects of variable evaporation or soil-moisture behaviour over time, and can only be used as a first approximation of drying behaviour.

10. There does not, however, appear to be any considerable experience of the evaporative dewatering of earthworks materials in the U.K. This may be due to reservations about climate, lack of experience, and a preconceived idea that the method can only be applied in warm climates.

EVAPORATIVE DEWATERING
Process Description

11. Water transport to the surface of bare soils under an evaporative flux occurs by two mechanisms, capillary action and diffusion. Capillary action is generally considered to be the predominant mechanism for granular soils whilst diffusion is generally considered to be the predominant mechanism for fine grained soils, and is described below.

12. A description of the drying process is fundamental to any investigation of evaporative drying. The mathematical description of the diffusive drying process is based upon the well known diffusion equation:-

$$\frac{dm}{dt} = \frac{d}{dz} (D \frac{dm}{dz})$$

Eq. 1

Where
m = volumetric moisture content (cm^3/cm^3)
t = time (days)
z = vertical co-ordinate (m)
d = soil moisture diffusivity (m^2/day)

13. An exact solution of Eq.1, which is non-linear, is not currently available. Diffusion problems are commonly solved using numerical approximations and the assumption of isothermal conditions which probably estimate total evaporation to accuracies of +/- 10% which is considered sufficiently accurate for earthworks purposes.

14. A typical soil drying curve for isothermal conditions is shown in Fig.1. Point A is the initial moisture content (Mo), point B the critical moisture content (Mcr) and point C represents the equilibrium moisture content (Me). Drying is characterised by two stages,the first stage A-B is called the constant rate drying period where drying occurs at a constant rate and is dependant upon the evaporative conditions. The second stage B-C is termed the falling rate drying period where the drying rate reduces asymptotically with time to some residual value (te). The two stages are separated by a critical point (tcr) when the surface moisture content becomes zero.

15. Drying is found to be a function of the potential evaporation at the soil surface, the initial moisture content, layer thickness and the soil moisture diffusivity. Differences in soil particle size or texture can strongly influence the soil-moisture properties and hence cumulative evaporation. Generally, coarse-textured soils will evaporate at a slower rate than fine-textured soils under both steady and cyclic evaporation, with evaporation from the finer soils approaching that of the potential rate. (refs. 13,14)

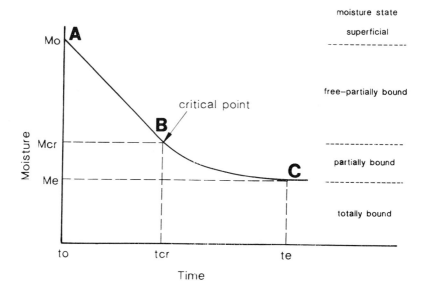

Fig 1.   Idealised rate of drying curve for isothermal conditions.

THE EVAPORATIVE DRYING MODEL
Development, Verification and Calibration
   16.   An approximate solution to the evaporative drying of thin
soil layers has been developed by Benson and Sill (ref.15). Their
model uses an integral method solution to the diffusion equation
to provide the total and profile moisture contents of the entire
soil layer for both the constant and the falling rate drying
period under isothermal conditions.
   17.   The use of material co-ordinates in the analyses greatly
simplifies the solution of the drying process with layer
thickness expressed as the thickness of solid particles only.
The layer thickness in material co-ordinates is given by

$$L=Ho/\,(1+Mo)$$ Eq. 2

Where L =   layer thickness in material co-ordinates
           Ho =   initial layer thickness (m)
           Mo =   initial volumetric content (cm^3/cm^3

   18.   Eq. 2 transforms the non-linear D(m) function into an
approximately constant coefficient per soil type, thereby
avoiding the time consuming determination of the soil-moisture
characteristics. The approximate moisture diffusivity and changes
in volumetric moisture content for thin layers can be easily
determined using laboratory drying experiments and Eqs. 3 and 4.

$$D\;=\;\frac{PEL}{3\;(Mcr-Me)}$$ Eq. 3

$$m\;=\;[\,\frac{W}{Wo}\,(1+\frac{Mo}{Gs})\,-1]\;Gs$$ Eq. 4

Where PE = potential evaporation (mm/month)
     Mcr, Me = critical, equilibrium volumetric
                    moisture content (cm³/cm³)
          Wo, W = initial, gravimetric moisture
                    content (%)
          Gs = specific gravity of soil

19. The maximum soil layer thickness which may be dried to the
base of the layer prior to crust formation (which inhibits
evaporative moisture losses) is termed the critical layer
thickness:

$$Lcr = 2\frac{D}{PE}(Mo - Me)$$

                                                      Eq. 5

Where Lcr = critical layer thickness in material co-
                    ordinates

A dimension drying intensity parameter, N (Eq. 6) is introduced
where N<2 for thin layers. If N>2 the layer is thick. The
mathematical description of the drying process for thick layers
is presented elsewhere (ref. 3).

$$N = PEL/MoD$$

                                                      Eq. 6

20. The Benson and Sill model was successfully calibrated
against sixty four 100mm diameter laboratory drying tests and
eleven field drying plots of plan area 1m² with Ho=0.25,0.50 and
0.75m with PE constant. Typical simulated and actual drying
curves are presented in Fig. 2 indicating that the model gives a
good description of the drying process under isothermal
conditions.
21. To enable use in real situations, the model was developed
to allow for variable PE (ref.3) with other input parameters
constant. The variable evaporation model was verified
mathematically prior to successful calibration against eight 1m²
field drying plots. The results indicate that scaling laws are
satisfied and the model can be used for simulation in the field
environment.

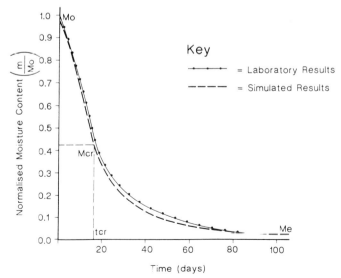

Fig 2. Actual and simulated drying curves

103

Application to Earthworks Dewatering

22. This section describes how the dewatering model may be used in practice. The example considers the case of a soil proposed for use as earthworks fill, which is unsuitable due to excessive moisture contents. The soil is a well graded clayey-silty sand with some gravel with a mean in-situ moisture content of 34%. The required moisture content loss (Wo-Wh) to improve soil properties is 20%. The commencement date for the earthworks is not known.

Answers to the following questions are required :-

   a) Can the soil be dewatered within a 'reasonable' time period and how long will drying take?
   b) What is the preferred dewatering procedure? (Layer thickness, handling etc).
   c) What are seasonal effects?

23. The input data required to perform the dewatering calculations is provided in Table 1 below. Potential evaporation can be determined from either simple pan evaporation tests or preferably from well established formulae based upon the Penman combination approach (ref.16) or by use of the Meteorological Office MORECS system,(ref.17).

Table 1. Dewatering Model Input Data

Required Information	Information Obtained from
Wo Initial moisture content	Site investigation data
Wcr Critical moisture content	{ Labortory
We Final moisture content	{ Classification
Wh Required moisture content for handling	{
BD Bulk density	{and
Gs Specific gravity	{ Drying tests
Ho Initial layer thickness	Defined by user
PE Potential evaporation	By calculation using the specific meterological data

24. Calculations were performed using the dewatering model with the following input data selected to model dewatering throughout the calender year:-

Initial moisture content   (Wo)   =   34%
Initial material height   (Ho)   =   100,250,500 mm
Potential evaporation   (PE)   =   Average monthly PE for dewatering starting at monthly intervals throughout the calendar year (MORECS data, Figure 3).

25. Output data is presented in three forms:-

a) plot of normalised moisture content vs time (the simulated drying curve, Figure 2)
b) "Checklist" summary of input data (Table 1)
c) Summary of drying parameters (Table 2)

Table 2 - Dewatering Model - Output Parameters

Symbol	Parameter description	Unit
Mo, to	Initial volumetric moisture content, initial time.	$(cm^3/cm^3$, days
Mcr, tcr	Critical volumetric moisture content, critical time.	$(cm^3/cm^3$, days
Me, te	Equilibrium volumetric moisture content, equilibrium time.	$(cm^3/cm^3$, days
m/Mo, t	normalised volumetric moisture content over time.	(days)
tf	Time for drying profie to reach bas of drying layer.	(days)
L	Layer thickness in material co-ordinates (Eq. 2).	-
Lcr	Critical layer thicknesss in material co-ordinates (Eq. 5).	-
D	Moisture diffusivity (Eq. 3).	(m^2/day)
N	Drying intensity parameter(Eq.6).	-

26. The output data was assessed using the simulated drying curves and Eq.4 to determine the predicated drying time to achieve the required moisture losses (Wo - Wh) for each initial height/month permutation. The results for the single material used in the example are summarised in Fig. 4. By inspection, the optimum initial layer thickness is about 250mm. This is confirmed from consideration of drying times, N values, initial material heights and L/Lcr ratio's, where L/Lcr = 1 is the maximum layer thickness for efficient dewatering. Caution is required in the interpretation of results. Although L/Lcr for Ho=500mm>L/Lcr for Ho=250mm (indicating increased drying performance) the drying times for the thicker layer becomes excessive as reflected by decreasing N values for increasing initail layer thickness (see Fig 5). In this case it may be beneficial to consider dewatering two layers at Ho=250mm.

Table 3. Initial material heights, L/Lcr ratios and N values

Ho (mm)	L	L/Lcr	N
100	0.0410	0.532	0.982
250	0.0794	0.689	0.611
500	0.1734	0.719	0.667

27.   Overall the results for this example calculation demonstrate that the quickest predicted drying times for a 250mm layer are in the summer months (April to August) where PE exceeds rainfall (see Fig. 3).   Reduced layer thickness may be required to achieve rapid dewatering between March to April and September to October. Evaporative drying times become excessive during the winter months (November to February) for all layer thicknesses as rainfall is high and PE is low.   It is emphasized that each set of results will be site specific.

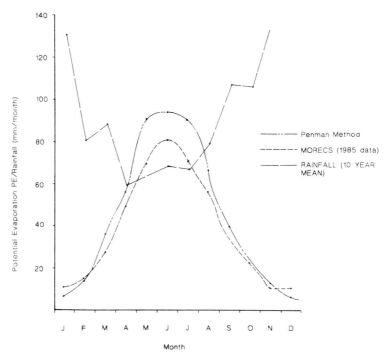

Fig 3. Potential evaporation and mean rainfall for field drying tests.

28.   Good management practices and quality control are essential during dewatering operations.   These include adequate monitoring of moisture content, selection of appropriate plant, the prevention of waterlogging and ponding by providing drainage measures and if necessary harrowing or ploughing of the soil. Surface drainage will be assisted by providing a 1:20 sloping crossfall and open drainage trenches across the drying area.

CONCLUDING REMARKS
29.   The application of evaporative dewatering techniques to enable the use of wet earthworks fill requires good management practices based upon a fundamental understanding of the dewatering process.
30.   An approximate solution to the evaporative drying of thin layers of earthwork fill is described. A variable evaporation dewatering model was developed and verified mathematically prior to calibration against laboratory and field drying experiments. The model uses easily determined soil-moisture properties offering great advantage and simplicity in process design calculations.   The model output and field dewatering experiments

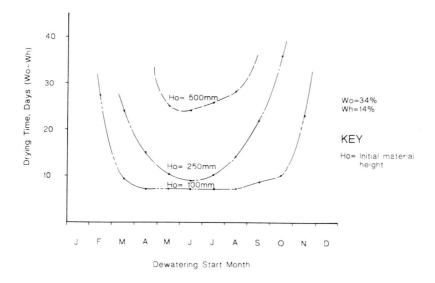

Fig. 4 Calculated drying times (Wo-Wh) for material spread at initial heights of 100,250 and 500mm.

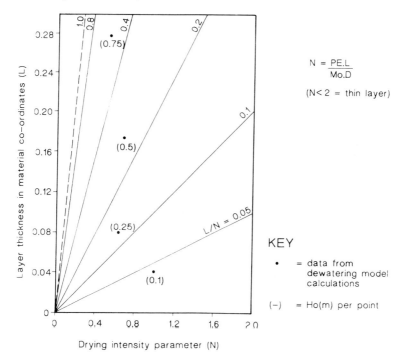

Fig. 5 L-N plot for initial material heights of 100-750mm

demonstrate that effective evaporative dewatering of wet
earthworks soils can be achieved. The approach has many
potential applications where the diffusive process operates
e.g. wastewater treatment sludges, mining waste slurries and
fine-grained dredgings, although large land areas may be required
for dewatering.

31. Correct application of the method, which will be site
specific, may allow the increased use of wet fills in earthworks
contracts.

REFERENCES

1. NORMAN, R. The effect of wet weather on the construction of
   earthworks. The Civil Engineering Research Association,
   ICE, London.1965.

2. FARRAR, D.M. A laboratory study of the use of wet fill in
   embankments. Department of the Environment, Road Research
   Laboratory, Crowthorne, 1971, Report LR406.

3. THOMAS, B.R. Clyde sediments : Physical conditioning in
   relation to use as a topsoil product for Land Reclamation.
   PhD Thesis, Department of Civil Engineering, University of
   Strathclyde, Glasgow. 1990.

4. GIBSON, R.E. & SHEFFORD, C. The efficiency of horizontal
   drainage layers for accelerating consolidation of clay
   embankments. Geotechnique,1968. vol 18, 327-335.

5. GRACE, H. & GREEN, P.A. The use of wet fills for the
   construction of embankment for motorways. Proceedings of
   the ICE Conference on Clay Fills, ICE, London,1979,113-118.

6. BELL, F.G. Stabilisation and treatment of clay soils with
   lime, Part 1 - Basic principles. Ground Engineering, 1988,
   vol 21, No.1, 10-15.

7. BELL, F.G. Stabilisation and treatment of clay soils with
   lime, Part 2 - Some applications. Ground Engineering,1988,
   vol 21, No.2, 22-30.

8. ANON. Specification for highway works, Part 2, Department
   of Transport, London. 1986.

9. RODIN, S. Earthworks - some potential aspects in the United
   Kingdom - Part 2. Muck Shift. Bulk Handler, 1964, May,
   34-41.

10. LIESZKOWSZKY, I.P. Fort Creek Dam - impervious clay core.
    Proceedings of the ICE Conference on Clay Fills, ICE,
    London, 1979, 157-164.

11. KUNO, G., SHINOKI, R., KONDO, T. & TSUCHIYA, C. On the
    construction methods of a motorway embankment by a
    sensitive volcanic clay. Proceedings of the ICE Conference
    on Clay Fills, ICE, London, 1979, 149-156.

12. REEVE, M.J. The selection and handling of soil materials
    for restoration of land to agriculture in the UK. From
    Planning and Engineering Geology, Geological Society
    Special Publication No.4, 1987, 561-567.

13. GARDENER, W.R. & HILLEL, D. The relation of external
    evaporation conditions to the drying of soils. Geophysical
    Research, 1962, vol 67, No 2, 4319-4325.

14. KEEY, R.B. Drying Principles and Practice. Pergamon Press,
    Oxford, 1972.

15. BENSON, R.E. & SILL, B.L. Modelling the drying of dredged
    material. Proceedings of the ASCE conference, Dredging and
    dredged material disposal, Florida, 1984, 866-875.

16. PENMAN, H.L. Natural evaporation from open water, bare soil
    and grass. Proceedings Royal Society London, A193, 1948,
    120-146.

17. THOMPSON, N., BARRIE, I.A. & AYLES, M. The Meteorological
    Office Rainfall and Evaporation Calculation System :
    MORECS. Meteorological Office, Bracknall, 1981,
    Hydrological Memorandum No.45

8. Forecasting the long-term acceptability potential of soils for earthworking

I. G. N. SMITH, J. OLIPHANT and S. G. WALLIS, Heriot-Watt University, UK, M. G. WINTER, Transport Research Laboratory, Scotland, and J. M. CROWTHER, University of Strathclyde, UK

SYNOPSIS. The acceptability of soil for earthworking may be characterised in terms of the Moisture Condition Value, which is related to moisture content. Annual variations in soil moisture content are affected by the features of the site and rainfall patterns. A technique has been developed, and implemented as a computer program, to predict these variations and forecast acceptability potential of soil for earthworking. The program uses an existing hydrogeological model to predict weekly changes of moisture content in the soil profile. Ground investigation and moisture condition data are combined with these moisture contents to forecast soil acceptability.

INTRODUCTION

1. In Scotland the Moisture Condition Apparatus (MCA) is routinely used for the assessment of soil acceptability for earthworking [1 & 2]. The MCA is used both in the laboratory and on site. Firstly, a laboratory determined calibration line, carried out at the ground investigation stage, gives a measure of the response of the Moisture Condition Value (MCV) to changes in moisture content. This response is described as sensitivity [1 & 2]. Secondly, at the time of earthworking, the MCA is used to give a rapid assessment of the acceptability of a soil. Typically, a soil is unacceptable if the MCV is below 8.5.

2. Figure 1 highlights the effect of a rise of 3% in the moisture content of a soil. At the particular sensitivity this change results in a decrease in MCV from 10.8 (acceptable) to 8.2 (unacceptable).

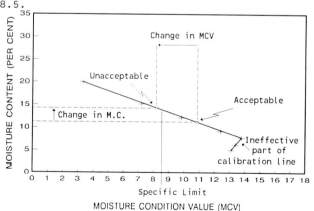

Figure 1. Changes in Moisture Content and MCV

3. Varying meteorological and hydrological conditions cause changes in the in-situ moisture content. The in-situ moisture content must be predicted to enable forecasts to be made of temporal variations in acceptability. Optimum times, or periods, for earthworking can thus be determined. A knowledge of future soil acceptability allows earthworks operations to be optimised. This leads to a more efficient use of plant, labour and materials with consequent cost savings.

4. The paper describes the development of such a predictive tool. A software package (Soil Acceptability in the Long-Term: SALT) has been devised to forecast MCV from predictions of soil moisture content. Predictions are based on historical meteorological conditions, ground investigation and hydrological data. These are achieved by using the hydrogeological model SWATRER (Soil Water Actual Transpiration Rate - Extended and Revised: [3 & 4]). The predicted changes in moisture content are then related to the MCV characteristics to yield the acceptability forecast.

SOIL ACCEPTABILITY AND RAINFALL

Soil and Water

5. The forecast of the acceptability of a soil for earthworking is considered initially in terms of meteorological data and in-situ moisture content. As meteorological conditions change, the soil-water regime is modified by the hydrological cycle. This results in changes in soil moisture content, fluctuations in the ground water flow and changes in the ground water level.

6. Figure 2 shows the types of subsurface water typically present. In the saturation zone the soil possesses a theoretical degree of saturation of 100%. Within this zone the voids are filled with water under hydrostatic pressure and as such cannot accept or hold any additional moisture.

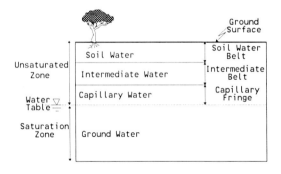

Figure 2. Types of Subsurface Water [5]

It therefore follows that the addition of water to a soil in this state will not result in a change in moisture content, unless a change in soil volume occurs. Instead, any additional water supplied in this situation results in a rise in the ground water table level (the extent will depend on the sub-soil drainage system present) and an increase in lateral ground water flow.

7. The unsaturated zone is commonly considered as comprising three smaller sub-zones as indicated in Figure 2.

110

Due to capillarity, water is drawn up above the water table (from the saturation zone) into the voids of the soil above. The region within which this occurs is the capillary fringe. The extent or height to which this water reaches depends on the material. However in general, the smaller the voids, the greater the suction pressures due to surface tension and the greater the capillary rise.

8. Above the capillary fringe is the intermediate belt. This is the belt within which some of the percolating rainfall is held within the soil mass by surface tension, adsorption, capillarity and chemical action. Excess water will move downward, due to gravity, toward the ground water table. The thickness of the intermediate belt varies with soil type and ground water level.

9. Although part of the unsaturated zone, the soil water belt may become temporarily saturated when excessive water reaches the ground surface as a result of, for example, high rainfall. The soil water belt extends from the ground surface down through the major root zone. Its thickness varies with soil type and vegetation.

Modelling in the Unsaturated Zone

10. If a slight suction (i.e., a water pressure slightly sub-atmospheric) is applied to a saturated soil, no net outflow of water from the pores is caused. However, as the suction is increased water starts to flow out of the larger pores within the soil matrix. As the suction is increased further, more water flows from the smaller pores until at some limit, corresponding to a very high suction, only the very narrow pores contain water. Increasing suction is thus associated with decreasing soil wetness or moisture content [6]. The amount of water remaining in the soil is a function of the pore sizes and volumes and hence a function of the matric suction. This function can be represented graphically as the soil-moisture retention curve.

11. Unsaturated flow processes are complicated and difficult to describe quantitatively since they often entail changes in the moisture content of the soil during flow. The most significant difference between flow in the saturated and unsaturated zones is the magnitude of the hydraulic conductivity (K). In a saturated soil all the pores are filled with water and the conductivity of the water is at a maximum. In an unsaturated soil however, some of the pores are filled with air and so the conductivity of the water is less. Furthermore, as suction develops, the first pores to empty are the large ones which are the most conductive. This in turn leaves only the smaller pores conducting the water. Thus as the conductivity of the soil changes, so does the moisture content. The most important physical soil properties for water movement in the unsaturated zone therefore, are the relations between the suction or matric head (h), the

111

volumetric moisture content (θ)[1] and the hydraulic conductivity (K).

Laboratory and In-Situ MCV Characteristics

12. If a natural MCV point for a non-granular material is plotted on the MCV/moisture content plane for the same material, it will in almost all cases fall above the calibration line (Figure 3). This is generally considered to arise from sample preparation. Each sample in the calibration test is dried out and reconstituted, whereas the natural moisture content sample is not. It appears, from a preliminary study, that the more granular the material the closer the natural point is to the calibration line. For a more clayey soil, the offset of the natural MCV point from the calibration line can be quite significant. Typically, this corresponds to 2 to 3% of moisture content. The relation between the offset of the natural point from the calibration line must therefore be taken into consideration.

13. In Figure 3 the in-situ sensitivity is plotted as a broken line through the natural moisture content MCV point and with the same slope as the laboratory calibration line. It is plain that the values of 'laboratory' and 'in-situ' MCVs are quite different for the same value of in-situ moisture content (MCV_2 versus MCV_1).

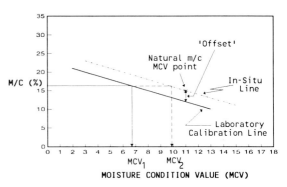

Figure 3. Laboratory and equivalent In-Situ MCV calibration line

While investigations into the relations between the in-situ MCVs and calibration lines are ongoing, parallelism of the two lines is assumed.

THE COMPUTER PROGRAM

14. The program has been developed to assist engineers with the planning of earthworking on trunk road contracts in Scotland. However, its potential application is much wider. Data input requirements have been minimised as far as possible and output screens designed to be clear and concise.

---

[1] The volumetric moisture content, θ is used rather than the mass, or gravimetric, moisture content as it is more adaptable to the computation of fluxes and water quantities added to the soil by rainfall, or removed by evaporation, root uptake and drainage.

These enable simple interpretations to be made of acceptability potential across a site at different times of the year. A comprehensive, context sensitive help facility is also present.

Input Data

15. The relations between rainfall and other factors to soil acceptability are shown in Figure 4. The input data are entered into the system and held in data files prior to analysis. A single analysis considers moisture content changes in a single soil profile (e.g., a trial pit or borehole). A series of analyses for several boreholes

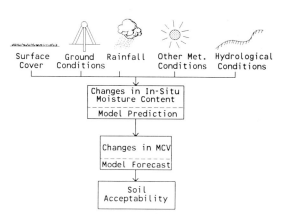

Figure 4. Relations between rainfall & other factors to soil acceptability

across the site are used to construct an overall picture of acceptability. Ground data are obtained from the ground investigation records and meteorological data from the Meteorological Office.

Soil Specific Data

16. The soil data required can be considered to comprise three categories; basic soil properties, soil suction, and MCV.

17. Basic properties such as in-situ moisture contents, dry density values and ground water level, determined from the ground investigation, enable an 'initial condition' state to be established. The analysis used in the model starts from the date of the ground investigation when these initial conditions were applicable.

18. Results from soil suction tests provide the required moisture retention characteristics. Although the suction test is uncommon in ground investigations, the characteristics are fundamental to the modelling processes. It may be that ground investigations in future will have to collect such data. However, as an alternative to the suction test, standard soil-moisture retention curves for several different types of soil have been prepared as part of this study. This allows the user to select the relevant curve, from a selection offered, for the soil type of interest. The appropriateness of this simpler alternative is being reviewed to assess its suitability.

19. The in-situ MCV line is used to forecast MCVs from the moisture contents predicted. These MCVs are then compared to

pre-set limits to give acceptability for earthworking (Figure 4).

Meteorological Data

20. The meteorological data required can be obtained by the end-user from the Meteorological Office in the form of MORECS (Meteorological Office Rainfall and Evaporation Calculation System) [7] output. This takes the form of weekly and fortnightly values of calculated potential evapotranspiration, actual evaporation and soil moisture deficit (SMD) for different surface covers on 40 x 40km grid squares across Great Britain.

21. Predictions of moisture content require values of anticipated weather conditions. This anticipated data takes the form of statistically adjusted historical meteorological data for the site under analysis. The data is used with the actual MORECS data to give meteorological data for the following 12 month period.

22. The use of MORECS is of three-fold benefit. Firstly, the MORECS model used by the Meteorological Office is widely accepted as producing accurate estimations of evaporation and SMD in periods of non-drought [8]. Secondly, MORECS output is compatible with the SWATRER model's meteorological data requirements. Finally, the MORECS output is a relatively inexpensive means of obtaining meteorological data. The cost of the individual items of data that are contained within a MORECS calculation greatly exceeds the cost of the MORECS output itself.

Surface Vegetation Data

23. The role of the surface vegetation within the soil-plant-atmosphere system is of major significance especially in the unsaturated zone. If rainfall fails to replenish the water supply in the soil, the soil moisture becomes depleted under the demands of the vegetation and leads to a soil moisture deficit. The soil moisture deficit is the amount of water required by the soil to bring it up to field capacity (the volumetric water content of the soil after the saturated soil has drained to equilibrium). Thus, the amount of water in the soil is dependent on the soil moisture deficit. This, in turn, is dependent on vegetation type and the particular point in time within the growing season. Information on the type of vegetation present is therefore required.

24. The vegetation data required is specific and includes rooting depth and leaf area index. These parameters are not normally used in the design or planning of civil engineering works. To this end, routines have been developed to simplify the data input.

Analysis and Forecasts

25. The nature of the analysis is indicated in Figure 4. The moisture contents are predicted using SWATRER, a one-dimensional numerical model for the unsaturated zone, which

considers water uptake by roots. It solves the Richard equation using an implicit finite difference technique [9 & 10]. The model has been modified slightly to make it more adaptable to the present needs. The forecasts of acceptability are made by relating the predicted moisture contents to the MCV.

26. All forecasts are based on weekly averages of moisture content and subsequent MCVs. Forecasts of the acceptability can be displayed in two formats. Firstly, as in-situ MCV (for any week after ground investigation) superimposed as a natural point on the laboratory determined calibration line. Secondly, as weekly MCVs for the 52 weeks following ground investigation displayed as a bar chart of MCV against week number. (An example of the latter is shown in Figure 5.)

27. The first format allows the user to relate in-situ MCV to the sensitivity of the soil. The user can also examine the trend of soil acceptability from week to week. If preferred, MCVs can be averaged over 2, 4 or 8 weeks. The MCV trend proves particularly useful if the material needs to be worked during a particular period of time and only the mean acceptability for the whole period is of interest. The second format could be used to assist in site planning. Particular periods of time when the material is most likely to be unacceptable become apparent from the display (Figure 5).

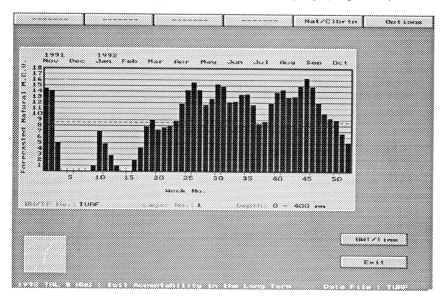

Figure 5. Screen display of weekly forecast MCVs for the 52 weeks following ground investigation

DISCUSSION

28. In its present state of development, the program can be used for planning earthworking contracts. At the time of

115

writing, full validation and calibration of the program is incomplete.

29. Initial validation was performed using test site data obtained over a period of four months. Weekly measurements of moisture content and MCV were taken and meteorological conditions monitored using an automatic weather station. The moisture contents and MCVs recorded are compared with those predicted and forecast by the program (Figure 6). The predicted moisture contents show good agreement with the measured values and predictions are generally within 2% of the in-situ moisture content. The accuracy of moisture content predictions obtained during initial validation work indicates that the SWATRER model shows considerable promise.

30. In contrast, the forecasts of MCV by the SALT model give rather poor agreement with the field measurements. However, the soil used in the initial studies has a sensitivity of 4.0 (MCV/% moisture content). This corresponds to 'very high' sensitivity [1].

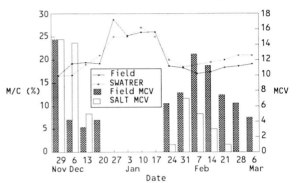

Figure 6. Comparison of forecast and measured results

Hence, even a small increase in moisture content causes a relatively large decrease in MCV. By the same token, for a small difference between predicted and measured moisture contents, there follows a significant difference between forecast and measured MCVs. This highlights the difficulty in forecasting long-term acceptability for highly sensitive soils. For soils of low and moderate sensitivity (less than 2.0) the difference is expected to be slight. Further validation will be carried out to assess the overall accuracy of the forecasts and to determine the limitations of the program. This will also enable routines to be devised to attach a statistical degree of confidence to the forecasts.

SUMMARY

31. A computer program for the forecast of soil acceptability has been developed. The program requires soil properties, surface vegetation and meteorological input data. Field water analysis is by the SWATRER model which gives predictions of soil moisture content. These predictions are then related to MCV to achieve a forecast of soil acceptability.

32. Initial validation work indicates that the program gives predictions to within 2% of moisture content. However even at such levels of accuracy, the forecast MCVs for

sensitive soils can be significantly at variance with measured field values. Further validation, in particular improvements to the input data, is expected to further increase the accuracy of the moisture content predictions and MCV forecasts. Even at current levels of accuracy, reasonable forecasts of MCV may be expected for relatively insensitive soils.

ACKNOWLEDGEMENTS

33. Crown Copyright 1993. The views expressed in this paper are not necessarily those of the Department of Transport or the Scottish Office Roads Directorate. Extracts from the text may be reproduced, except for commercial purposes, provided the source is acknowledged.

34. The work described in this paper forms part of a Scottish Office Roads Directorate research programme conducted by the Transport Research Laboratory, Scotland, (Director, Dr. G.D. Matheson), and the paper is published by permission of the Scottish Office Roads Directorate and the Chief Executive of TRL.

REFERENCES

1. SCOTTISH DEVELOPMENT DEPARTMENT. The use and application of the moisture condition apparatus in testing soil suitability for earthworking. Scottish Development Department, Edinburgh, 1989.

2. MATHESON G.D. and OLIPHANT J. Suitability and acceptability for earthworking with reference to MCA testing of quaternary soils in Scotland. in Quaternary Engineering Geology, A. Forster et al. (Eds.), Engineering Geology Special Publication 7, Geological Society, London, 1991.

3. FEDDES R.A., KOWALIK P.J. and ZARADNY H. Simulation of Field Water Use and Crop Yield. Pudoc, Wageningen, 1988.

4. DIERCKX J., BELMANS C. and PAUWELS P. SWATRER: A computer package for modelling the field water balance, Reference manual. Laboratory of Field and Water Engineering, Catholic University of Leuven, Belgium, 1986.

5. SMITH G.N. Elements of Soil Mechanics, Sixth Edition, BSP Professional Books, 1990.

6. HILLEL D. Soil and Water: Physical Principles and Processes, Academic Press, 1971.

7. THOMPSON N., BARRIE I.A. and AYLES M. The Meteorological Office Rainfall and Evaporation Calculation System: MORECS, Hydrological Memorandum No. 45, Meteorological Office, 1981.

8. GARDNER C.M.K. and FIELD M. An evaluation of the success of MORECS, a meteorological model, in estimating soil moisture deficits, Agricultural Meteorology, 1983, vol. 29, 269-284.

9. FEDDES R.A., BRESLER E. and NEUMAN S.P. Field test of a modified numerical model for water uptake by root systems, Water Resources Research, 1974, vol. 10, pt. 6, 1199-1206.

10. BELMANS C., WESSELING J.G. and FEDDES R.A. Simulation model of the water balance of a cropped soil: SWATRE, Journal of Hydrology, 1983, vol. 63, 271-286.

9. The swelling behaviour of compacted clayey backfill

K. O'CONNOR and R. N. TAYLOR, City University, London, UK

ABSTRACT: Several different types of laboratory experiments are described which were designed to investigate the key factors affecting the development of swelling pressures in compacted clayey soils. The different apparatus used included a stress path cell, an oedometer and a geotechnical centrifuge. The results demonstrated the importance of determining swelling pressure under comparable conditions and constraints as would apply in the full scale situation.

INTRODUCTION

1. On the majority of construction sites where earthworks are involved unsuitable clayey soil is removed and granular backfill imported. This common practice is becoming more costly as the availability of granular fill reduces and the costs of transport increases. The unsuitability of compacted cohesive fills arises from their behaviour which can be difficult to characterise. As the moisture content of the compacted soil changes after construction the soil may tend to swell or compress leading to significant changes in stress on, for example, adjacent structures. An example of the detrimental effects of such adverse swelling pressures on a retaining structure was reported by Symons and Murray (ref. 1).

2. The current specification in the U.K. (ref. 2) requires that when clayey (often referred to as cohesive) soils are to be used for backfill adjacent to structures the properties and state of the soil must lie within certain limits. The soil should have a liquid limit less than 45% and plasticity index less than 25% Also, at the end of compaction the dry density of the soil must be at least 100% of the maximum dry density (as specified by the Proctor compaction test B.S.1377) or less then 5% air voids - which ever gives the lower dry density at the field moisture content. The placement moisture content generally has an upper limit of 1.1 PL to allow trafficability by compaction plant and a lower limit of PL-4 below which collapse and settlement of the compacted soil structure becomes significant as the in situ moisture content increases. Figure 1 shows the constraints within which the soil can lie.

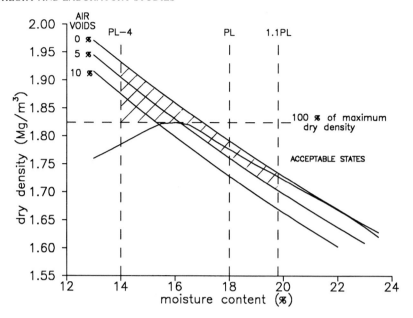

Fig. 1. Current U.K. specification for compaction
of backfill class 7A. Standard Proctor compaction
with 2.5 kg rammer. Soil: Brickearth.

3.   This specification restricts the types of material that
can be used as backfill. A costing study (ref. 3) indicated that
substantial reductions in the cost of backfill could be made if
the specification was widened.

4.   The research described is aimed at investigating the
factors affecting the build up of swelling pressure in compacted
clayey fills. A variety of experimental techniques have been
used to assess the swelling behaviour of three different soils
under different initial conditions.

EXPERIMENTAL INVESTIGATIONS

5.   Swelling pressure is defined as the pressure required to
hold the soil at constant volume when water is added. There are
however a number of different methods for arriving at this value
(ref. 4). Each satisfies the definition but gives quite
different swelling pressures (ref. 5). Methods for determining
swelling pressure have generally involved variations of the one-
dimensional consolidometer test where the soil is subjected to
constraints in the radial and some times the axial direction.
On adding water, measurements of pressures and strains are made.
The magnitude of pressures obtained in this way vary with the
constraints on the soil during the experiment. A larger degree
of confinement produces greater swell pressures.

6.   The experimental investigations have aimed to examine the
differences in swelling pressures determined in tests which have
a high degree of confinement and tests that reflect the

conditions that exist behind a stiff retaining wall. (The purpose of modelling a stiff retaining wall is that the swelling pressures would be greater than for a flexible wall). The test data may then be used to develop a relationship between the different types of test and form a predictive method for determining the swelling potential of clayey fills in such circumstances.

SOIL TESTED

7.   The three soils used in this investigation are described in Table 1. London Clay is a well known soil and was chosen for its high plasticity and known swelling potential. It was also used by the Transport Research Laboratory (TRL) in a pilot study investigating the development of swelling pressure (ref. 6). Brickearth is a silty clay with a plasticity which lies towards the limit of acceptability for clayey backfill. Wadhurst clay is a natural clay with a plasticity which lies between Brickearth and London Clay and just outside the current Department of Transport recommendations.

	LL	PL	PI	G_s
London Clay	78	29	50	2.78
Wadhurst Clay	57	24	33	2.70
Brickearth	38	18	20	2.72

Table 1. Classification parameters of soils tested

LABORATORY STUDIES

8.   A number of different experimental techniques have been used to measure and compare the swelling pressure developed by the clays described above.

9.   100 mm Stress Path Apparatus. Tests on 100mm diameter samples were performed using a computer controlled stress path apparatus which allowed compacted soil samples to be confined with close to zero lateral displacement, while pore suctions induced during compaction were allowed to dissipate by a gradual water inflow. A schematic diagram of the stress path cell is shown in Figure 2.

10.   Water under a small pressure equivalent to approximately 1m head could flow into the sample using a central porous pipe. The sample was permitted to swell or consolidate vertically under a constant axial stress of 50kPa, corresponding to a total stress level of approximately 2-3m of overburden. Any tendency to deform radially was monitored by local instrumentation and was counteracted by the control program which adjusted the radial stress (cell pressure). These tests were designed to simulate the condition that would be experienced by an element of soil behind a stiff retaining wall.

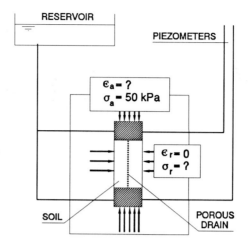

Fig. 2.   Schematic diagram of 100mm stress path

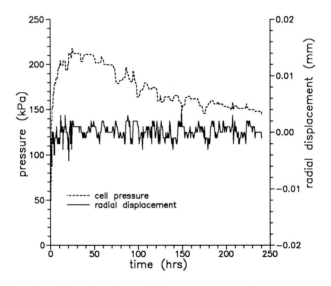

Fig. 3.   Result from 100mm stress path apparatus

11.   Data from a typical test for London Clay are presented in Figure 3. After reaching the initial isotropic confining pressure of 50kPa the computer control maintained the axial stress constant at 50kPa while adjusting the radial stress in order to maintain the average radial strain near to zero. The typical test response was for a rapid rise in the swelling pressure which reached a maximum and then reduced as the soil continued to soften and strain in the axial direction. The same type of behaviour was recorded in a pilot scale study performed at TRL (ref. 6).

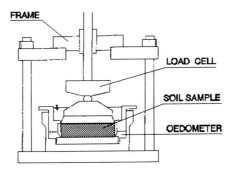

Fig. 4.   Schematic diagram of modified oedometer apparatus

12.   Modified Oedometer.   A standard oedometer cell was mounted in a relatively stiff frame as shown in Figure 4. Samples of soil compacted to the same initial states as for the 100mm stress path cell tests were flooded with water and the development of swelling pressure recorded on a load cell between the top platen of the oedometer and the surrounding frame.

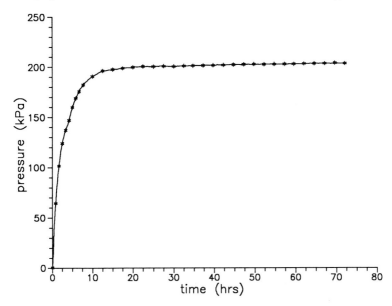

Fig. 5.   Test result from modified oedometer

13.   The experiment commenced from a state of near zero confining vertical stress. The purpose of these tests was to establish a correlation between the controlled stress path test designed to mimic field conditions and the relatively simple modified oedometer test in which the soil was confined in all directions. The pressure time relationship for an oedometer test

is shown on Figure 5. On adding water to the soil the confining
pressure increased to a constant maximum value.

14.   Computer Controlled Oedometer.   Complementary to the
above, the behaviour of compacted fills at low dry densities and
at varying initial conditions was examined using the same
oedometer cell but placed in a computer controlled apparatus
where it was possible to measure and control the axial stresses
and strains throughout the wetting process.

15. The first series of experiments investigated the influence
of dry density on the change in volume of the soil during
wetting. Each of the samples was compacted to a different
initial dry density (at constant moisture content) and was then
compressed to the same initial vertical stress of 50kPa. Then,
while keeping the vertical stress constant, the change in axial
strain was measured as water was added. Figure 6 presents data
for a series of tests on Wadhurst clay with a constant vertical
stress of 50kPa. By interpolation it is possible to determine
a dry density at which the clay, would not undergo a change in
volume when wetted. The implication of this graph is that there
is a combination of dry density and initial stress state at
which the soil will undergo a zero volume change and will not
exhibit a swelling pressure as initial suctions are allowed to
dissipate.

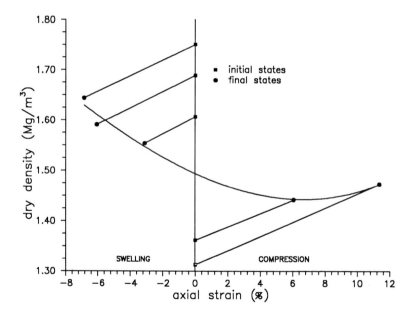

Fig. 6.   Variation of strain with initial dry
density at same vertical stress of 50 kPa. Soil:
Wadhurst clay, compaction m_c=20.3%

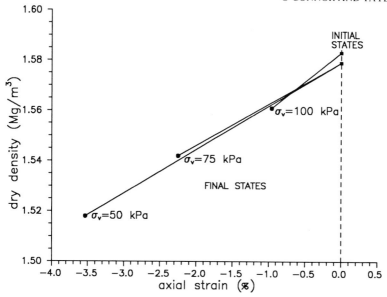

Fig. 7. Variation of strain with initial dry density at different vertical stresses. Soil: Wadhurst clay, compaction m_c=20.3%

16. A second series of experiments used five soil samples at a similar initial dry density (at constant moisture content). These were compressed to different initial pressures prior to wetting. This simulates the conditions that would apply at different horizons behind, for example, a stiff retaining wall. Figure 7 suggests that for a particular dry density and moisture content there is a vertical stress at which there would be no volume change and hence no swelling pressure developed.

17. <u>Centrifuge Experiments</u>. A series of centrifuge model tests was designed to study the development of swelling pressure on idealised retaining walls. Since the dissipation of pore pressures following compaction will not be uniform with depth, it is important to investigate any stress redistribution that occurs and to correlate the development of swelling pressures on correctly modelled retaining walls with those determined from the single element tests.

18. A stiff wall was used to indicate an upper limit on swelling pressure. The wall comprised 4 horizontal segments each 40 mm high and supported by 3 load cells (Figure 8). Figure 9 shows the arrangement of the model wall and the strong box. After manual compaction of the soil behind the wall, wick drains were inserted into the soil to reduce the drainage path lengths and speed up the swelling process. The package was taken to the test acceleration and water was then introduced at the surface of the soil. The ensuing horizontal swelling pressure was monitored by the load cells. As in prototype retaining walls the soil was allowed to strain vertically during swelling and this was recorded by a number of displacement transducers.

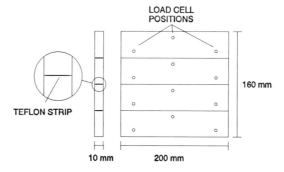

Fig. 8.   Model retaining wall

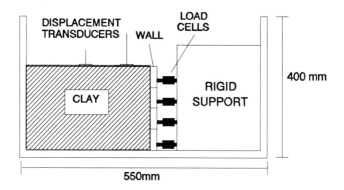

Fig. 9.   Centrifuge model configuration

19.   The variation of swelling pressure with time is shown on Figure 10. During the compaction of the clay the wall was supported as it was considered necessary to protect the load cells from any excessive force that could damage them. The values of average pressure on the wall are the changes in pressure that occurred from commencement of the swelling stage when water was first added to the surface of the model. All panels of the wall detected rises in swelling pressure which reached maximum values after about 10-20 hours. The greatest swelling pressures were measured at the top and bottom panels which were closest to the source of water. The swelling pressures tended to reduce after reaching maximum values, particularly at the top panel (section 1). At this high level, the in situ stresses were relatively low and substantial softening of the clay fill could occur.

20.   From the scaling relationships that apply when using centrifuge testing techniques it can be shown that linear dimensions scale from model to prototype as 1:n where n is the multiple of Earth's gravity used for the test acceleration. Pressure has a scale factor of 1:1 and the time dependant seepage or diffusion of water through soil has a scale factor of $1:n^2$. This implies that the swelling pressures obtained in a

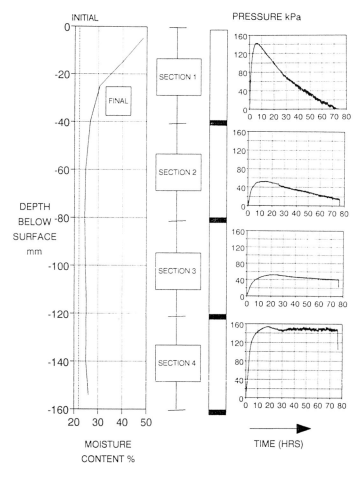

Fig.10. Variation of swelling pressure with time and change in moisture content during centrifuge test. Soil: London Clay, compaction m_c=22%

centrifuge experiment at 38g lasting for 3 days on a model wall 160mm high represents the development of swelling pressures on a 6m high prototype retaining wall over a period of 12 years.

DISCUSSION

21. By collating the results from the 100mm stress path apparatus and modified oedometer a number of relationships can be seen. Figure 11 shows the variation of swelling pressure with initial dry density for the tests performed on London Clay. The value of swelling pressure shown is the swelling pressure in excess of the initial confining stress. Also marked on the figure is the range of swelling pressures that were exerted on the centrifuge model retaining wall. A similar relationship exists between the swelling pressure and initial moisture content.

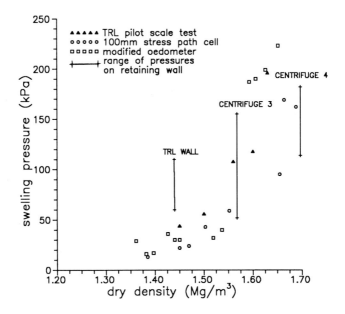

Fig.11.   Swelling pressure vs. dry density for London Clay tests

22.   These graphs indicate that it is possible to use the relatively quick and easy oedometer experiment to give a good indication of the swelling tendency of a potential backfill material.

23.   Figures 6 and 7 suggest that for the case of a stiff retaining wall there is an initial stress state (corresponding to a depth) below which compacted fill with a known dry density and moisture content will neither swell nor compress. The implication is that by controlling dry density and moisture content of the fill as it is compacted behind a retaining wall the development of swelling pressures can be minimised. However it is recognised that these requirements may be difficult to achieve in practice.

CONCLUSIONS

24.   The swelling pressure developed by a compacted clayey fill is a function of the dry density, moisture content, soil type, initial stress state and degree of confinement placed on the soil. Until recently, the measurement of swelling pressures arising from such soils has been directed at determining the potential of a soil to swell. The constraints applied on the soil in each testing method can alter the value of the swelling pressure (ref. 5). The more constraints applied to the soil the greater the magnitude of the swelling pressure. For design purposes it is necessary to choose the test method that is most appropriate to the conditions of the prototype situation.

25. The use of a geotechnical centrifuge in the analysis of the behaviour of compacted clayey fills has proved successful and can give data of swelling pressures over a very long equivalent prototype time period.

26. The time dependent development of swelling pressure behind retaining structures has been demonstrated clearly in both the 100mm stress path apparatus and centrifuge experiments. The swelling pressure rises quickly and then reduces with time as softening and associated vertical straining occurs. This augments previous work (ref. 6 and ref. 7) where pilot scale retaining wall studies were performed on different clayey soils.

27. The aim of the work was to investigate the possibility of widening the range of acceptable plasticity of soils used as backfill material. It has been shown that for higher plasticity soils, there may be combinations of dry density, water content and initial stress state which would minimise the development of swelling pressures. However complex problems such as the control of the field placement conditions and the potential for long term movements would need to be resolved before any change in current practice could be considered.

ACKNOWLEDGEMENTS

The work described in this paper forms part of a research project carried out at City University, London. It was funded by a SERC CASE award in conjunction with the Transport Research Laboratory. The authors are grateful for the help, support and encouragement received from Dr IF Symons of TRL throughout the course of the research. The views expressed in this paper are not necessarily those of the Department of Transport.

REFERENCES

1. Symons IF and Murray RT (1988). Conventional retaining walls : Pilot and full-scale studies. Proceedings of the Institution of Civil Engineers, Part 1,84,June,pp 519-538.
2. Department of Transport (1986). Specification for highway works, Part 2, HMSO, London.
3. Naish NG (1988). A cost survey of backfill around bridge abutments and retaining walls in South-East England. Department of Transport, TRRL, Contractor Report 112, Crowthorne.
4. Johnson LD (1989), Horizontal and vertical swell pressures from a triaxial test- a feasibility study. Geotechnical Testing Journal. Vol 12, No 1,pp 87-92.
5. Sridharan A, Rao AS and Sivapulliaih PV (1986). Swelling pressures of clays. Geotechnical Testing Journal, Vol 9, No 1, March 1986, pp. 24-33.
6. Symons IF, Clayton CRI and Darley P (1989). Earth Pressures against an experimental retaining wall backfilled with heavy clay. Department of Transport, TRRL, Research Report RR 192, Crowthorne.
7. Carder DR, Murray RT and Krawczyk JV (1980). Earth pressure against an experimental retaining wall backfilled with silty clay. Department of Transport, TRRL, Laboratory Report LR 946, Crowthorne.

10. Expansive behaviour of unsaturated fine-grained soils in compacted canal embankments

K. GARBULEWSKI, W. WOLSKI and S. ZAKOWICZ, Warsaw
Agricultural University, Poland, and I. KARIM AL-HELO, University of
Technology, Warsaw, Poland

SYNOPSIS. This paper presents selected results of a wide program of
investigations conducted to explain the causes of damages to canal
embankments, which are the main earth structures of a valley irrigation
system situated in a sub-tropical region. Damage was sustained mainly by
canals protected with concrete lining. Results had shown beyond doubt that
the main reason for damages to earth structures was the expansive behaviour
of local soils used to build the canal embankments. Besides the traditional
methods used to assess expansive soils and their properties (the constant soil
volume method), the investigation of swelling capacity of soils was
performed on selected samples taken from the destroyed canals using
suction measurements in a modified Proctor apparatus. Results of tests
permitted a dependence between suction pressure and swelling potential,
which is useful in forecasting soil swelling.

INTRODUCTION
1. Water in irrigation systems in a subtropical region is delivered to
cultivated fields using a system of canals, consisting usually of a main
canal, distributaries, water courses and field channels. The canals are
usually built in special embankments using local cohesive soils from vicinity
of the embankments. The embankments are built using the generally known
principles of soil mechanics and foundation engineering, including i.a. soil
compaction, required to attain the appropriate strength and filtration
properties of soil in-built into the embankments. Many of the reclamation
structures in subtropical regions are built under difficult and complex soil
conditions, one of which - as concluded from this paper - is the occurrence
of expansive soils. The expansive properties of soils used in construction of
canal embankments are not necessarily reported during reconnaissance
investigations, and thus are omitted in the design stage, and may lead to
serious damage to land reclamation structures (refs 1-2). The embankments
may undergo complete destruction over a short period of time under adverse
atmospheric conditions (temperature, wind, precipitation, etc.) and
continuously changing water content of soil in the embankments, caused by
a cycle of canal filling and emptying. In consequence, the problems
associated with appropriate and easy recognition of expansive behaviour of
soils become important also in respect to canal embankments.
2. Results analyzed in this paper have been obtained for canal
embankments (Fig. 1) situated in an irrigation system of a river valley,
located in a subtropical region (ref. 1). Soils in this object are of fluvial
origin; the surface layer down to 5-10 m consists most often of silty clays,
silts and mainly thin sand layers. The soils are salinated in a varying
degree; a certain portion of the area is only slightly saline, other areas have
high salination level.

Engineered fills. Thomas Telford, London, 1993

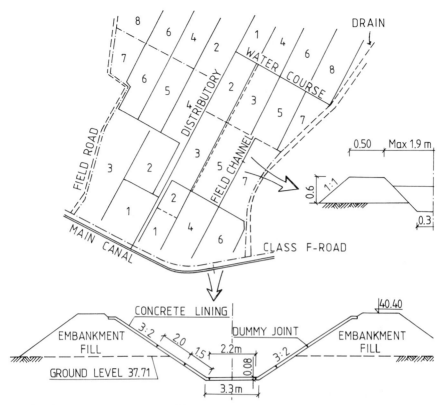

Fig. 1   Example layout of irrigation system and typical cross-sections
of canal embankments

CANAL EMBANKMENT CONSTRUCTION

3.   Canal embankments in the analyzed irrigation project (Fig. 1) were
built using the local cohesive soils situated in surface borrow areas. The soil
was scraped towards the canal site and built in layers 0.2 m thick into the
embankment. Each layer was compacted using a vibration roller to attain a
relative compaction index above 95% of the standard Proctor compaction.
Soil compaction was controlled on regular basis in a field laboratory.
Samples for compaction testing were taken from each layer as a cross-
section, every 100-150 m. Results proved that the compaction technology
applied guaranteed easy attainment of the required compaction parameters.
Only 5% of the results showed compaction below the desired level. In those
cases, compaction was repeated until the required parameters were attained.

4.   Upon completion of the compacted embankment, a cutting was made
using RAHCO equipment and concrete lining was inserted. Embankment
slopes and bottom of the cutting were previously moistened with water to
attain a smooth surface and protect the soil against excessive drying.
Thickness of the concrete lining was 7 cm for canals with flow rate of Q <
2.5 m³/s and 8 cm for Q > 2.5 m³/s. The concrete lining was protected
against drying using chemical agents, mainly sodium silicate. Compressive
strength of concrete was regularly controlled in the field laboratory,
attaining values exceeding the required 20 MPa after 28 days. The lining
was divided into panels using dummy joints. Transverse joints were spaced

every 3.5 m, lengthwise joints - about every 3.0 m, with the lowest joint some 0.5 m from the canal bottom. Joints depth was 1/3 of the panel thickness. Joints were filled with mastic. Concrete lining was used for the main canals and water courses. Field channels were made without protecting the embankments. Typical cross-sections of the canal embankments have been presented in Fig. 1.

CHARACTERISTICS OF COMPACTED FINE-GRAINED SOILS

5. Construction of canals was performed using cohesive soils classified according to the USCS mainly as CL varieties, and sometimes the CH ones, containing (Fig. 2) up to 20% of sand fraction, about 60% of silt and 20-25% of clay (<0.002 mm). This soil is characterized by the following plastic properties (Fig. 3): liquid limit, w_L = 25-50% (sometimes up to 60%), plastic limit, w_P = 15-25%, plasticity index, I_p = 7-25% (sometimes up to 35%), activity, A = 0.3-1.5.

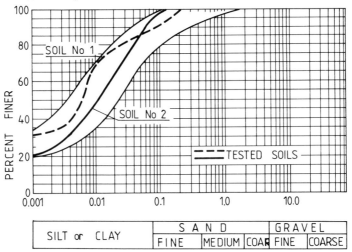

Fig. 2   Grain size distribution of soils used in canal embankment construction

6. The soil tests in the Proctor device with normal energy usually gave compactness parameters in the following range (Fig. 4):

maximum dry unit weight, γ_{dmax} 16.0-17.5 kN/m^3
water content, w_{opt} 15.0-25.0 %

Soil in-built into the embankment had water content similar to optimal. Distribution of water in the embankment showed (Fig. 5) that shortly after construction of the embankment soil water content underwent considerable reduction, particularly in the embankment slopes. Distribution of water content shows that soil in the embankment is partially saturated.

7. X-ray diffraction experiments showed that all soil samples had similar mineral composition : smectite, illite, kaolinite and chlorite, as well as mixed layer minerals, such as illite-smectite, illite-chlorite. The following regularity has been reported in quantitative composition of illite (I) kaolinite (K) and smectite (S) in the clay : I > S > K. A more detailed quantitative analysis was difficult because peak intensity was affected by a variety of factors, including chemical composition, structure and crystallites, etc.

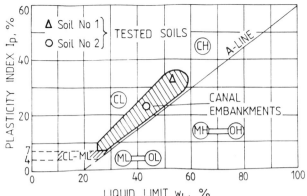

Fig. 3   Plasticity behaviour of soils used for canal embankments

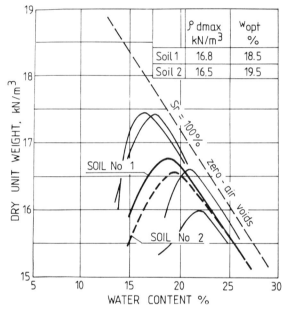

Fig. 4   Standard compaction curves for soils used in construction of canal embankments

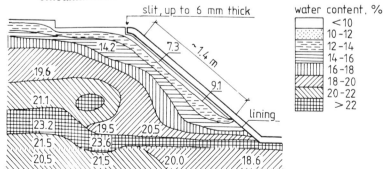

Fig. 5   Water content distribution under side slope lining of canal embankment (refs 1-2)

DAMAGES TO THE IRRIGATION CANALS

8. Cracking of the concrete lining was reported in various periods after placing, ranging from several weeks to several months. This was particularly true for main canals and certain distributaries, as well as in the water courses. Cracking was reported most frequently in panels located in lower portions of the canal embankment slopes, but some cracking was seen also in canal bottom, and in the upper parts of the lining. Crack dimensions varied from very thin, hardly detectable, to fully open gaps several millimeters wide, accompanied by displacement of the lining. A change of side slope inclination above the crack can be observed in the cracks, as well as in the broken dummy joints. Upper portion of the slope, above the crack, was usually more flat, which indicates a deformation of the earth slope of the canal. Generally, the following crack types may be distinguished: - small fissures, usually in the bottom part, - longitudinal cracks at the bottom, about 0.5-1.0 m above the dummy joint, several mm wide; - cracks along the bottom , near the longitudinal axis of the canal, - cracks of the upper parts with a heave along the dummy joint, - panel displacement with gap along the dummy joint. A typical distribution of the fissures in a cracked water course lining has been shown in Fig. 6.

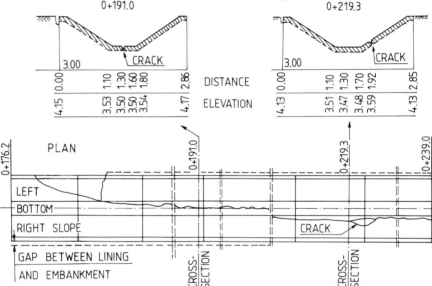

Fig. 6   Crack location and cross - sections of canal embankment (watercourse)

IDENTIFICATION OF SOIL EXPANSIVE BEHAVIOUR

9. Index property methods. In order to identity the expansive potential of the soils in canal embankment construction, the following index properties were used (ref. 1): - shrinkage limit w_s, - liquid limit w_L, - plasticity index I_p, - clay content and percentage passing No. 200 sieve, - activity, A. Pursuant to the classification (refs 3-5) proposed by Holtz and Gibbs (1956), Seed, Woodward and Lundgren (1962), Van der Merwe (1964) it may be concluded that the analyzed soil is expansive, with the real degree of expansiveness impossible to assess. Depending on the classification, in which different properties were used, the soil may be

considered as weakly or strongly swelling. Most of the soils may be classified as having medium expansiveness (Fig. 7).

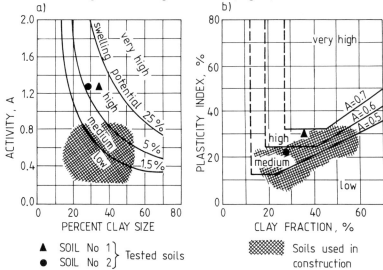

Fig. 7   Identification of expansive soils; a) Seed, Woodward and Lundgren diagram (1962), b) Van der Merwe diagram (1964)

Soils selected for more detailed investigation of swelling and suction properties have been marked in Fig. 7 as soil No. 1 and soil No. 2. Using Seed' et al classification (ref. 4) the tested soils should be included into the group of strongly swelling soils, while the Van der Merwe classification (ref. 5) shows soil no. 1 to be strongly swelling, while soil no. 2 as having medium swelling properties.

10.   Swelling test.   Swelling investigations were performed for two selected soils, the properties of which were presented in preceding sections on Figs 2-7. Samples were take from the canal embankments in which damage to the concrete lining was reported. The constant soil volume method was applied using a consolidometer base-mounted in triaxial apparatus (Fig. 8). Soil samples were prepared in a standard Proctor instrument using normal compacting energy of 60 Ncm/cm³. Soil samples 60 mm in diameter and 20 mm high were cut out of the Proctor compacted mould. Before moistening with water in the consolidometer, the sample was loaded with a pressure of 7 kPa to attain good contact between individual components of the instrument. The increasing swelling pressure during moistening was measured using a tensometric transducer placed between the consolidometer piston and frame of the triaxial apparatus (Fig. 8). Swelling pressure measurements were made every 30 minutes for the first 5 hours and then, after 18, 24 and 48 minutes. The maximum swelling was attained usually at under 20 hours of moistening. The tests were performed on soil samples with 15-30% water content.

11.   Results presented in Fig. 9 show that both soils have a similar swelling pressure of about 50 kPa. These results confirmed the well known dependence of distinct lowering in swelling pressure with increase in water content of the compacted soil. Tested soils compacted on the "dry" side of the compaction curve show almost twice higher swelling pressure as compared with soils compacted on the "wet" side. The test also proved the soils' expansive force to diminish together with compaction increase on the

"dry" side. Instead, "wet" side compacted soils display a reverse property, i.e. the expansive force value increases together with growth in compaction (Fig. 9).

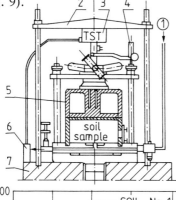

Fig. 8 Scheme of device for swelling test; 1-supply of water, 2-frame, 3 - tensometric converter, 4-displacement gauge, 5-consolidometer, 6-measuring bridge, 7-base

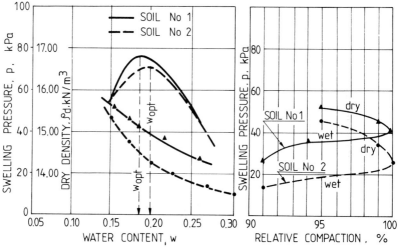

Fig. 9   Results of swelling tests

12. <u>Suction test</u>. The difficulties encountered in laboratory testing of soil swelling force and the duration of such tests induce many investigators to search for more precise and more rapid methods. Proposals have been made recently to use suction measurements to assess swelling pressure of soils (refs 6-8). Laboratory and field methods of measuring soil suctions are based on direct, mechanical testing procedures which incorporate the principles of thermodynamics (refs 8-10). However, the dependence between the suction pressure and swelling force of soils has not been determined so far from direct measurements of those values.

Suction tests for the two soils investigated here were performed in a modified Proctor instrument, where a fine-porous tensiometric ceramics element connected by a rigid conduit to the pressure gauge was assembled into the cylinder (Fig. 10). A Polish PS-100 tensiometric sensor was used for measuring suction pressures up to 100 kPa accurate to 0.1 kPa. The ceramic element assembled into the Proctor cylinder has an air-inflow limit of over 140 kPa. A 1000 cm³ soil sample was compacted in the Proctor apparatus at different initial water contents, covered with plastic film and a metal disk to protect the soil against loss of water and volume change.

Suction pressure was measured every 15 minutes for 32 hours, and then after 48 hours.

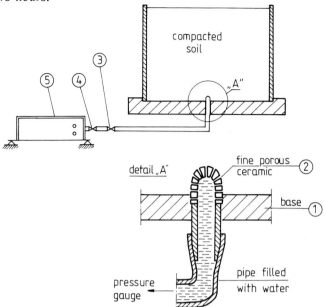

Fig. 10   Scheme of measurement unit for suction tests; 1-base, 2-fine porous ceramic, 3,4-valves, 5-transducer

13. Results have been presented in Figs 11-12 as a dependence between suction pressure and soil water content (Fig. 11) and between suction pressure, void ratio and saturation degree (Fig. 12). Same as for soil swelling pressure, suction pressure decreases visibly with increase in soil water content. Both soils exhibited similar values of suction pressures. Determinations of swelling force and suction pressures for the same soil sample were used to establish a direct dependence between those parameters (Fig. 13). Results show that the dependence between suction pressure, h, and swelling pressure, p, is exponential and may be presented as the following equation:

$$p = a\,(h)^n \qquad\qquad (1)$$

where a and n determined using the least squares method are :

for soil No. 1 $a = 12.71, \quad n = 0.32$
for soil No. 2 $a = \ \ 3.85, \quad n = 0.58$

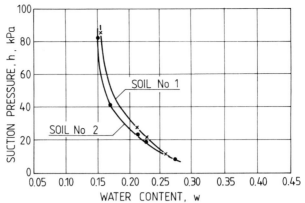

Fig. 11   Relationship of suction pressure to soil water content

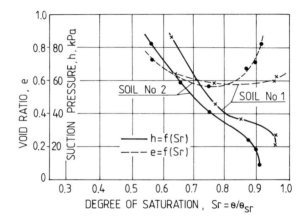

Fig. 12   Relationship of suction pressure and void ratio to saturation degree

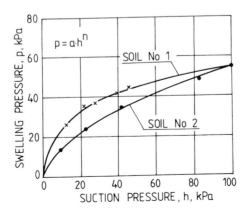

Fig. 13   Relationship between swelling pressure and suction pressure

CONCLUSIONS

The analysis of expansive properties of soils which mainly caused damagesof canal embankments with concrete lining and experimental results of soil swelling potential presented in this paper permit to formulate the following basic conclusions :

14. The scope of reconnaissance testing to assess properties of soils, also for design of canal embankments in subtropical regions should be wide enough to assess the possible expansive behaviour. Omission of those properties in the project may lead to significant damages to light concrete lining structures and other earth objects.

15. Nomographs and classifications widely presented in the literature should be used with caution and should not be the sole basis for assessing expansive properties of soils. Different classifications lead to divergent opinions on expansive properties of soils, there are situations where soil may be classified as strongly swelling in one system and weakly swelling according to another.

16. The method based on measuring soil suction gives relatively rapid and meaningful assessment of soil swelling force. The method should be developed to attain a more precise dependence between the swelling pressure and suction pressure of various soils. Results of experiments presented here show this dependence to be exponential. Soil suction measurements performed under field conditions (tensiometers) may give a rapid assessment of soil swelling parameters necessary for development of project design·s.

REFERENCES

1. Wolski W. (editor) Report on cracks in linings of Irrigation Project. Department of Geotechnics, Warsaw Agricultural University, 1986 (Interim report).
2. Garbulewski K. and Wolski W. Damages to irrigation embankment canals constructed with expansive soils. Proceedings of 15th International Congress of ICID, The Hague, 1993 (to be published).
3. Holtz W.G. and Gibbs H.J. Engineering properties of expansive clays. ASCE Transaction Paper No. 2814, Vol. 121, 1956.
4. Seed H.B., Woodward R.J. and Lundgren R. Prediction of swelling potential for compacted clays. Journal of SMFD ASCE, Vol.88, No SM3, 1962.
5. Van der Merwe D.H. The prediction of heave from the plasticity index and the percentage clay fraction. Trans. S.A. Inst.Civ. Engrs, Vol.6, No 6, 1964.
6. Komornik A., Livneh N. and Smucha S. Shear strength and swelling of clays under suction. Proc. of the Fourth Int. Conf. on Expansive Soils, Denver, Vol.I, 1980.
7. Acar Y.B. and Nyeretse P. Total suction of artificial mixtures of soil compacted at optimum water content. Geotechnical Testing Journal, GTJODJ, Vol.15, No 1, 1992.
8. Stannard D.I. Tensiometers - theory, construction and use. Geotechnical Testing Journal, GTJODJ, Vol.15, No 1, 1992.
9. Swartzendruber D. The flow of water in unsaturated soils, in: Flow through porous media. ed. R.J.M. de Wiest, Academic Press, New York, 1969.
10. Żakowicz S. The basis of water management in light textured soils on the irrigated areas. D.Sc.Thesis, Faculty of Land Reclamation and Environmental Engineering, SGGW, Warszawa, 1978 (in Polish).

11. Dynamic earth pressure and dynamic water pressure of fills behind quaywalls

M. KAZAMA and T. INATOMI, Ministry of Transport, Japan

SYNOPSIS. The behaviour of the dynamic earth pressure and dynamic water pressure of fills behind quaywalls are discussed in relationship to the shaking table test and analytical studies. Fundamental information concerning the dynamic earth pressure developed by dry sandy soil was obtained by experiment. It was found that it was necessary to estimate the dynamic earth pressure by considering the dynamic interaction between the quaywall and the backfill. The behaviour of dynamic pressures of saturated fills due to forced vibration was analyzed by using the thin layered element method. It was found that permeability and loading frequency can affect considerably the response of dynamic pressures and its distribution.

INTRODUCTION
1. Quaywalls backfilled with engineered fills are subjected to additional dynamic earth pressure and dynamic water pressure during earthquakes. These pressures sometimes cause damage to quaywall structures. For example, Photo 1 shows the damage suffered by the gravity quaywall in Kushiro Port following the Kushiro-oki earthquake (M=7.8) in January 1993. The increase in earth pressure was one of the main causes of the damage.

Photo 1. Damaged gravity-type quaywall (-7m Caisson type) in Kushiro Port at 1993 Kushiro-oki earthquake

Engineered fills. Thomas Telford, London, 1993

2. The estimation of dynamic pressures produced during earthquakes has been the subject of engineering research into quaywall earthquake proof design. At present, seismic stability assessment is based on pseudo-static calculations. In this method the earth pressure is estimated using the Mononobe-Okabe equation (M-O equation) which depends on the properties of the backfill materials and a seismic coefficient. The dynamic interaction between the structure and the backfill is not taken into account, although this represents a major departure between the assumed and actual conditions. The soil-structure interaction effect of gravity quaywalls, subjected to seismic activity, has been investigated by a number of authors using the shaking-table (refs. 1 and 2).

3. The construction of quaywalls used as waterfront structures present another problem due to water-soil-structure interaction in which hydrodynamic pressure is induced both in front and behind the quaywall. It can be concluded that the permeability of the backfill and the loading rate of the dynamic force can significantly affect the behaviour of any dynamic pressures acting on a quaywall. When a backfill material has an extremely small permeability, pore water movement relative to the soil skeleton can be ignored. It is normally assumed during the analysis of saturated ground in earthquake conditions that fills can be assumed to exhibit undrained conditions. However, there are a few guidelines as to how small the permeability has to be for this to be a valid assumption. The validity of various simplifying assumptions used to describe soils subjected to 1-dimensional vertical loading has been analyzed by Zienkiewics et al (ref. 3).

4. The authors have studied the behaviour of dynamic earth pressure and dynamic water pressures using the shaking table test and have conducted a number of numerical analyses using the thin layered element method. In the numerical study, the main focus was on the behaviour of the dynamic water pressure subjected to lateral loading. It is important to note that this paper does not cover liquifaction. (It is accepted that critical conditions occur when the soil liquifies, at which condition the soil exhibits a unit weight of approximately $2t/m^3$.)

SHAKING TABLE TESTS COVERING DYNAMIC EARTH PRESSURE

The Model
5. Shaking table tests have been undertaken to study the dynamic earth pressure acting on rigid caisson type quaywalls backfilled with dry Akita port sand. Figure 1 shows details of the model caisson used to measure the dynamic earth pressures and the arrangement of transducers. The dimensions of the model caisson were 800mm in height, 478mm in length and 500mm in width including three aluminium plates. Each plate was supported by three bi-axial load cells and had several earth pressure cells located on the centre line. The resultant dynamic earth pressure force was calculated by superimposing the time histories of the three load cells and eliminating the inertia force of the plate.

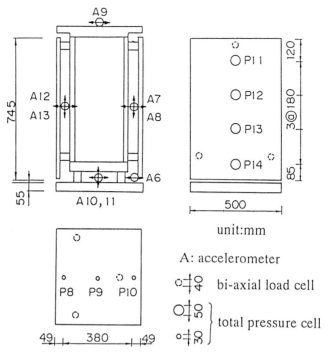

Fig. 1.   The model caisson

6.   Fig. 2 shows a perspective view of the model tests
which were performed. The model was made up of three
caissons, including two dummy caisson placed side by side
across a rectangular testing box (3m long by 1.5m wide by
1.2m deep). Small gaps between the model caissons, the dummy
caisson and the side of the box were provided to prevent any
forces from adjacent sections. These gaps were covered with
bellows-shaped vinyl sheets in order to prevent any spillage
of sand fill.

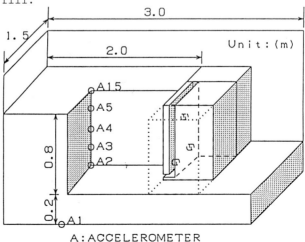

A:ACCELEROMETER

Fig. 2.   The bird's-eye view of the model

7.   The backfill was placed by raining the sand from a
height of 200mm.   After placement it was compacted by
vibration at a frequency of 30Hz, and an amplitude of 200Gal
for a duration of 30 seconds.   Table 1 shows the physical
properties of the backfill and the static earth pressure
following placement of the fill.   Table 1, Case-1 represents
the test using the caisson filled with sand, and Case-2
represents the test using the caisson without fill.   Both
tests were performed to study the effects of the weight of
the caisson on dynamic earth pressures.

Table 1.   Physical properties of the model

Item	Case-1	Case-2
Unit Weight of Backfill (kN/m^3)	15.30	15.25
Apparent unit weight of caisson (kN/m^3)	15.43	10.68
Relative Density (%)	53	52
S-wave Velocity (m/s)	108	108
Coefficient of the Static earth pressure	0.359	0.270
Height of the Action Point (H : Caisson Height)	0.475 H	0.443H

8.   An imput horizontal acceleration in the form of a
sinusoidal wave with a frequency varying from 1Hz to 50Hz was
used to represent the resonant frequency of the backfill.
The magnitude used ranged from 20, 50 and 100Gal.   In this
case the dynamic earth pressure was defined as the change
from the static earth pressure as illustrated in Fig. 3.

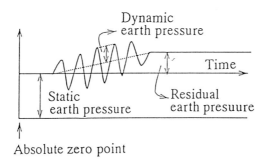

Fig. 3.   Illustration of the earth pressure time history

Results of the experiment
9.   Fig. 4 shows the response curve of the backfill to the
acceleration.   It is seen that the resonant frequency of the
backfill was about 28Hz, and, as the input acceleration
increased the maximum peak value decreased.   However, there
was no change in the frequency at which the peak value
appeared.

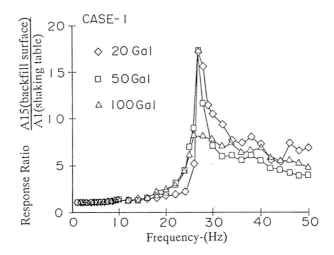

Fig. 4.   Response curve of the backfill

10.   Fig. 5 shows the variation of the earth pressure coefficient K with frequency. The coefficient K was calculated using the following equation.

$$K = 2P/(\gamma H^2) \tag{1}$$

Where; P is the magnitude of the earth pressure per unit of depth; g is the unit weight of the backfill, and H is the height of the caisson.

11.   It can be seen that the dynamic earth pressure was amplified close to the resonant frequency of the backfill. As the input acceleration increased, the peak value developed at a lower frequency. Thus, the non-linear aspect of the dynamic earth pressure is different from that of the backfill above. The reason why the dynamic earth pressure varied with the frequency of the input motion is that the response of backfill varied with the frequency as shown in Fig. 4. Therefore, the experimental relationship of the magnitude of the earth pressure and the input motion, as compared with M-O equation, considers the average acceleration response of the backfill. Fig. 6 shows the variation of the resultant force developed by the seismic earth pressure with the backfill response at mid-depth. Here, it is possible to consider the lateral acceleration axis as the strain axis developed in the backfill. In addition, if the earth pressure obtained from the M-O equation represents this condition of large strain developments in backfill, then the values obtained from the M-O equation are identified at each end of Fig. 6. Obviously, the behaviour of the dynamic earth pressure is different from that obtained from the M-O equation.

12.   Another important feature which can be seen in Fig. 5, is that the effect of the weight of the caisson influences the dynamic earth pressure. The value of the dynamic earth pressure developed with a heavy caisson (Case-1) is much

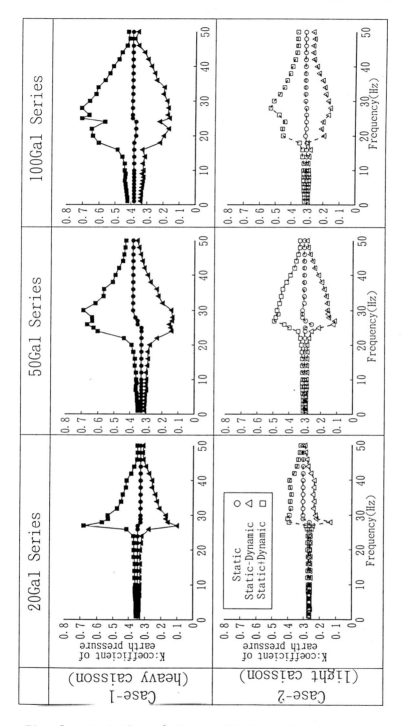

Fig. 5. Variation of the coefficient of seismic
earth pressure with frequency

larger than that developed with light caisson (Case-2). This shows that the caisson weight affects the dynamic earth pressure and this result cannot be explained by the M-O equation. It is necessary when estimating the earth pressure to consider the dynamic interaction between the structure and the backfill identified as the coupled vibration effect.

Ka=Coefficient of seismic active earth pressure by M-O equation
Kp=Coefficient of seismic passive earth pressure by M-O equation
($\phi=40°, \delta=7°$)
k=Seismic coefficient

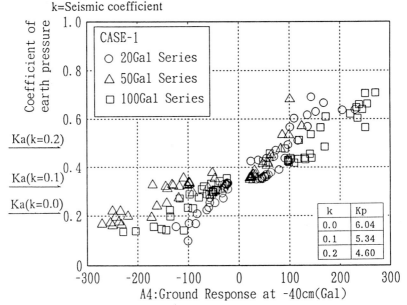

Fig. 6. The coefficient of seismic earth pressure versus the backfill

13. Fig. 7 shows the distribution of the acceleration, displacement and dynamic earth pressure recorded at the time when the bottom frictional force indicated a maximum value. As a steady sinusoidal wave was used in the study, the absolute displacement was easily obtained from the acceleration records. It was found that the dynamic earth pressure distribution was consistent with the relative displacement between the backfill and the caisson. The proportion of the resultant external forces at any time is shown on Fig.7. There is a phase difference between the inertia force of the caisson and the resultant force of the dynamic earth pressure.

i) Case-1 50Gal 10Hz (at P_{BX} max.)

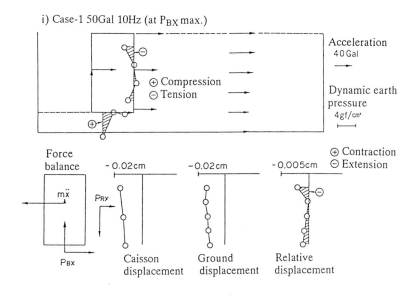

ii) Case-1 50Gal 30Hz (at P_{BX} max.)

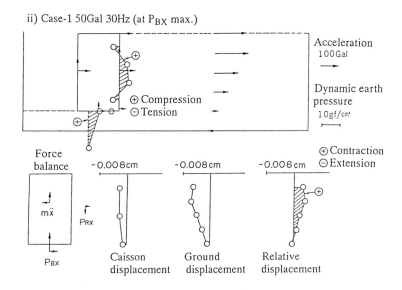

Fig. 7. Relation between the dynamic earth pressure distribution and relative displacement of the caisson and backfill at the time when the bottom friction indicates maximum

14. Fig. 8 shows the distribution of the maximum dynamic earth pressure recorded by the total earth pressure cells. The shape of the dynamic earth pressure distribution was not triangular and it appeared that the recorded distribution corresponded to the rotational motion of the caisson.

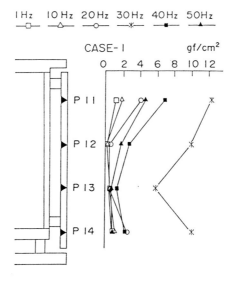

Fig. 8. Distribution of the maximum dynamic earth
pressure

DYNAMIC WATER PRESSURE BEHIND QUAYWALLS
15. Geotechnical engineering problems associated with
waterfront structures are often related to the behaviour of
the pore water. Typical examples of this type of problem
include the consolication of soft clay and the liquefaction
of loose sand. When considering liquefaction, the
relationship between the permeability and loading rate is
acknowledged as being a critical factor in the control of
excess pore pressure behaviour, including both the build up
of pressure and the accumulation diffusion process. The two
phase nature of the material is very important. When
studying the dynamic water pressure generated behind
quaywalls, it is also necessary to consider the two-phase
nature of the ground.

Numerical analysis
16. The behaviour of the dynamic pressures behind
quaywalls subjected to dynamic loading have been analysed
using the thin layered element method. The forumulation was
based on Biot's (ref. 4) equation describing the dynamic
behaviour of a saturated elasto-porous medium. Details of
the formulation of the method have been given by Nogami and
Kazama (ref. 5).
17. The condition considered is shown in Fig. 9, the
submerged soil, with a water level above the ground surface,
was subjected to lateral harmonic motion in the x = 0 plane.
The semi-infinite ground was divided into 11 layers in order
to consider the distribution with depth, Fig. 9. Stresses,
including the total stress (σ), effective stress (σ') and
pore water pressure (π) were computed at the middle depth of
each layer. The constants used in the calculation are shown
in Fig. 9.

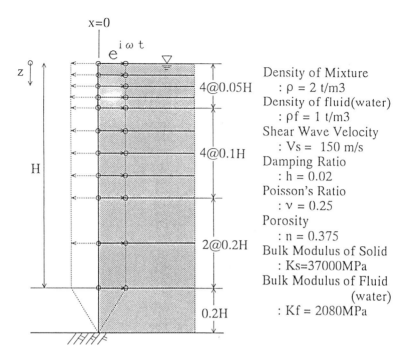

Density of Mixture
  : ρ = 2 t/m3
Density of fluid(water)
  : ρf = 1 t/m3
Shear Wave Velocity
  : Vs = 150 m/s
Damping Ratio
  : h = 0.02
Poisson's Ratio
  : ν = 0.25
Porosity
  : n = 0.375
Bulk Modulus of Solid
  : Ks=37000MPa
Bulk Modulus of Fluid
            (water)
  : Kf = 2080MPa

Fig. 9. Geometry and physical constants used in the
calculation

18. The computed stress ratio ($\sigma x'/\sigma x$) and ($\pi/\sigma x$) at x = 0
plane are shown in Fig. 10 for a frequency = 1Hz and various
permeabilities of the soil. The dotted line represents the
results under undrained conditions. The lower the
permeability, the higher the stress ratio ($\pi/\sigma x$), due to poor
diffusion of the porewater. Classically the stress ratio ($\sigma
x'/\sigma x$) behaves to the contrary. It can be seen that ($\sigma x'/\sigma
x$), plus ($\pi/\sigma x$) is not equal to one because of the phase
difference between $\sigma x'$ and π. In addition, any free drainage
at the ground surface affects the distribution of the stress
ratio ($\pi/\sigma x$). The pore water was far stiffer compared with
the soil skeleton stiffness and thus the soil response along
the depth was affected by the difference in the permeability
as shown in Fig. 10.
19. When comparing the stress ratio between the undrained
condition and the drained condition, the stress ratio ($\pi/\sigma x$)
for k = 10^{-4}m/s and 10^{-5}m/s are very little different from that
for the undrained condition. Thus, the soil with a
permeability less than k = 10^{-4}m/s at a frequency = 1Hz can be
considered to behave as a soil under undrained conditions.
Conversely, the stress ratio ($\sigma x'/\sigma x$) for k = 10^{-1}m/s at a
frequency - 1Hz is about 1, in this case the soil needs to be
considered as a two phase material. It was also found that
the distribution of the pore water pressure is similar to
that estimated by Westergaard's equation.

149

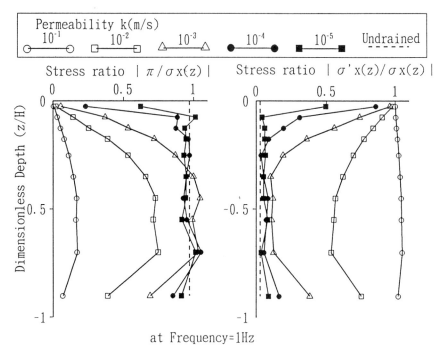

at Frequency=1Hz

Fig. 10.   Distribution of the stress ratio (σx'/σx) and
(π/σx) at frequency 1Hz

20.   The features described above are summarized in Fig.
11, where, $(\pi/\sigma x)_{total}$ is obtained by summing (π/σx) along the
depth in accordance with the equation ;

$$(\pi/\sigma x)_{total} = \{ \sum_{i=1}^{10} (\pi/\sigma x)i\ Hi \}/H \qquad (2)$$

and where Hi is the length of i-th layer.
21.   The assumption of undrainded conditions becomes more
valid when the permeability of the soil is decreased and the
loading frequency increased. In this condition the total
stress in the lateral direction can be considered as the
dynamic pressure. Also when the permeability is relatively
small in respect of the loading frequency, the dynamic
pressure can be separated into two parts. The first is the
dynamic water pressure induced by pore water, the second is
the dynamic effective stress of the soil skeleton in the
lateral direction.
22.   A second feature found in Fig. 11 was that the
resonancy of the surface ground affected the variation of the
stress ratio. In this case, the first resonant frequency of
the shear wave was Vs/(4*1.2H), then ωH/Vs=1.31. The nature
of the stress ratio in respect of the resonant frequency of
the ground was shown to be complex.
23.   In practice, the dominant frequency range of a strong
earthquake is approximately 0.5Hz to several Hz, and the

permeability of the backfill material (large gravel) is about 10^{-2}m/s to 10^{0}m/s. This range of values is intermediate between the undrained state and full the drained state shown in Fig. 11.

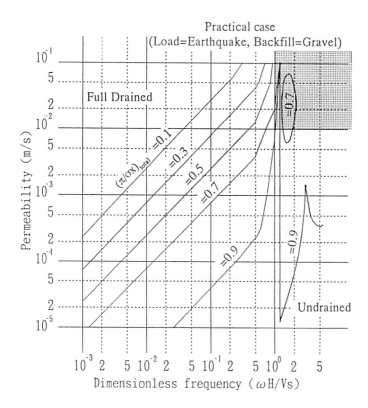

Fig. 11. Contour line of the stress ratio $(\pi/\sigma x)$ in permeability - dimension-less frequency $(\omega H/Vs)$ plane

CONCLUSIONS

24. The behaviour of the dynamic earth pressure and the dynamic water pressure behind quaywalls have been studied using a shaking table and by analysis. It was found that the dynamic earth pressure during an earthquake was consistent with the relative displacement between the quaywall and the backfill. Thus, when considering the dynamic earth pressure acting on a quaywall, it is necessary to consider the dynamic response of the total system including the structure and the backfill. It was also found that the permeability and loading frequency considerably affects the behaviour of the dynamic water pressure behind quaywalls. For practical earthquake proof design, further information about the proportion of the dynamic water pressure to the total dynamic lateral stress and about the phase difference between the dynamic water pressure and the effective dynamic lateral stress, is required.

REFERENCES
1. KAZAMA, M. and INATOMI, T. Model vibration tests for the seismic earth pressure acting on the rigid caisson foundation, Proc. of JSCE, No.416/I-13, pp.419-428, 1990.
2. INATOMI, T., KAZAMA, M. and MURAKAMI, Y. Shaking table tests for the seismic stability of caisson type quaywalls both in ordinary 1g field and in centrifugal field, Proc. of the 10th World Conf. on Earthquake Engineering, Vol.5, pp.2801-2806, 1992.
3. ZIENKIEWICZ, O.C., CHANG, C.T. and BETTESS, P. Drained, undrained, consolidation and dynamic behaviour assumptions in soils, Geotechnique, Vo.30, No.4, pp.385-395, 1980.
4. BIOT, M. A : Mechanics of deformation and acoustic propagation in porous media, J. of Applied Physics, Vol.33, No.4, pp.1482-1498, 1962.
5. NOGAMI, T. and KAZAMA, M. Dynamic response analysis of submerged soil by thin layer element method, Soil dynamics and earthquake engineering, Vol.11, No.1, pp.17-26, 1992.

12. The deformation behaviour of partially saturated granular fill at low stress levels

W. F. ANDERSON, I. C. PYRAH, A. McKINLAY and T. H.
SALMON, University of Sheffield, UK, and A. K. GOODWIN, Scott
Wilson Kirkpatrick and Partners, UK

SYNOPSIS
To obtain a better understanding of the behaviour of
unsaturated granular soils at low stress levels a
comprehensive programme of one dimensional compression tests
and triaxial compression tests has been carried out on
limestone backfills with gradings satisfying the DTp
specifications for granular sub bases and pipe bedding
material.

Two types of cell (1-D and triaxial), both of which allow
the separate measurement/control of pore air and pore water
pressures, have been used to examine the effect of suction
on soil behaviour, particularly compressibility and
stiffness. Significant strains were observed on inundation,
the magnitude being dependent upon the magnitude of the
suction prior to the inundation.

INTRODUCTION
   1. A considerable amount of construction activity is
carried out at relatively shallow depths where soil is
likely to be partially saturated and subjected to low
stresses. In particular the installation, repair and
replacement of service pipes takes place in the top five
metres or so of the ground. Although the use of trenchless
technology is increasing, the majority of service pipes are
installed in trenches which have to be backfilled with the
excavated material or imported fill. Post construction
settlements of reinstatements across highways are often of
such a magnitude that repair to the road surface has to be
carried out some time after initial reinstatement. When
trenchless technology methods are used displacements due to
pipe bursting or moling may cause damage to neighbouring
services and/or movements at the ground surface. For pipe
replacement the largest deformations will almost always
occur in previously compacted fill. Current practice in
predicting movements at relatively shallow depths is usually
to ignore the unsaturated nature of the soils because of the
lack of understanding of their behaviour at low stress
levels.

2. Previous research (eg ref. 1) into the behaviour of partially saturated soils has shown that the suctions induced by non-saturation decrease as the particle size, and hence void size, of the soils increase, and most research into partially saturated soils has concentrated on the behaviour of unsaturated silts and clays in which the suctions may be very large. The influence of the lower suctions induced in granular soils has largely been ignored because these suctions are expected to be considerably smaller than the stress levels imposed on the soil by self weight or construction for most engineering activities. However, for shallow depth activities such as service pipe replacement, the stress levels will usually be less than 100kPa and even the suctions induced in granular soils are likely to have an effect on the soil behaviour.

3. To obtain a better understanding of the behaviour of partially saturated granular soils at low stress levels a comprehensive programme of one dimensional compression tests and triaxial compression tests has been carried out on typical limestone backfills, using equipment which was specially designed to overcome many of the problems associated with laboratory testing of unsaturated soils. The paper describes the equipment developed for testing the partially saturated granular soils, reports the results of tests to assess the influence of partial saturation on compressibility and stiffness and discusses the implications for engineering practice.

STRESSES IN PARTIALLY SATURATED SOILS

4. The behaviour of partially saturated soils is more complicated than that of saturated soils because of the additional presence of air in the void spaces. If the degree of saturation is high the air may be in the form of discrete bubbles in the pore water which is continuous throughout the void spaces. If the degree of saturation is low then the air will be continuous throughout the void spaces and the water will be held at intergranular contacts by surface tension. At intermediate degrees of saturation both the air and the water may be continuous within the void spaces. Because of surface tension effects the pore air pressure, u_a, is higher than the pore water pressure, u_w. Fredlund (Ref. 2) has shown that the behaviour of partially saturated soils may best be described in terms of independent stress variables, $\sigma - u_a$ and $u_a - u_w$, the latter being called the matrix suction. The deformations measured in one dimensional compression tests and triaxial tests are examined in terms of these two variables.

TEST MATERIALS

5. A quarried limestone with the properties listed in Table 1 was used throughout the test programme. The material was supplied air dry and bagged in four

Property	Description/Value
Grain Shape	Flaky or equidimensional
Angularity	Angular to sub-angular
Aggregate Crushing Value	22
Aggregate Impact Value	18
Specific Gravity	2.71
Aggregate Absorption - coarse	0.8%
Aggregate Absorption - fine	1.4%

Table 1. Properties of Limestone Aggregate

industrially sieved size ranges. These were batched in appropriate proportions to produce the gradings shown in Figure 1. The fine grading complies with Clause 503 of the Department of Transport Specification for Highway Works (ref. 3) and is representative of material that may be used in a reinstatement for the bedding, laying and surrounding of pipes. The coarse grading satisfies Clauses 803 and 804 of the above Specification, which refer to granular sub-base materials Types 1 and 2. The uniform grading and the gap grading are representative of those found in natural materials which may be excavated and reused as trench backfill.

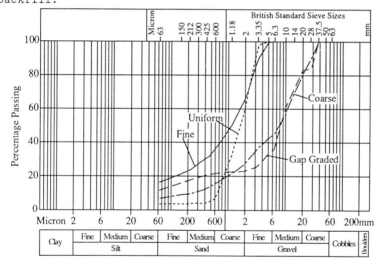

Fig. 1. Gradings used during test programme.

6. The limestone aggregate absorbed water into the particle interstices and during the research it was important to differentiate between the total water content and the free water content, the latter being calculated on the basis of the free water being the difference between the total water and the absorbed water found from British Standard aggregate absorption tests. For both the one dimensional compression tests and the triaxial tests compaction tests were carried out using the compaction

procedure to be used for specimen preparation. This enabled
an optimum free water content to be determined for each
material and each type of test.

ONE DIMENSIONAL COMPRESSION TESTS
Equipment and test procedures

7. Two new 250mm diameter consolidation cells, (Figure 2)
were designed and manufactured for this research. The novel
features are separate drainage facilities for the pore air
and the pore water, and the direct measurement of the
applied stress on the specimen by load cells incorporated in
a two part top platen. This arrangement eliminates the
platen friction effects, which can be significant when
dealing with low stresses. The conventional Rowe Cell
method of taking the applied stress to be equal to the water
pressure in the rubber membrane leads to significant errors.
All tests used a constant air pressure in the test specimens
and, for the water phase, drained or undrained conditions
could be maintained independently at each platen. Pore
pressure measurements are made behind the high air entry
stones.

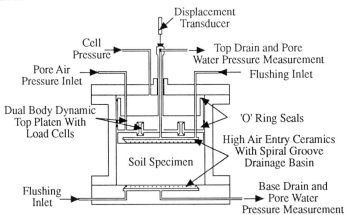

Fig. 2. 1-D compression cell for testing partially
saturated soils

8. Specimens were prepared at a predetermined water
content (usually lower than the optimum free water content)
by compaction with a vibrating hammer in the compression
cell. Most test series were carried out on well compacted
material using a technique similar to the British Standard
Vibrating Hammer Test, and only these tests are reported in
this paper. The range of water contents used gave initial
degrees of saturation which were normally below 50%; this
ensured that the specimens had continuous air voids (ref.4).

9. When the consolidation cell had been assembled a small
bedding pressure (<5kPa) was applied. The vertical stress

and air pressure were raised together to 100 kPa; an
elevated air pressure of 100kPa was used in all tests to
ensure that the measured pore water pressures were positive.
The specimen was then left until pore pressures had
equalised, generally 24 hours, and the initial suction could
be determined. After equalisation the specimens were loaded
incrementally and the suctions and strains monitored.
Maximum vertical stresses ranged from a nominal 75kPa to
150kPa. In some series of tests the specimens were unloaded
in stages and the rebound monitored , whilst in others the
specimens were inundated while under a vertical stress, and
the subsequent strain monitored.

10. One series of tests was carried out allowing water
drainage as well as air drainage, but water volume change
measurements indicated that for the range of initial free
water contents used negligible water drainage occurred. The
results of this series could therefore be compared with
other series which involved specimens with similar initial
free water contents but with no water drainage.

Matrix suctions
11. Matrix suctions were found to be higher at the top of
the specimen than at the bottom of the specimen. Suctions
under the nominal bedding load varied considerably,
particularly for the coarse grading. More consistent
suctions were measured when the initial vertical stress of
10kPa had been applied and the specimen left to attain
equilibrium. These values are considered to be more
realistic values for the initial suction. The suction was
found to increase with decreasing initial free water
content, and maximum initial suction values measured at the
top of the specimen for each of the soil types tested are
given in Table 2. Also shown are the initial free water
content and degree of saturation of the specimens in which
these suctions were measured. The maximum dry densities and
optimum moisture contents, determined using the vibrating
hammer method in the compression cell, are reported for
three of the gradings in Table 2. Compaction of the uniform
graded limestone at different water contents did not give
results which allowed a normal compaction curve, with
clearly identifiable maximum dry density and optimum
moisture content, to be drawn.

12. During the compression tests the suction values at
the base of the specimens remained fairly constant but the
values at the top decreased with increasing vertical stress.
At the end of each test the water content distribution
within the specimen was determined and, not unexpectedly,
higher water contents were found at the base of the
specimens than at the top. The difference between top and
bottom water contents increased with initial free water
content, implying that even in partially saturated specimens

with continuous air voids downward flow of water may occur
even if the water in the void spaces is discontinuous. For
all gradings this flow appeared to become significant at
initial free water contents of 3% to 4%.

Soil Grading	Optimum (free) Moisture Content %	Maximum Dry Density Mg/m^3	Initial Free Water Content %	Initial Degree of Saturation %	Initial Suction kPa
Coarse	6.4	2.28	3.4	33.3	38
Fine	8.5	1.95	2.4	12.9	90
Gap Graded	5.5	2.05	2.4	12.6	50
Uniform	*	*	1.6	5.5	17

Table 2. Compaction characteristics and maximum initial
suctions from one dimensional compression tests.

13. The differences between suctions measured at the top
of coarse grained and fine grained specimens was not as
large as may have been expected. A possible explanation for
this is that small congregations of finer grain particles at
the contacts between coarse particles give rise to higher
suctions than may be expected for coarse soils. The lowest
initial suctions were measured with the uniformly graded
limestone which, as can be seen in Figure 1, had little fine
material to congregate round the largest particles in this
grading.

Effect of initial water content on one dimensional
compression under incremental vertical stress

14. The effect of initial free water content on
compressibility may be appreciated from Figure 3 which shows
the total strain which occurs after the application of a
stress increment plotted against initial free water content
for the series of tests on coarse graded samples. Similar
trends were found for the other gradings although the water
content at which strains peaked varied.

15. This behaviour is attributed to the influence of soil
suctions. Starting at a free water content of about 1%, the
high localised suctions at particle contacts with a
congregation of fines, together with some possibly dry
surface contact points increasing the average friction
coefficient between the particles, are believed to
counteract the low dry density of the mass sufficiently for
the soil to have a low compressibility. As the free water
content increases, the localised suctions reduce the
internal stability of the mass at a greater rate than the
increasing dry density tends to increase this stability.
Hence the compressibility increases. As the water content
increases further still beyond a critical value of about 3%
the increasing dry density increasingly dominates and the

compressibility decreases once more.

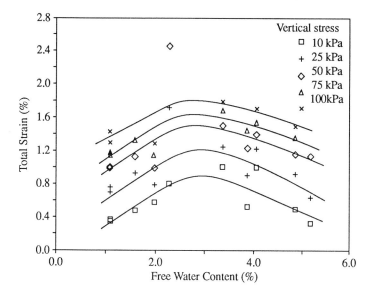

Fig. 3. Effect of initial free water content on
compressibility of coarse grading under increasing
vertical stress

16. For specimens of different gradings compacted in a
similar manner the peak axial strains under any particular
vertical stress were fairly similar. This similarity exists
despite the higher dry density, relative density and broader
grading, all of which would be expected to decrease the
compressibility of a dry material. It is suggested that the
higher suction stresses developed internally within the fine
grading and within the finer elements of the uniform and gap
graded materials may explain these results by virtue of
their stiffening influence.

17. For granular soils settlements are often considered
instantaneous but case records examined by Burland and
Burbridge (ref. 5) show that foundations on sands and
gravels exhibit time dependent settlements. In the current
tests Goodwin (ref. 6) has shown that there is a time
dependent component of the settlement which becomes an
increasing proportion of the total settlement at very low
stresses. However, the total settlements under these very
low stresses are unlikely to be significant in field
situations and at higher stresses some 90% of the
compression occurs within 30 minutes of loading and the time
dependent component will again be insignificant in field
terms.

Effect of inundation on one dimensional compression

18.   Series of tests have been carried out in which the specimen has been incrementally loaded in the normal way (usually to a nominal vertical stress (σ_v-u_a) of 75 kPa) and then, maintaining this stress, water has been allowed to percolate slowly into the specimen through the top porous stone while monitoring any additional settlement.

19.   Figure 4 shows the results of inundation tests, carried out on two coarse grained specimens prepared in an identical way with an initial free water content of 1.1%. The similarity of the curves demonstrates the repeatability of the one dimensional compression tests, and it can be seen that even in a well compacted material, which has already been subjected to a vertical stress of 75 kPa, further strains occur on inundation. This straining is almost certainly due to the destruction of the suction, as monitored by changing pore water pressure during inundation. The relationship between inundation strains and suction would appear to be borne out by Figure 5 which shows a trend of increasing inundation strain with suction.

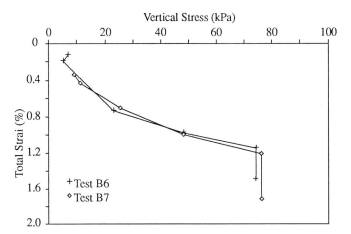

Fig. 4.   Effect of inundation on compressibility of coarse grading.

20.   The other materials showed similar trends.   The maximum strain caused by inundation was about 2.3% which was measured in a gap graded specimen prepared with an initial free water content of 1.5% under a vertical stress of 120 kPa.

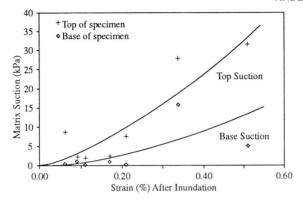

Fig. 5. Relationship between matrix suction and inundation strain.

TRIAXIAL TESTS
Equipment and Test Procedures

21. During triaxial tests on saturated soil specimens volume changes are found by monitoring the volume of water going into or out of the specimen. Because of the volume changes which occur in the gas phase this method cannot be used for partially saturated soils. For these soils specimen volume changes are found by monitoring the flow of water into or out of the cell, rather than into or out of the specimen. For short term tests this, together with the use of conventional equipment may give a reasonable estimate of specimen volume change, but for long term tests the permeability of the perspex cell body and the expansion or contraction of the perspex when cell pressures change are likely to lead to inaccuracies. Wheeler (ref 7) has demonstrated how the use of a double wall cell, with the inner and outer cells being kept at the same pressure, can allow accurate assessment of specimen volume changes by monitoring the flow of water into or out of the inner cell.

22. A new double wall triaxial cell (Figure 6), suitable for testing 150 mm diameter by 375 mm high specimens, has been designed and constructed for testing partially saturated granular soils at low stress levels. In order to improve the accuracy of the measurements, particularly during the shearing stage of the test, the cell has facilities for internal measurement of the vertical strain over the central portion of the specimen and the radial strain at the mid-height of the specimen. The vertical strains are measured over the central 150mm of the specimen by three equally spaced LVDTs fixed in a positioning ring attached to the specimen membrane, with the LVDT spindles resting on targets attached to the membrane. The LVDTs are suspended from springs to ensure that they do not impose any load on the specimen. The radial strains are monitored by

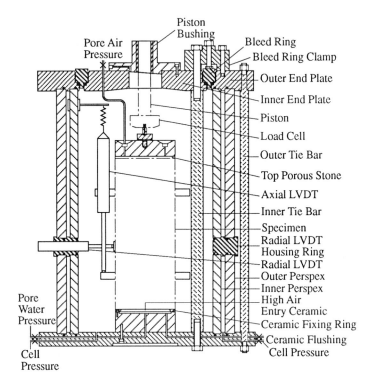

Fig. 6. Triaxial cell for testing partially saturated
soils.

three equally spaced horizontal LVDTs fitted into a metal
ring in the double wall cell so that they rest on targets at
the mid height of the specimen at the start of the test.

23. Triaxial specimens of the fine grading (Figure 1)
have been prepared with an initial free water content of
2.7%. The specimens are compacted in layers by a vibrating
hammer inside the triaxial test membrane within a split
metal mould. Using this technique a maximum dry density of
2.21 Mg/m^3 was achieved at an optimum free moisture content
of 6.9%. After compaction the top platen with a
conventional porous stone is placed on the specimen and the
membrane attached to it before the specimen is slid
carefully onto the triaxial cell base plate which is fitted
with a high air entry ceramic stone. The split mould is
removed and, there being sufficient suction for the specimen
to retain its shape without application of a vacuum, a
second membrane is placed around it and sealed to the end
platens in case the inner membrane has been damaged during
compaction. The internal strain measuring transducers are
put in position and the cell assembled and filled with

162

water.

24.   Consolidated undrained and consolidated drained tests
have been carried out. The initial stage of the
consolidated undrained tests is similar to the one
dimensional compression tests in that the cell pressure and
air pressure, which is applied to the specimen through the
top platen, are raised simultaneously to a value that will
ensure that measured pore water pressures are always
positive. The specimen is left for pore pressure
equalisation to occur and an initial suction measured.
After equalisation the cell pressure is raised to the
desired value and the specimen allowed to consolidate. At
the end of consolidation the specimen is sheared with
drainage of air but no drainage of water. Suctions are
monitored during the shearing stage.

25.   Consolidated drained tests were carried out with a
predetermined suction ranging from 25 kPa to 75 kPa. For
most tests the procedure was speeded up by combining the
equalisation and consolidation stages. During the shearing
stage both air and water were allowed to drain with the
suction being maintained at its predetermined value.

Matrix suctions
26.   Because pore air pressure was controlled through the
top porous stone the pore water pressure could only be
measured at the base of the specimen and initial suctions
determined for the base. For the fine grading, well
compacted at an initial free water content of 2.7%, values
as high as 38 kPa were found. This may be compared with a
maximum initial suction of 90 kPa found at the top of a fine
grading specimen with an initial free water content of 2.4%
in the one dimensional compression tests. The lower initial
suction found in the triaxial tests is partly due to the
higher initial free water content in these specimens and the
non uniform distribution of water found in the 1-D
compression cell specimens.

27.   During undrained shearing the suction changed as the
specimen tried to contract or dilate with a maximum suction
of 46 kPa being measured during a shearing stage. This
suction determined at the base of the specimen may not be
the suction at the shearing zone in the middle of the
specimen since the rate of strain of 0.001%/min may not have
been slow enough to ensure equilibrium of pore water
conditions throughout the specimen.

Strain measurement and volume changes
28.   The design of the triaxial cell allowed comparison of
axial strains measured externally by piston movement and
internally over the centre portion of the specimen using
LVDTs. Figure 7 shows a comparison of these two

measurements. It can be seen that the external transducer may be indicating axial movement at the start of the shearing stage, but the internal transducers monitoring specimen strain do not start to move for some time due to bedding errors. The three axial strain devices around the specimen did not indicate initial movement at the same time indicating non uniform bedding, but having started the straining appeared to be uniform up to approximately 0.5% axial strain. At this stage the specimen began to show signs of failure with either significant barrelling or the formation of a distinct shear plane. Failure generally occurred at about 0.9% strain and therefore the internal strain measuring devices should not be used when examining failure conditions.

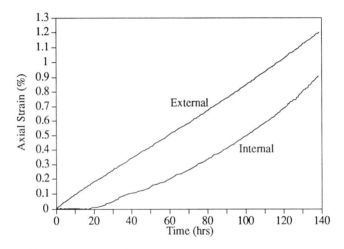

Fig. 7. Comparison of axial strains measured externally and internally.

29. The radial strain transducers usually indicated movement before axial strain was noted on the internal axial strain transducers. This early response is probably due to small lateral movements of the specimen during initial take-up of load rather than the development of strains in the centre portion of the specimen.

30. From the measurements recorded by the internal strain measuring devices it was possible to estimate the specimen volume change by assuming the specimen remained cylindrical. Figure 8 shows a comparison of this calculated volume change with the volume change deduced from the volume of water draining into or out of the inner cell during the shearing stage of one of the drained tests. It can be seen that although the trends are similar there is a discrepancy in the values. The apparent contraction of the specimen during the early stages of shearing, as indicated by the inner cell volume change, may be due to water flowing into the cell

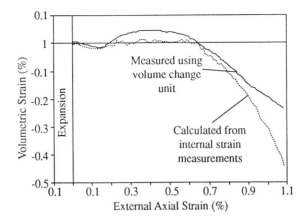

Fig. 8. Comparison between measured volume change and
volume change deduced from internal strain measurement.

during bedding in of the platens and a more accurate measure
of volume change, particularly at low strains when the
specimen remains cylindrical, is obtained from the internal
transducers. The divergence of the curves at around 1%
axial strain is probably due to the specimen losing its
cylindrical shape.

Stiffness

31. The drained elastic modulus of the soil may be
expressed as the slope of the deviator stress - axial strain
curve during the shearing stage of the triaxial test. Since
the deviator stress - strain relationship is curved, the
value of the modulus will depend on the stress or strain
range over which it is measured. For comparison purposes
the modulus values in this work have been calculated as
secant values over a deviator stress range from 0 to 30% of
the deviator stress at failure, i.e. a stress range likely
to be experienced in the field. The strains from the
internal transducers have been used in calculating modulus
values.

32. Figure 9 shows how the derived elastic modulus varies
with cell pressure and with suction. The soil stiffness
increases, not only with confining pressure, but also with
soil suction.

DISCUSSION

33. The results show that for a partially saturated
coarse grain soil at low stress levels both the
compressibility measured in 1-D compression tests and

165

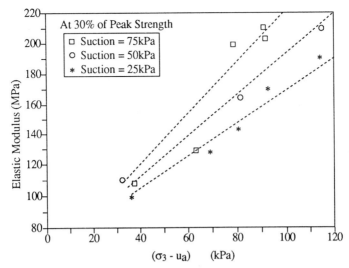

Fig. 9. Influence of suction on elastic modulus.

stiffness measured in triaxial tests are a function of
suction; the higher the suction the lower the
compressibility and the higher the stiffness. A reduction
in the suction, by inundation for example, is likely
therefore to lead to strains resulting in movements, eg
ground settlement of a reinstated trench backfill, or to a
redistribution of stress and possible associated ground
movements, eg around a pipe where pipe bursting has taken
place. If predictions are to be made of their effect then
the analysis should include suction as one of the variables.

34. A practical implication of the reduction in volume
caused by elimination of the suction during inundation is
that if post-construction settlements are to be expected due
to a change in groundwater conditions, these settlements may
be reduced by inducing such movements during construction.

35. The tests, in addition to highlighting the
significance of suction and hence inundation on soil
behaviour, provide a more accurate picture of the stress-
strain behaviour for the type of soil tested than has been
available previously. Of particular value for further
predictions of ground movement are the local strain
measurements in the triaxial tests.

CONCLUSIONS
36. New equipment for examining the behaviour of
partially saturated granular soils at low stress levels has
been developed. The equipment includes a 250mm diameter one
dimensional compression cell capable of applying and
accurately measuring stresses in the range 5 - 150 kPa, and
a double walled triaxial cell for testing specimens 150mm

166

diameter by 375mm high. Both cells allow the measurement and/or control of air and water pressures and the triaxial cell allows the local measurement of axial and radial strain.

37. The tests have highlighted the significance of matrix suction even for the coarse grained material used in this study. As suction increases the compressibility of the soil decreases and the stiffness increases.

38. The inundation of a partially saturated coarse grained soil induces strains which are dependent upon the suction present prior to inundation.

ACKNOWLEDGEMENTS

40. This work has been carried out with the support of the Science and Engineering Research Council and British Gas plc.

REFERENCES

1. CRONEY, D. AND COLEMAN, J.A. Soil thermodynamics applied to the movement of moisture in road foundations. Proc. 7th Int. Conf. for App.Mech., 1948, Vol 3, 163-177.
2. FREDLUND, D.G. Volume change behaviour of unsaturated soils. Ph.D. Thesis, Univ. of Alberta, 1973.
3. DEPARTMENT OF TRANSPORT(1992) "Specification for Highway Works". HMSO, London.
4. FREDLUND, D.G. Appropriate concepts and technology for unsaturated soils. Can. Geot.J., 1979, Vol 16, 121-139.
5. BURLAND, J.B.and BURBIDGE, M.C. Settlement of foundations on sand and gravel. Proc. I.C.E., Part 1, 1985, Vol 78, 1325-1381.
6. GOODWIN, A.K. One dimensional compression behaviour of unsaturated granular soils at low stress levels. Ph.D Thesis, University of Sheffield, 1991.
7. WHEELER, S.J. The undrained shear strength of soils containing large gas bubbles. Geotechnique,1988, Vol 38, No 3, pp 399-413.

13. A study of the stress–strain behaviour of a compacted soil from Nova Avanhandava Dam

F. SCHNAID, Federal University of Rio Grande do Sul, Brazil, and
L. M. COSTA FILHO, Fluminense Federal University, Brazil

SYNOPSIS. The stress-strain behaviour of a coluvionar compacted soil from Nova Avanhandava Dam, Brasil, was studied. Nineteen triaxial tests were carried out, eleven of which under controlled strain conditions (UU) and eight under controlled stress conditions (stress path $K = \sigma_h/\sigma_v = $ constant). All tests in this work were carried out in partially saturated undisturbed samples, with water pressure measured at the base and centre of the triaxial specimens and radial strains measured internally to the triaxial cell. Reconstitution of the unconsolidated-undrained tests and prediction of the K stress tests were made on the basis of a hyperbolic stress-strain relationship used in conjunction with an exponential formulation to determine Poisson's ratio. The accuracy of the predictions is assessed and the need of incorporating the elastic cross-anisotropic formulation in the behaviour of compacted soils is discussed. This analysis aims at the development of a simple model for prediction of displacements in dams, where a K stress path is expected to occur. The accuracy of predictions of constant K stress path tests in the laboratory encourages the use of such approach in the field.

INTRODUCTION

1. For studies of stress, strain and displacement distributions in embankment dams during and after construction, a stress-strain model for the soil(s) involved is required. This model should be able to predict correctly the behaviour of the soils along the stress-paths occuring in the different points of the dam. Previous studies both for homogeneous (Lacerda and Mahler, 1973; Eisenstein, 1979; Charles, 1976) and zoned dams (Costa Filho et al, 1982; Azevedo Filho, 1990) have indicated that for most of the dam typical stress-paths during construction are essentially linear, along $K = \sigma_3/\sigma_1 = $ constant lines. Values of K vary between 0.4 and 0.7. This occurs irrespective of soil parameters, dam geometry and initial (K_o) state of stress after compaction (Costa Filho, 1985).

2. The Nova Avanhandava Dam is located in the Tiete River approximately 450 Km from the City of São Paulo. It was built to increase the electrical capacity of CESP (Companhia

Engineered fills. Thomas Telford, London, 1993

Energética de São Paulo) by 300 MW. Figure 1 shows a typical cross-section of the dam.

3. Most of the dam was built using a colluvial soil which occurs in the borrow areas overlying a residual soil derived from basalt.

4. A series of laboratory tests was carried out to investigate the accuracy of a simple stress-strain model to reproduce the behaviour of the colluvial soil from Nova Avanhandava Dam along K = constant stress-paths.

5. The purpose of this paper is to present the results of this investigation.

MODELLING

6. Compacted soils are partially saturated and its stress-strain-volumetric behaviour is a function of different factors including water content, structure, strain conditions, stress levels and energy of compaction. Several attempts have been previously made to model this behaviour by making use of different types of stress-strain models with varying degrees of complexity. Despite its limitations, the non-linear elastic model proposed by Duncan and Chang (1970) using the hyperbolic formulation has been widely used for stress-strain analyses of dams up to the present. The probable reason for this lies in its relative simplicity and incorporation of some of the important aspects of observed soil behaviour (non linearity, influence of stress-level, failure, etc).

7. Several developments and evolutions of the model were presented (eg., Orgler, 1983; Duncan et al, 1980; Eisenstein, 1979), particularly as related to the calculation of volumetric strains (associated with Poisson's ratio determination).

8. In the present work the conventional hyperbolic formulation for the tangent Young's modulus E_t was used, for which the parameters K, n, R_f, C and Ø have to be determined.

9. For the Poisson's ratio determination, studies carried out by Lade (1972), Orgler (1983), among others, as well as in the present work, have indicated that the exponential formulation proposed by Lade (1972) gives a better approximation than the original hyperbolic one. The values of Poisson's ratios are computed according to the equation

$$\nu_t = L \cdot m \cdot (\frac{\sigma_3}{P_a})^q \cdot \varepsilon_a^{m-1}$$

where the parameters L, m and q have to be determined experimentally.

10. An alternative approach was suggested by Duncan et al (1980), using the bulk modulus B instead of the Poisson's ratio in the elastic stress-strain relationships. In this formulation the bulk modulus is computed by the equation

$$B = K_b \, P_a \, (\frac{\sigma_3}{P_a})^s$$

169

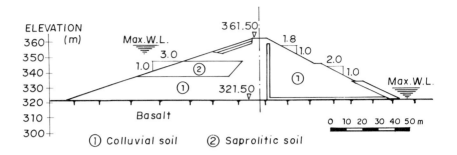

Fig. 1.   Main cross-section–Nova Avanhandava Dam

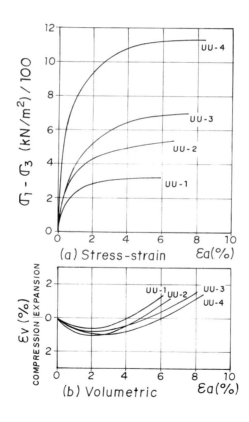

Fig. 2.   Experimental results – UU tests

where the parameters K_b and s have to be determined experimentally. The values of Poisson's ratio can be obtained from B and E_t.

11. Both approaches were used in this study.

TESTING PROGRAM

12. Undisturbed block samples were obtained from a location in the central part of Nova Avanhandava Dam immediately after a layer placement. The soil is a silty clay with a liquid limit of 62,5% and a plasticity index of 33%.

13. Compaction in the field yields values of dry density of approximately 16 kN/m^3 and water content of 26% (-0.9% to +2.7% variation), which is about the optimum water content obtained from laboratory compaction. The degree of saturation is high, ranging from 85% to 90%. The immediate effect of such compaction conditions is that the soil is partially saturated and the air in the voids of a soil element is likely not to be continuously connected throughout the specimen but constrained in the form of bubbles.

14. In unsaturated soils, the pore fluid pressure is known to have two components, pore water pressure U_w and pore air pressure U_a. These components must be combined in two independent stress parameters in order to describe the mechanical behaviour of unsaturated soils. The two parameters normally selected are $\sigma - U_a$ (referred as net stress) and $u_a - u_w$ (referred as suction). Various equations were proposed to assess the shear strength and deformation of partially saturated soils using these concepts but their usefullness was limited. It was just recently that a constitutive relationship was developed based on critical state framework to predict both volumetric and stress strain for any stress path and any drainage condition (Alonso et al, 1990). This approach involves nine constants which cannot be assessed from conventional triaxial techniques and its applicability is still restricted to research developments.

15. In the present work the testing equipment allowed the measurement of pore water pressures but without an independent control of suction. Since air pressure, and therefore suction, was not monitored, interpretation of triaxial data could only be made in terms of effective stress by adopting assumptions regarding fluid pressures, or simply in terms of total stresses.

16. For the high degrees of saturation presented by the samples, neglecting the pore air pressures would not lead to a significant error in interpretation of test results directly in simple conventional effective stress terms. However, the total stress approach has been adopted in combination to the simple hyperbolic model. Such methodology has been used because it has been considered desirable to formalize a simple stress-strain model for incorporating in a finite element program.

17. Furthermore, undrained conditions were assumed for the dam construction, which enables the use of a total stress analysis.

18. Whether undrained conditions, even aproximately, occur during earth dam construction is subject to considerable dispute, not only due to the possibility of air drainage in the initial stages of loading but also for the relatively high coefficients of consolidation of brazilian compacted soils, associated with the times involved in construction. However, the consideration of partially drained conditions would require a complex model, which is outside the aim of this investigation.

19. All tests were carried out in conventional triaxial cells either under strain controlled axial compression (UU) or under stress controlled axial compression (constant K stress path), using samples with 101,6 mm diameter.

20. Axial loads were monitored by internal load cells and confining pressures measured by electrical transducers. Pore pressures were measured simultaneously at the base and centre of the triaxial cell by two independent transducers. Axial deformation were measured by mechanical gauges corrected for load cell compliance.

21. The influence of Poisson's ratio parameters in the stress distribution and displacements (particularly horizontal ones) has been stressed by Lacerda and Mahler (1973) and Costa Filho (1985) among others. Monitoring radial strains is an important factor for determining accurately Poisson's ratio parameters and K stress path. A common procedure often adopted in triaxial tests is to infer radial deformations from axial and volumetric strains. The error involved in this procedure in tests carried out in saturated samples was discussed by Costa Filho (1980). Assessment of radial strains is even more complex in partially saturated soils where volume change is not only a function of water drainage but also air compressibility. Accurate measurements of radial deformations therefore require a direct monitoring of radial variation at the boundaries of the specimen, internally to the triaxial cell. A strain-gauged sensor measuring device to do so was then developed to achieve the necessary precision on radial strain measurements (Schnaid et al, 1985).

LABORATORY RESULTS

22. The stress-strain-volumetric behaviour and failure envelope of unconsolidated-undrained tests are presented in Figures 2 and 3, respectively. Two identical tests were carried out at each stress level; four stress levels were tested corresponding to iniᵗial mean stresses of 100 kPa, 300 kPa, 500 kPa and 900 kPa. The curves represent the average between the two tests.

23. Nine tests were carried out under stress controlled conditions (K stress path), also under undrained conditions, with K values of 0.4, 0.6 and 1.0. Three identical tests

172

were performed at each stress path to check the influence of
compaction variability in the field. Volumetric and stress
behaviour is presented in Figure 4, corresponding to average
of the 3 curves.

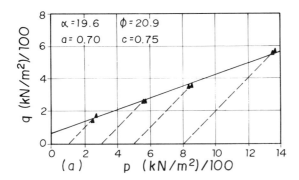

(a)

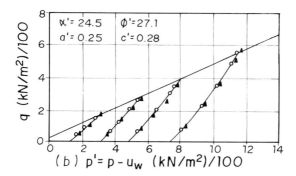

(b)

Fig. 3.   Failure envelopes - UU Tests
          (a) Total stresses
          (b) Total - pore water pressures

MODEL ADJUSTMENTS

24. The empirical parameters for the different formulations were determined from the UU triaxial tests, using the points corresponding to 1 and 5% axial strains in the experimental curves. These values of axial strains were chosen because they correspond more closely to the range of axial strains observed in dams built with the same material (Orgler, 1983).

25. The following parameters were obtained:

- Young's modulus $\qquad K = 807$
  (hyperbolic formulation) $\qquad n = 0,35$
  $\qquad R_f = 0,90$
  $\qquad C = 75$ kPa (total stresses)
  $\qquad 0 = 21°$ (total stresses)

- Poisson's ratio $\qquad L = 0,225$
  (exponential formulation) $\qquad q = -0,072$
  $\qquad m = 1,58$

- Bulk modulus $\qquad K_b = 170$
  $\qquad s = 0,37$

26. Comparison of reconstituted and experimentally determined stress-strain-volumetric curves for the UU tests is shown in Figure 5. Calculated values were in general agreement to experimental data, the diference being smaller than the dispersion between test results. Besides 1 and 5% axial strain, other different pairs of values were used, but the overall adjustment was not improved. The conventional proposal by Duncan and Chang (1970) at 70 and 95% of maximum shear stress did not reproduce satisfactorily the experimental curves at small strains, grossly underestimating the initial moduli.

27. Predicted and observed behaviour of K stress-path tests are presented in Figure 6 for Poisson's ratio formulation and in Figure 7 for bulk modulus formulation.

28. Good agreement was obtained for the stress-strain curves, using both formulations. Again the difference was smaller than the scatter of test results.

29. For the volumetric strain curves, predictions were fair for K stress ratios of 0,6 and 1,0. Predictions for K = 0,4 tests were very poor, since they did not consider the dilatant behaviour of the soil. The above occurred irrespective of the formulation used.

30. An attempt was made of using elastic cross-anisotropy to better represent the volumetric strain behaviour of the compacted soil, as already proposed by Eisenstein and Law (1979). The volumetric strain curves presented in Figures 6 and 7 indicate predicted values greater than the observed ones. Since the stress-strain curves were correctly predicted, the difference must lie in the radial strains. The consideration of anisotropy might explain the observed difference, considering the compacted soil as an overconsoli-

dated material with a greater stiffness in the horizontal direction.

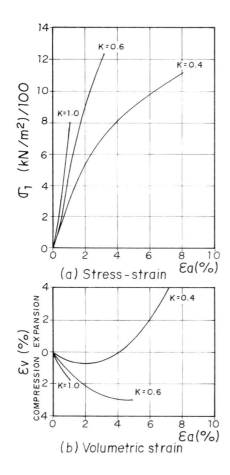

Fig. 4.   K stress path tests results

31. The consideration of cross-anisotropy in elasticity requires the use of 5 parameters to correlate stresses and strains: E_a, E_r, ν_{rr}, ν_{ar} and G_{ra} in which **r** and **a** are the radial and axial axis, respectively. Since the conventional triaxial tests carried out do not enable the determination of all five constants, in the present analysis radial deformations were predicted with the simplifications $\nu_{rr} = \nu_{ar}$ and $\varepsilon_r = 2\varepsilon_a$ (Eisenstein and Law, 1979).

32. An indication for the use of $E_r = 2 E_a$ (was also given in the measured values of E_r and E_a during isotropic compression (K - 1,0) tests, which gave $E_a \sim 2E_r$. The parameter G_{ra}, while important in the analyses of dams, is not required for the prediction of the present laboratory

tests. Non-linearity was considered using the same parameters as above.

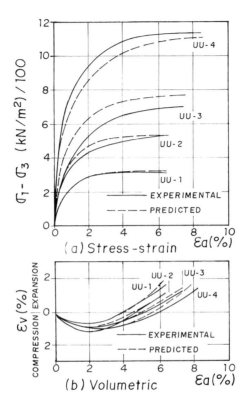

Fig. 5.   UU - Tests experimental and reconstituted curves

33. Figure 8 shows the new volumetric strain curves computed using elastic anisotropy. A better agreement was obtained, including a dilatant behaviour, but still with a significant difference for K = 0,4. The stress-strain curves remained essentially unnaffected.

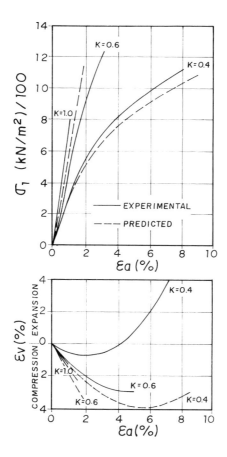

Fig. 6. K stress-path tests/Poisson's ratio formulation-
experimental and precited curves

CONCLUSIONS

34. The simple hyperbolic model used to predict the stress-
strain-volumetric behaviour of the compacted soil of Nova
Avanhandava Dam yielded good results for the stress-strain
curve observed along different stress-paths with $K = \sigma_3/\sigma_1 =$
constant. The prediction of volumetric strains was fair to
poor.

35. It is likely therefore that vertical displacements
should be better predicted in enbankment dams than horizontal
ones.

36. The limited amount of investigation carried out suggests that the introduction of elastic anisotropy significantly improves the prediction of volumetric strains. Further
work is clearly necessary in this respect.

177

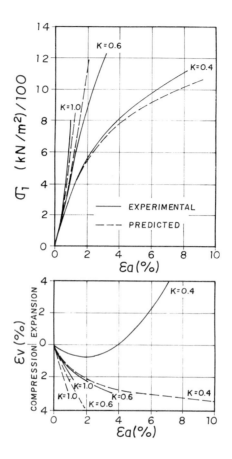

Fig. 7. K stress-path tests/bulk modulus formulation –
experimental and predicted

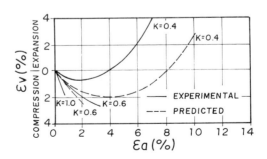

Figure 8. K stress-paths/volumetric curves –
elastic anisotropy

37. Alternatively, more complex elastic (Veiga Pinto, 1983) or elastoplastic (Lade, 1977) models such as used by Azevedo Filho (1990) could be employed but with loss of the desired simplicity.

38. It seems that the best approach to be used in conjunction with the simple hyperbolic model would be to determine the basic parameters for Young's modulus and Poisson's ratio (or bulk modulus) from conventional triaxial compression tests and adjust them so that a good prediction is made of K = constant triaxial tests with different K values.

39. A similar approach was used by Celestino and Marechal (1975) in relation to the prediction of end-of-construction settlements for Ilha Solteira Dam, in a similar soil to Nova Avanhandava Dam. The predicted settlements based on parameters obtained from K = constant tests compared well with the measured ones.

40. Finally it should be mentioned that oedometer tests could be used for the same purpose but only under drained conditions (e.g. for transitions and rockfill). For compacted clayey soils, even under partly drained-partly undrained conditions, triaxial K = constant tests should be used.

REFERENCES

1. LACERDA W. and MAHLER C. (1973). Some comments on earth dam deformations during and after construction, obtained by the finite element method - IX Brazilian Seminar on Large Dams, Rio de Janeiro.

2. EISENSTEIN Z. (1979). The role of constitutive laws in analysis of embankments - 3rd Int. Conf. on Numerical Methods in Geomechanics, vol. 4, 1413-1430, Aachen, AA. Balkema.

3. CHARLES J.A. (1976). The use of one-dimensional compression tests and elastic theory in predicting deformations of rockfill embankments - Canadian Geotechnical Journal, vol. 13, nr. 3, 189-200.

4. COSTA FILHO L.M.; ORGLER, B. and CRUZ, P.T. (1982). Some considerations on the prediction of end-of-construction pore-pressures in dams using laboratory tests - VII Brazilian Congress on Soil Mechanics and Foundation Engineering, vol.4, 11-27, Recife.

5. AZEVEDO FILHO R.N. (1990). Analyses of dam behaviour by the finite element method using an elasto-plastic model - M.Sc. Thesis, Catholic University of Rio de Janeiro.

6. COSTA FILHO (1985). Panel discussion, Session 4, Deformations of Dams, XVI Brazilian Seminar on Large Dams, vol. 2, Belo Horizonte.

7. SCHNAID F.; COSTA FILHO L.M. and MEDEIROS L.V. (1985). An equipment for the measurement of radial displacements in triaxial tests. Soils and Rocks, Journal of Brazilian Society of Soil Mechanics, vol. 6, nr. 3, 35-40.

8. DUNCAN J.M. and CHANG C.Y. (1970). Non-linear analysis of stress and strain in soils. Journal of the Soil Mechanics and Foundation Division, ASCE, vol.96, nr.SM5, 1655-1682.

179

9. DUNCAN J.M.; BYRNE P.; WONG, K.S. and MABRI P. (1980). Strength, stress-strain and bulk modulus parameters for finite element analyses of stresses and movements in soil masses - Geotechnical Engineering Report nr. 80-01, Univ. of California, Berkeley.

10. LADE P.W. (1972). The drained stress-strain and strength characteristics of cohesionless soil. Ph.D. dissertation, University of California, Berkeley.

11. ORGLER B.L. (1983). Stresses and displacements in earth and rockfill dams during construction - M.Sc. Thesis, Catholic University of Rio de Janeiro.

12. ALONSO E.E.; GAUS A. and JOSA A. (1990). The constitutive model for partially saturated soils. Geotechnique, vol. 40, nr. 3, 405-430.

13. COSTA FILHO L.M. (1980). A laboratory investigation of the small-strain behaviour of London Clay - Ph.D. Thesis, University of London.

14. EISENSTEIN Z. and LAW S.T.C. (1979). Influence of anisotropy on stresses and displacements in embankments - 3rd Int. Conf. of Numerical Methods in Geomechanics, vol. 2, Aachen, 709-716, AA. Balkema.

15. VEIGA PINTO, A.A. (1983). Prediction of structural behaviour of rockfill dams - Specialist thesis, LNEC, Lisboa.

16. LADE P.V. (1977). Elasto-plastic stress-strain theory for cohesionless soil with curved yield surfaces. International Journal of Solids and Structures, vol. 13, nr. 11, 1019-1035.

17. CELESTINO T. and MARECHAL L.A. (1975). Stresses and strains in the Ilha Solteira earth dam - 5th Pan-American Conference on Soil Mechanics and Foundation Engineering, vol. 2, Buenos Aires, 189-198.

14. Compaction characteristics of some arid swelling soils

M. A. EL-SOHBY, Al-Azhar University, Egypt, and
A. M. ELLEBOUDY, Banha University, Egypt

SYNOPSIS. The paper presents the results of an experimental
programme conducted to characterize some desert clayey soils
to be used for road construction. Alluvial clay Nile deposits
were included in the research for comparison. All soils tend
to have high swelling pressures. Soil samples were compacted
at different moisture contents and compactive efforts. Swelling pressure and percent were determined for natural and compacted samples, using the oedometer, to study the effect of each
factor on swelling behaviour. CBR test was also used to evaluate the swelling characteristics. Suitability of the classification systems for identifying the expansive soils was
discussed.

INTRODUCTION
1. In arid and semi-arid regions, vast areas of rather dry
clayey soils are usually encountered. In Egypt, wide areas are
covered with potentially expansive soils specially in the desert.
Detrimental heave of light weight structures and road pavement
have been a major concern of civil engineers in the last two
decades, specially in the locations of the new communities. In
planning and design of roads and highways, the available soils
have to be used as subgrades and embankment material, because
in most cases functionally and economically, it is impossible
to replace these soils by more stable soils.
2. Soil reconnaissance in a newly developed area, located in
the desert north east of Cairo, detected the presence of potentially swelling soils. These soils exist in an arid environment
where the natural moisture content is quite low and the water
table is very deep. Since the new land use involves erection
of networks of water and sewage pipes to serve the new community,
the inevitable break down of water supply and sewage systems,
in addition to surface drainage and irrigation, will cause the
formation of perched water table and temporary water horizons
which will alter the natural moisture equilibrium and change
the degree of saturation.
3. Experimental research programme was conducted to characterize three types of arid desert clayey soils, obtained from
the above mentioned area, to be used as subgrade and road embankments. Samples from partially saturated alluvial Nile

clayey deposits in Cairo area, taken from the desiccated layers
above the water table, were included in the research for
comparison.

PHYSICAL PROPERTIES

4. The stiffness of soil and the low level of water table
furnished suitable conditions for obtaining undisturbed block
samples from open pits. The block samples were immediately
transported to the laboratory in their natural condition.

5. Classification tests were carried out on representative
samples including natural moisture content, mechanical analysis,
Atterberg limits, specific gravity, and bulk density. The tested soils were classified according to the American Association
for State Highways and Transportation Officials (AASHTO), and
the Federal Aviation Administration (F.A.A.) classification
systems. The desert soil samples were denoted as D1,D2 and D3,
and the alluvial soils were A1 and A2. It was observed that
the desert clayey soils have low natural moisture contents and
high dry unit weights, compared with the alluvial soils, due to
their old geological history and environmental conditions. The
alluvial clayey soils were less active, and possessed higher
natural moisture contents and moderate dry densities, since
they are considered relatively recent deposits. All the physical properties are shown in Fig. 1 and Table 1.

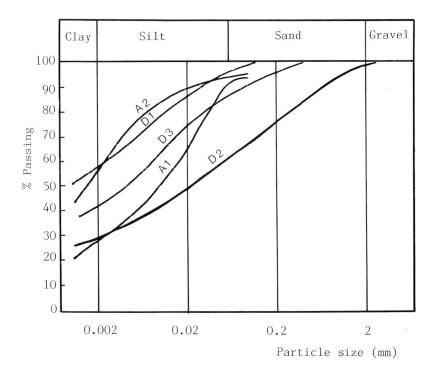

Fig. 1. Grain size distribution of tested soils

Table 1. Properties of tested soils

Sample No.	D1	D2	D3	A1	A2
Depth of sample (m)	1.25	1.50	1.50	1.50	1.25
Natural water content %	13.0	8.50	4.42	24.0	36.40
Dry unit weight kN/m^3	18.50	20.0	19.0	15.4	12.8
Specific gravity	2.73	2.97	2.68	2.76	2.72
Liquid limit %	80	73	50	63	74
Plastic limit %	28	34	25	25	34
Shrinkage limit %	15.80	22.80	16.05	12.00	10.13
Plasticity index %	52	39	25	38	40
Activity (Skempton)	0.93	1.35	0.926	0.91	0.71
Activity (El-Sohby)	0.75	1.51	0.69	0.60	0.66
Free swell (Holtz) %	100	150	40	70	50
Free swell (< 2μm) %	150	250	70	110	100
AASHTO classification	A-7-6	A-7-5	A-7-6	A-7-6	A-7-5
F.A.A. classification	E-12	E-11	E-8	E-10	E-11

MINERALOGICAL COMPOSITION

6. The mineralogical composition of the clay fraction of the five types of soils were determined using the x-ray diffraction and the differential thermal analysis. The main clay mineral in all soils was the montimorillonite. Soils D1 and A1 contained montimorillonite and kaolinite, soil D3 and A2 contained montimorillonite, kaolinite and illite, and soil D2 contained montimorillonite, kaolinite, and vermiculite (2,3). In soils D1,D3,A1 and A2 the exchangeable cation is calcium, but it is sodium for soil D2.

ACTIVITY AND FREE SWELL

7. Skempton (5) defined the activity of soil as the ratio of the plasticity index to the abundance of the clay fraction (less than 2 μm) as percent of the dry weight of the sample. El-Sohby (1) modified the activity to be the plasticity index of the clay particles (less than 2 μm) divided by 100. Van Der Merwe (6) suggested a guide to estimate the degree of expansion in relation with Skempton's activity. According to his activity chart, soils D1,D2,D3 and A2 are expected to have a high degree of expansion, but soil A1 is highly expansive. Activity of soils are shown in Table 1.

8. The free swell test results (Table 1) showed that soil D2 possessed the highest swelling percent using the procedure suggested by Holts and Gibbs (4). This result agrees with the

measure activities since soil D2 has the highest activity number. It is important to mention that the simple free swell test measures the actual tendency of soil to swell without the influence of the boundary conditions imposed in the other sophisticated swelling tests.

COMPACTION TESTS

9. Samples from each soil were compacted according to AASHTO modified Proctor test. The compaction curves are shown in Fig. 2. In order to study the influence of compaction moisture content on the swelling properties of soils, compacted samples were prepared at 95 % and 105 % of the optimum moisture content. In addition, samples from each soil were prepared using higher compactive effort by increasing of blows per layer to 35 blows, and another set of samples were compacted using less energy by compacting the samples according to the standard Proctor test. All the compacted samples were used in the swelling tests.

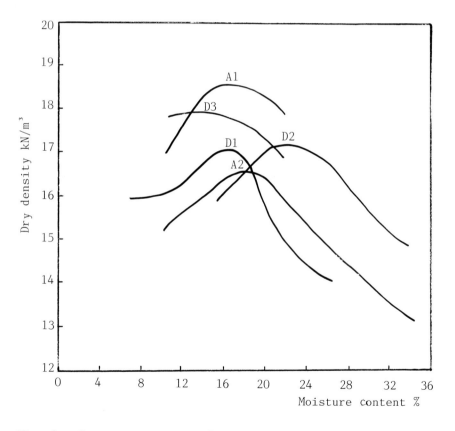

Fig. 2. Compaction curves of tested soils

SWELLING BEHAVIOUR

10. The swelling pressure and swelling percent of soils were
evaluated by the oedometer test. Undisturbed specimens, carefully carved from the block samples, were used to evaluate the
swelling properties in the natural condition. The first set of
prepared samples were used to derive a relationship between the
compaction moisture content, and the swelling pressure and the
swelling percent. It was evident that the swelling percentages
of the alluvial clayey soils are generally greater than those
of the desert soils. The swelling percent of the desert soils
were more influenced by the amount of compaction water content
than that of the alluvial soils. This observation emphasizes
the advandage of compacting the active clayey desert soils on
the wet side of the optimum. Values of swelling pressure, however, for both types of soils, were very close within the used
range of compaction moisture content. Table 2 illustrates the
results of these tests.

11. Comparing the swelling pressure and swelling percent of
the natural soil samples and the remoulded samples, it was found
that compaction increased both the swelling pressure and swelling percent. It became obvious that disturbance of the structure of these expansive soils increased their swelling potential.

12. Both swelling pressure and swelling percent increased
with increasing the compaction effort as shown in Table 3. Thus,
it is recommended to use the equivalent of the standard compaction effort when compacting an expansive soil in the field,
and decrease the requirement of percentage of field compaction
to 85 % to 90 % to reduce the expected swelling.

CBR TEST

13. The swelling percent of the samples compacted at different compactive efforts was measured using the standard CBR procedure. The CBR test results do not agree with the swelling
percent measured by the oedometer test. The swelling percentages measured by the CBR test were much less than the values
determined by the oedometer test. It seemed that the high side
friction in the CBR test attributed to the decrease in the swelling percent values. Table 4 gives an example of the comparison
between the values of swelling percent determined by both tests
using samples from each soil, compacted by the modified compactive effort at the optimum moisture content. The disagreement
in the results between the oedometer and CBR tests questioned
the reliability of the CBR test for evaluating the swelling behaviour of the active clayey soils.

Table 4. Swelling percent from CBR and oedometer tests

Sample No.	D1	D2	D3	A1	A2
Swelling percent (CBR test)	2.60	3.70	6.30	1.40	6.40
Swelling percent (oedometer test)	21.60	14.80	19.60	11.40	22.50

Table 2. Effect of compaction moisture content on swelling characteristics

Soil No.	Soil properties	Compaction moisture content %		
		95% O.M.C.	100% O.M.C.	105% O.M.C.
	Compaction moisture content %	16.15	17.00	17.85
D1	Swelling %	22.00	21.60	20.60
	Swelling pressure kN/m²	1060	830	790
	Compaction moisture content %	20.90	22.00	32.10
D2	Swelling %	16.80	14.80	11.10
	Swelling pressure kN/m²	1460	1230	1159
	Compaction moisture content %	13.30	14.00	14.70
D3	Swelling %	26.50	22.59	20.80
	Swelling pressure kN/m²	1459	1280	1249
	Compaction moisture content %	15.20	16.00	16.80
A1	Swelling %	13.20	11.40	10.90
	Swelling pressure kN/m²	390	359	340
	Compaction moisture content %	17.58	18.50	19.43
A2	Swelling %	22.70	19.60	17.10
	Swelling pressure kN/m²	780	700	640

Table 3. Effect of compaction effort on swelling
characteristics

Soil No.	Soil properties	Type of compaction		
		Standard	Modified	High
	O.M.C. %	18	17	16
	$\gamma_{d\ max.}$ kN/m^3	15.4	17.0	17.6
D1	Swelling %	18.20	21.60	26.4
	Swelling pressure kN/m^2	300	830	1200
	O.M.C. %	22.5	22.0	20.0
	$\gamma_{d\ max.}$ kN/m^3	15.2	17.2	17.4
D2	Swelling %	14.6	16.2	28.6
	Swelling pressure kN/m^2	360	1230	1600
	O.M.C. %	21	14	13
	$\gamma_{d\ max.}$ kN/m^3	15.6	18.1	18.3
D3	Swelling %	6.5	22.4	23.0
	Swelling pressure kN/m^2	140	1270	1450
	O.M.C. %	19	16	15.6
	$\gamma_{d\ max.}$ kN/m^3	16.2	18.6	20.3
A1	Swelling %	6.0	10.2	13.7
	Swelling pressure kN/m^2	190	400	790
	O.M.C. %	28	18.5	14
	$\gamma_{d\ max.}$ kN/m^3	13.8	16.5	18.3
A2	Swelling %	4.0	19.6	38.8
	Swelling pressure kN/m^2	100	780	2180

CLASSIFICATION SYSTEMS AND SWELLING PREDICTION

14. The soils included in this research were classified according to the AASHTO classification system; the most widely used in highway engineering. They were also classified according to the F.A.A. classification system which used for air fields. The classification of the five soils is shown in Table 1. Both systems were based on the grain size distribution and Atterberg limits. They ignored totally the swelling characteristics of soils and treated them as other normal types of clays. They have even treated silt and clay as one portion (passing 200 sieve) inspite of the fact that the clay fraction is the most effective part in expansive soils. Since Atterberg limits are not uniquely related to expansiveness, it is suggested to add a rather simple test such as the free swell test, plus the determination of the activity number to both classification systems to accomplish a proper identification of expansive soils.

CONCLUSIONS

15. In the light of the preceding results, we can conclude the following:

(a) Both the swelling pressure and swelling percent decrease with increasing the compaction moisture content. This confirms the importance of compacting the active clayey soils on the wet side of the optimum.

(b) Both the swelling pressure and swelling percent increase with the increase of compaction effort. Thus, it is recommended to compact the expansive soils at 85 % to 90 % field compaction, using the standard compactive effort, to reduce its tendency to swelling.

(c) The evaluation of swelling of soils by the oedometer test was more reliable than the CBR test, since the swelling percent measured by the CBR test was much lower than the values measured by oedometer test. This was attributed to the high side friction between the sample and the walls of the mould during the test. Thus, it is not recommended to use this test for evaluation of swelling characteristics of expansive soils.

(d) Remoulding has an essential influence on the swelling percentage. Disturbing the natural structure of the tested expansive soils increased their swelling potential.

(e) The AASHTO and F.A.A. classification systems do not differentiate between expansive clays and less active clays, and do not detect their tendency to swelling. For more accurate classification, it is recommended to calculate the activity number of soil, and to measure the free swell to identify the expansive soils.

REFERENCES
1. EL-SOHBY M.A. Activity of soils. Proc. 10th Int. Con. on Soil Mech. & Found. Eng., Vol. 1, 587-591, Stockholm 1981.
2. EL-SOHBY M.A. and MAZEN S.O. Mineralogy and swelling of

expansive clayey soils. Geotechnical Engineering, South-east Asia, 1983, Vol. 14, 79-87.

3. HABIB S.A. Behaviour of some swelling soils under light weight structures. M.Sc. Thesis presented to Al-Azhar University, Cairo, Egypt, 1984.

4. Holtz W.G. and Gibbs H.J. Engineering properties of expansive clays. Transaction ASCE, 1956, Vol. 121, 641-677.

5. Skempton A.W. The colloidal activity of clays. Proc. 3rd Int. Conf. on Soil Mech. & Found. Eng., Vol 1, 57-61, Zurich 1953.

6. VAN DER MERWE C.P. and AHRONOVI M. The behaviour of flexible pavement on expansive soils in Rhodesia. Proc. 3rd Int. Conf. on Expansive Soils. 267-276, Haifa 1973

15. The behaviour of a laterite fill subjected to drying and wetting, including the effects of crack jacking

R. J. JARDINE, O. OYEWUMNI and E. ARMAH, Imperial College, UK, H. GRACE, Henry Grace and Partners, UK, and D. W. HIGHT, Geotechnical Consulting Group, UK

SYNOPSIS

Field observations of laterite based low–volume roads in tropical environments suggest that their performance is improved by allowing drying and wetting weathering cycles between being placed and any subsequent bituminous surfacing. This paper describes a related laboratory investigation which involved a variety of tests on a compacted plastic laterite from Kenya. The soil was subjected to drying and re–wetting, with and without the (substantial) shrinkage cracks being infilled; the results are used to explain and quantify the effects of weathering cycles on laterite pavement fills.

INTRODUCTION

Background

The suitability of tropical laterite for pavement construction in low–volume roads has been studied over the past twenty years. Notable contributions include the field work described by Grace et al (1988) and the laboratory research reported by Toll (1988), Grace and Toll (1987) and Hight, Toll and Grace (1988). Their work shows that laterite road bases can perform well, even when they fail to conform with standard specifications. Good compaction appears to be the key; exposing the laterite base course to weather and traffic before surfacing with bitumen gives additional benefits.

Investigations in Kenya and Malawi have shown that laterite base courses undergo extensive cracking within a few days of their initial compaction in place, due to the drying out of plastic fines. Cracking even occurs during wet seasons, as sunny intervals invariably develop in between the rain storms. Surface water and traffic act subsequently to wash sand, silt and clay into the cracks. Initially the crack widths fluctuate with the weather, but the tendency to close up in wet intervals and re–open in dry periods reduces with each successive cycle. This process leads to an overall drying of the laterite, with water contents typically falling to about half the Proctor optimum by the end of the wet season. Combined with the kneading action of traffic, the wetting and drying cycles eventually cause the cracks to heal. Furthermore, the traffic and weathering render

Engineered fills. Thomas Telford, London, 1993

the initially smooth compacted laterite road surface into a rough mosaic of hard laterite nodules which are held firmly in position by a well compacted and dried—out soil mortar.

After sweeping away any loose fine material, a thin bituminous surface dressing is applied, which adheres well to the weathered laterite. Evaporation is greatly reduced by the surface dressing and the relatively dry base course draws in water from the surrounding soil mass until the suctions in the road base are in equilibrium with the local piezometric conditions. Wetting—up is relatively rapid at the road edges, but two years may be required to reach final equilibrium suctions and stable water contents on the centre—line.

The road locations studied by Grace and Toll had their formation levels at least several metres above the regional water table giving, with plastic laterites, final soil suctions of around 100 kPa, or more. Field measurements in Kenya and Malawi suggest that the final equilibrium sub—base water contents are 10 to 20% below those operating at the time of construction. The weathering and re—wetting process appear to give substantial improvements in fill properties at depths of up to 1.0m.

The present study

This paper describes experiments that add to Toll's earlier work. The behaviour of one plastic laterite was studied during drying and wetting; a wider range of suctions was considered than was possible previously and further attempts were made to investigate crack infilling. The laterite was sampled by Grace in 1982 from the Cheronge borrow pit, near to the Nyeri North West 7 route; see Figure 1. The experiments were performed at Imperial College by Oyewumni (1992) and Armah (1992).

EXPERIMENTS ON THE CHERONGE LATERITE

Sample classification, preparation and compaction testing

The Cheronge laterite is a deep red—brown, well graded soil. Particle size distributions and Atterberg limits were determined on four large bulk samples, giving the results shown in Figure 2 and in Table 1. The limits all plot just above the Casagrande 'A—line', classifying the sub 0.5mm fractions as high plasticity clays. Although about half of the as—dug material was coarser than 0.5mm, this fraction consisted mainly of friable concretions that could be broken by hand, were easily pulverised and would be broken down to some extent by excavation and field compaction.

The laboratory study required a relatively large volume of stable soil that contained no coarse particles and a mixture was

191

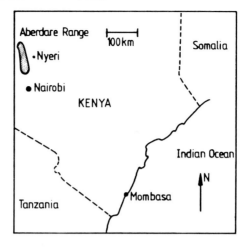

1. Location of laterite sampling area

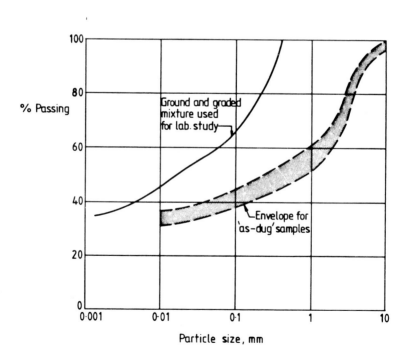

2. Grading curves for Cheronge laterite

made with 50% sieved as-dug soil and 50% ground-down coarse
material. This composite soil had virtually identical index
properties to the mean of the intact samples (see Table 1). Its
modified grading curve is shown in Figure 2; Table 2 gives other
relevant index data.

Sample Type	Liquid Limit %	Plastic Limit %	Plasticity Index %
As dug, means	60.8	28.3	32.5
Test mixture	59.0	26.7	32.2

Table 1. Atterberg limits for Cheronge laterite

Compaction curves for the mixture and the sieved unground soil
are presented on Figure 3, showing that adding the ground
component increased γ_d and decreased the optimum water content.
The laboratory samples were prepared by compaction following
either (i) the BS heavy (4.5kg) procedure in either standard
compaction or CBR test moulds, or (ii) static loading combined
with dynamic pounding to achieve the same densities in a special
mould made for standard oedometer test samples.

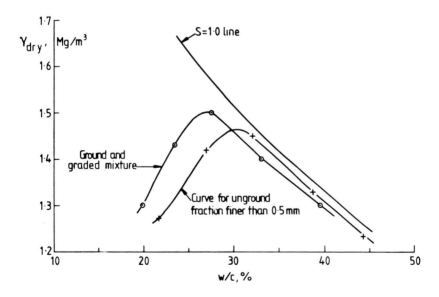

3. Compaction curves for Cheronge laterite

Shrinkage Limit (@ 21°C)	20.0%
Shrinkage Limit (@ 40°C)	21.4%
Specific Gravity	2.78
Linear Shrinkage	15.4%
Activity	0.575
Clay Fraction	35%
Optimum w/c (BS heavy comp.)	27.6%
Maximum γ_d (BS heavy comp.)	1.49Mg/m³

Table 2. Index data for test mixture

The use of laboratory compacted samples is supported by fabric studies by Toll, Shaw and Hight (1985), which indicated reasonable agreement between field and laboratory compacted laterites. Most of the laboratory testing involved samples compacted wet of optimum, with γ_d falling within 95% of the maximum BS heavy compaction dry density. Field compaction appears to achieve a similar range of dry density (typically 92 to 95% of heavy compaction optimum).

Suctions and total radial stresses after compaction

The matrix suctions developed by Cheronge samples when compacted (in standard Proctor, CBR and oedometer moulds) to different water contents were determined following the filter paper procedures and calibrations proposed by Chandler, Crilly and Montgomery-Smith (1992). As shown in Figure 4, suctions of up to 1 Mpa apply after standard compaction at the BS heavy optimum water content, with initial suctions of 20 to 100 kPa being developed in the wetter samples that were used for most of the study.

The Cheronge data are broadly comparable to Toll's results for other laterites. His laboratory experiments gave post-compaction matrix suctions in the range 50 to 700 kPa, depending on the initial water content and the compactive effort. Toll also measured the lateral stresses that were locked-in by laboratory compaction in a rigid mould, showing that they were of comparable magnitude to the matrix suctions.

It is not clear how large the suctions and lateral stresses set up by field compaction are, but smaller values probably pertain (for any given water content) in the field than in the laboratory, because the degree of lateral and basal confinement is lower.

194

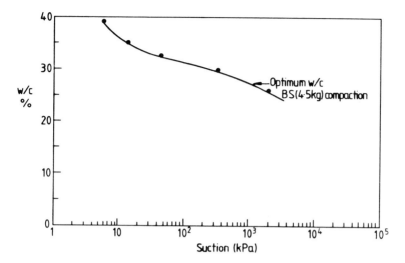

4. Suction water-content relationship after BS heavy compaction for Cheronge laterite

Drying tests

Samples of Cheronge laterite were compacted at a range of water contents in the oedometer and CBR moulds and then dried out completely in either (i) a temperature controlled laboratory (at 21°c with typically 40% relative humidity), or (ii) in a heated cabinet (with relative humidity ≈ 25% and temperature ≈ 40°C). The samples were dried from both ends; their weights and dimensions were measured at regular intervals.

Evaporation was roughly ten times faster in the heated cabinet and it was found that weight loss rates were hardly affected by the initial water content. Figure 5 shows some typical data obtained in the heated cabinet, giving curves of average water content against time for (i) a 76mm diameter, 19.5mm thick 'oedometer' sample and (ii) a 152mm diameter, 128mm thick 'CBR' specimen. Both curves are linear to well beyond the shrinkage limit. To a first approximation, the linear portions follow the expression:

$$\Delta w/c = A. t. (1m/H)^{1.2} \qquad Eq. 1$$

Where t is the time in seconds, H is the sample half thickness, 1m is the one metre equivalent distance (in the units used for H) and A is 7.8×10^{-8} s^{-1} for drying at 40°C.

All samples developed pronounced drying cracks, with radial stresses dropping quickly to zero. Thin radial cracks formed first but, as drying continued, the soil separated from the confining rings, the radial cracks healed and lateral shrinkage was accommodated by a circumferential boundary gap. The ultimate radial shrinkage strains (for samples compacted at 32% water content) amounted to around 5% in both 'oedometer' and 'CBR' sample types, with corresponding gaps of about 2 and 4mm respectively. Further insights into the drying process are given in Figures 6 and 7.

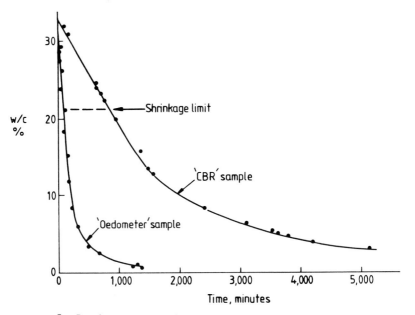

5. Drying curves for compacted Cheronge laterite

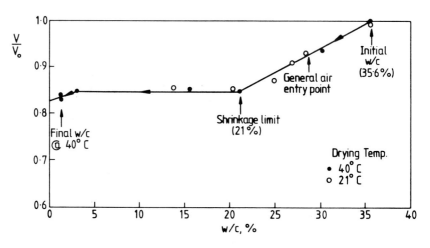

6. Volume change water–content relationship for Cheronge laterite during drying

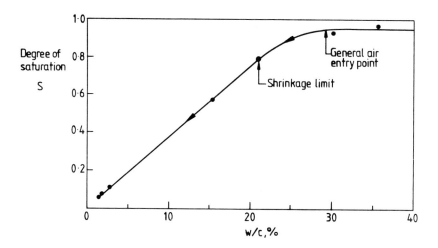

7. Variations in degree of saturation with water content for compacted Cheronge laterite during drying

Figure 6 shows how the volumes (V) of 'oedometer' and 'CBR' samples compacted at ≈ 35% water content changed during drying at either 20° or 40°C. A unique, tri-linear, relationship was found; drying to the shrinkage limit involved ≈ 15% volume change, with no additional strain developing until the final 2 to 3% of water evaporated. Figure 7 reports the corresponding changes in the degree of saturation (S). When compacted wet of optimum, the Cheronge laterite was 95% saturated. The degree of saturation began to decrease markedly after the water content had reduced by about 5% (at the 'general air entry point' proposed by Marinho 1992). S fell to 80% at the shrinkage limit; a linear relationship was found at lower water contents that projected back to the origin.

Following from the continuous drying tests, ten identical oedometer sized samples were compacted at 33% water content and allowed to dry until a specified weight change had occurred. The samples were sealed against further evaporation and left for 24 hours at 20°. Matrix suction and water content determinations were then made, giving the results plotted on Figure 8. Drying to the general air entry point generated a suction of ≈ 4Mpa, which increased to 10MPa when the shrinkage limit was reached. Suctions of up to 50MPa were indicated at low water contents.

The reduction of water content to half the BS optimum seen in field tests under drying pavements would take the Cheronge laterite to about 14%, which is well below its shrinkage limit High suctions and considerable vertical cracking would certainly develop; drying to lower water contents would increase the suctions, but have no effect on void ratio or crack widths. We can also use Equation 1 to estimate that a 1m thick layer of

197

Cheronge laterite could easily dry to below its shrinkage limit within a few weeks under dry tropical conditions.

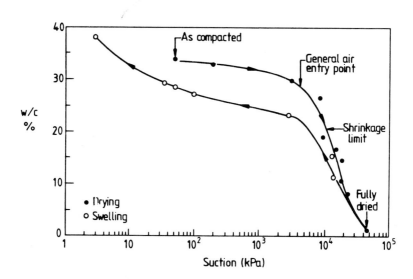

8. Variations in matrix suction for compacted Cheronge laterite during drying and re-wetting

Behaviour on re-wetting

It is well known that soil drying and wetting is hysteretic. The points of interest are how hysteresis affects engineering behaviour and how the processes are modified by crack infilling. Toll's instrumented CBR mould tests indicated that laterites compacted to suctions between 20 and 600 kPa experienced little overall change in their locked-in compaction total radial stress as a result of soaking, whereas partial drying and re-wetting (without effective crack infilling) caused the radial total stress to drop dramatically. The studies by Oyewumni (1992) and Armah (1992) extend from Toll's work. Firstly, samples were dried completely, so that substantial cracks developed that could be infilled more effectively. Secondly, single stage and incremental tests were performed that covered a far wider range of suctions than was considered previously.

Suction-water content relationships

The main set of re-wetting experiments involved oedometer sized samples which were first compacted at about 35% water content then dried to just 1 to 2%. The 2mm wide circumferential gaps between the dried soil and confining rings were then infilled, by rodding in soil from the laterite's 63μm and finer fraction. A graduated amount of water was sprayed over the surface of each sample, which was then sealed and allowed to reach an internal equilibrium at 20°C over 24 hours. The sample's matrix suction and water content were then determined. Although the changes in sample height were not measured, the shape of the wetting curve and visual observations suggest that volume changes were small until a 'swelling limit' was exceeded, when water contents rose above 20% and suctions had fallen below 2 to 3 MPa.

The complete drying and wetting curve is shown in Figure 8. The relationship is hysteretic and, provided the final equilibrium suctions do not fall below about 10 kPa, the overall process involves a net densification. For the final suction range expected in the field (>100 kPa) the final equilibrium water content is similar to that at the BS heavy optimum water content, and is at least 7% dryer than when compacted.

Radial total stress changes

Two incremental swelling tests were carried out on larger 'CBR' samples using the mould designed by Toll (1988), which incorporates thinned strain gauged sections to measure average total lateral stress. Calibrations involved applying known internal air pressures to a closed cylindrical section.

CBR samples were compacted in the rings (at initial water contents of about 29%) before being dried out fully and having base plates fitted to prevent leakage. The 3 to 4mm wide circumferential shrinkage gaps were then infilled. In the first test laterite fines were used, but it proved difficult to compact this infill satisfactorily. A rounded medium sand (0.3 to 0.6mm diameter) was used for the second test, giving a tighter packing. A graduated amount of water was added to the top of the samples at the start of each increment, a cover plate

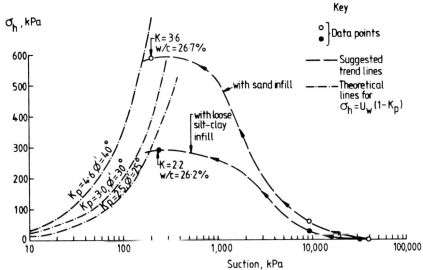

9. Radial total stresses during swelling for dried Cheronge laterite; laboratory measurements and theoretical upper limits

199

was placed to prevent evaporation, and the system was left to equalise, with progress being monitored via the radial stress measuring circuits. At least 24 hours elapsed between increments and the swelling tests were stopped when the soil reached the BS heavy optimum water content.

The swelling increments were not completely monotonic. Rapid swelling occurred at the boundary when the soil was temporarily in contact with free water, followed by drying as moisture was drawn towards the higher suction central portion of the sample. As a result of the soil's hysteretic characteristics, water contents determined after the end of the first test were 3% higher than the average (27%) at the top, and 3% lower at the base. Recharging through the sand infill used in the second test halved the final centre-line water content variations.

The average radial total stresses measured at four water contents for each sample are presented in Figure 9 in a plot of σ_r against suction; the suctions were estimated by referring the average water contents to the independently derived curve shown in Figure 8. Figure 9 also shows theoretical lines relating suction to σ_r assuming: (i) the stress conditions cannot violate a passive failure criterion with $c' = 0$ and (ii) that the samples were nearly saturated at the ends of the tests, with the effective stresses being close to $\sigma - u_w$. For these conditions Equation 2 holds, where K_p is the Rankine Passive

$$\sigma_r = u_w (1 - K_p) \qquad \text{Eq 2}$$

Pressure coefficient. From Figure 9 we may note further:

1. The radial stresses remained relatively small in both tests until the water content exceeded the supposed swelling limit (at 2 to 3 MPa) and the soil started to expand against the confining ring.

2. Towards the end of the tests the radial total stresses were approaching the passive failure lines corresponding to relatively high ϕ' values (Toll measured critical state angles of $\approx 40°$ in comparable soils).

3. Once on a passive failure line, the suctions and radial total stresses are inter-linked by Equation 2 and are dependent only on the equilibrium suction. However, the effectiveness of the infill both controls σ_r up to this point, and fixes the suction value at which passive failure starts. If no infill is placed the gap may not be able to close before reaching the 'field' final suction.

Oedometer loading tests

Five oedometer loading tests were performed on the Cheronge laterite. Figure 10 shows the void ratio-effective vertical

stress diagram of a standard test performed on soil compacted at
31.5% water content, along with data obtained from a test
designed to model the 'weathering' process. The 'weathered'
sample was prepared by compacting the laterite at the same water
content, then drying it fully before filling the shrinkage gap
with the medium sand. The vertical swelling pressure was then
determined (≈ 850 kPa, compared with the 'dry' soil's pre-
testing matrix suction of ≈ 50 MPa) and the sample was swelled
back before being recompressed. As anticipated from the data
shown in Figure 8, the 'weathered' sample was markedly denser
than the freshly compacted material. The 0.08 difference in void
ratio persisted during swelling and when relatively high
stresses were imposed. As an aid to interpretation the Intrinsic
Compression Line (ICL) expected for the laterite when compressed
from a reconstituted slurry has also been added; this was
projected following the relationships with liquid limit given by
Burland (1990).

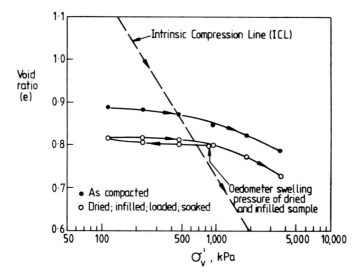

10. Oedometer tests on compacted and dried and infilled Cheronge
    laterite

Figure 11 shows data from three further tests on dried compacted
laterite; the projected ICL discussed above is also shown. In
one test a dried laterite was flooded, without any infill being
placed, at a stress 100 kPa below the previously determined
swelling pressure for an infilled sample. An immediate collapse
was seen, involving ≈ 1.5% strain, and the sample was marginally
softer during subsequent tests than the infilled equivalent
shown in Figure 10. The two other samples were flooded at 3,600
kPa – one without any infill, the other with silt and clay
rodded into the gap. Both samples collapsed, but the infilled
sample experienced far smaller deformations. Overall, the
oedometer tests emphasise the beneficial effects of crack
infilling.

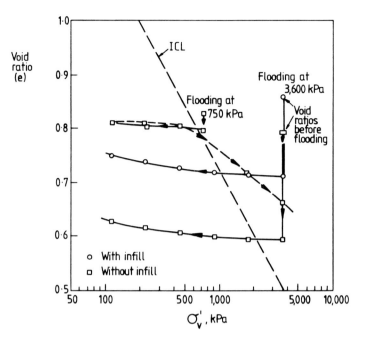

11. Oedometer tests on dried Cheronge laterite, showing collapses when flooded due to excessive vertical stress or absence of crack infilling.

ENGINEERING IMPLICATIONS

The Cheronge laterite experiments illustrate the stress and suction changes that a compacted laterite road pavement might experience, giving results that are broadly compatible with Grace and Toll's field observations.

The effects of weathering before bituminous sealing

In addition to preparing the road surface for surface dressing, weathering appears to improve the performance of the laterite road-base. The potential benefits depend on the change in matrix suction between compaction and final equilibrium.

Provided that there is no large net reduction in suction, and that shrinkage cracks are infilled, the weathered laterite is likely to densify. Grace and Toll's field measurements and the present tests suggest that Cheronge laterite would experience a 0.08 overall reduction in void ratio through weathering. If the laterite was surfaced without undergoing the drying process it would experience a small net increase in water content with time. Assuming that the critical state undrained shear strength varies with effective stress according to a curve parallel to the ICL (for which $C_c^* \approx 0.4$) suggests that the final saturated

202

undrained shear strength, Cu, of the weathered soil should be at least 60% higher than that of the same soil if sealed without being weathered. A pro-rata gain in static pavement bearing capacity would be expected.

Soil shear stiffness is also affected by the weathering process. Triaxial experiments performed by Toll (1988) prove that the shear stiffness of saturated laterite increases with both soil dry density and mean effective stress, p'. A relatively dry weathered laterite which swells from a high suction with all cracks closed is likely to tend to the passive stress condition described by Equation 2, where the long term p' is $(1 + 2K_p)/3$ times the equilibrium suction, u_w. If the laterite was surfaced without being pre-weathered it would be less likely to reach full passive failure during swelling and, for the same value of u_w, a smaller p' would apply. The combined effects of higher p', smaller void ratio and the passive stress path followed during wetting up could increase shear stiffnesses by a factor of two or more, so considerably stiffening the road foundations. Reducing the shear strains experienced under traffic loading is likely to lead to a greatly improved fatigue life for the road base under traffic loading (see Brown and Selig 1991).

The role of crack infilling

Effective crack infilling is vital if the laterite is to develop a high K_0 when swelled back to the equilibrium suction. Without this lateral restraint, little improvement may be achieved in γ_d, Cu, p', or shear stiffness, G.

CONCLUSIONS

1. The laboratory research helps to explain and quantify the potential benefits of allowing a laterite road to weather between fill placement and bituminous sealing.

2. The potential benefits of the process depend critically on effective crack infilling and on achieving high long-term suctions.

3. Practical measures might be considered to optimise the cracking and infilling processes, such as placing crack initiators and using sand infill.

4. Conventional laterite pavements may be ineffective in cases where the regional water table is less than a few metres below formation level. Alternative measures should be considered in such cases.

ACKNOWLEDGEMENTS

The practical help provided by Dr A Ridley in the laboratory and the financial support of the Rees Jefferies Road Fund are gratefully acknowledged. Mr Oyewumni was supported at Imperial College by an EC study grant.

REFERENCES

Armah, E (1992)
Report on 'An experimental investigation of drying, swelling and crack infilling on a tropical soil', Imperial College, London.

Brown, S F and Selig, E T (1992)
The design of pavement and rail track foundations. Cyclic loading of soils. Blackie, Glasgow, pp 249–305.

Burland, J B (1990)
On the compressibility and shear strength of natural clays. 30th Rankine Lecture, Geotechnique, Vol 40, No 3, p329–378.

Chandler, R J, Crilly, M S and Montgomery-Smith, G (1992)
A low cost method of assessing clay dessication for low-rise buildings. Proc. ICE, Vol 92, May, pp 82–89.

Grace, H, Scott Wilson Kirkpatrick and Partners and Imperial College. (1988) An investigation into the use of laterite instead of crushed stone or stabilised material as a base course for bituminous roads. Report for Republic of Malawi, World Bank, SERC and African Development Bank; Volumes I to III.

Grace, H and Toll, D G (1987)
Recent investigations into the use of laterites as bases for bituminous surfaced low volume roads. Proc. 4th Int. Conf. on low volume roads. Transportation Research record 1106, Vol 1, pp 80–88.

Hight, D W, Toll, D G and Grace, H. (1988)
Naturally occurring gravels for road construction. Proc. 2nd Int. Conf. on Geomechanics in Tropical Soils, Singapore, Vol 1, pp 405–412.

Marinho, F (1992)
Personal Communication

Oyewumni, O L (1992)
A study of a lateritic soil used for low cost road construction in Africa. MSc Dissertation, Imperial College, London.

Toll, D G, Hight, D W and Shaw, H F (1985)
The rôle of soil fabric in determining the engineering behaviour of compacted lateritic and quartzitic gravels. Int. Conf. on Geomorphology, Manchester. Published in Zeitschrift fur Geomorphologie, 1987.

Toll D G (1988)
The behaviour of unsaturated compacted naturally occurring gravel. PhD Thesis, University of London (Imperial College).

Toll D G (1990)
A framework for unsaturated soil behaviour. Geotechnique, 40, 1, pp 31–44.

204

16. Compactive measure of cohesionless soils

N. MOROTO, Hachinohe Institute of Technology, Japan, and
S. SHIMOBE, Nihon University, Japan

SYNOPSIS : The authors first proposes a new classification
diagram for cohesionless soils. Based on this diagram, the
strength characteristics of sands and gravels can be easily
determined according to their particle properties. Under
normal confining pressure, the introduction of relative
compaction for cohesionless materials becomes more effective
and useful for design and execution management of engineered
fills. In higher pressure ranges, due to the reduction in
strength by particle breakage, we suggest that well graded
rounded materials have superior engineering properties and
are more suitable for use in high fill dams.

INTRODUCTION
1.   Cohesionless materials are extensively used in
earthwork construction, for example in highway embankments
and rockfill dams. The engineering properties of compacted
fills that are most relevant to civil engineers are
compressibility, shear strength, and permeability. These
properties are mosly related to the compaction
characteristics of fills, an increase in density changes the
properties. Therefore, granular soils as fill materials have
been densely compacted.
2.   In this paper, the authors propose a new classification
diagram to correlate the particle characteristics with the
mechanical properties of cohesionless soils such as sands and
gravels. This diagram is drawn in terms of the minimum void
ratio e_{min} and Mogami's strength coefficient k obtained from
the limit density test and shear tests, respectively. After
detailed analysis of the k-e_{min} relationship, a simple method
to evaluate the angle of internal friction ϕ introducing the
relative compaction C_f is suggested. (Further, some
compactive measure of cohesionless materials also are
discussed.)

BASIC PROPERTIES OF k-e_{min} RELATIONSHIP FOR COHESIONLESS SOILS
3.   Mogami's formula is often referred to (ref.1) :

$$\sin\phi = \frac{k}{1 + e} \qquad (1)$$

Where, e : void ratio, k : Mogami's strength coefficient

The strength coefficient, k, can be determined from drained shear tests (mostly triaxial compression tests) on cohesionless soil specimens which have different initial void ratios e_o. Thus,

$$k = (1 + e_o) \sin\phi \qquad (2)$$

From Eq.2, the consistency of k and its average value $\bar{k}$ can be examined and calculated for each material.

    4. Moroto (ref.1) showed that the consistency of the coefficient k was reasonable for different granular soils and that the dependence of k on the particle properties of material was notable. Further, he concluded that k was a material constant closely related to the minimum void ratio e_{min} and that this relationship increased linearly with an increase in e_{min}. This could be represented as :

$$k = a\, e_{min} + b \qquad (3)$$

where, a and b are constants depending on particle shape and grading. This feature of Eq.3 can be observed for a confining pressure range of (σ_3 50~500 or 700 KN/m²).

    5. Shimobe and Moroto (ref.2), applying the k-e_{min} relationship of Eq.3 to data obtained by many researchers from tests on sands and gravels, investigated the effects of particle characteristics such as roundness, R, and coefficient of uniformity, U_c, on the constants a and b. It was found that :

(a) For 23 kinds of sands with a uniform grading ($U_c \sim 2$) and different particle shape, the following equation was obtained :

$$\bar{k} = 1.290\, e_{min} + 0.314 \qquad (4)$$

    The coefficient of correlation for linear regression analysis is 0.953. The 'A-line' given by Eq.4 for uniform samples represents a particular characteristic line and the position of each material on the A-line depends on the degree of roundness (also from Fig.1).

(b) For granular materials with similar roundness but with a variation in grading, the $\bar{k}$-e_{min} relationship for each soil is linear. The lines are at an angle and to the left of the A-line and are parallel to one another as shown in Figs. 1 and 2. These lines can be expressed by :

$$\bar{k} = 0.260\, e_{min} + b_2 \qquad (5)$$

    Where, b2 is constant depending on the roundness. Eq.5 was determined by linear regression analysis (γ = 0.987) from tests on Kakogawa sands by Fukumoto (ref.3).

(c) In order to relate the constant b2 as defined by Eq.5 to roundness R, the $\bar{k}$-e_{min} relationship for uniform quartz sand (U_c = 1.09) with varying roundness (R = 0.2~0.7) based on tests by Norris (ref.4) was determined. The b2-R relationship obtained from this data together with other supplementary data from tests

206

on sands with the same roundness and varying grading
was established by the linear regression analysis as
(γ = 0.915) :

$$b_2 = 1.114 - 0.544R \qquad (6)$$

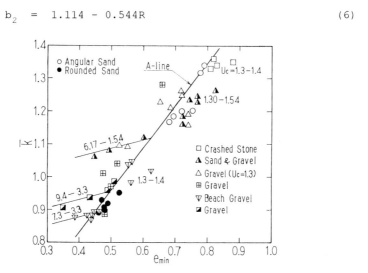

Fig. 1. $\bar{k}$-e_{min} relationship of sands and gravels

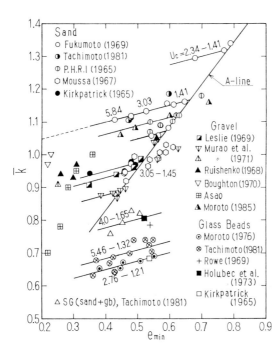

Fig. 2. k-e_{min} relationship of coarse grained
materials

207

CHARACTERISTICS OF k-e_{min} DIAGRAM

6. Moroto (ref.1) proposed that the k-e_{min} diagram could be used to classify engineering properties of granular soils. It should be possible to overlay lines of equal internal friction on the k-e_{min} diagram. Consider ϕ_M, which is defined as the angle of friction at e_{min}, that it is the maximum internal friction angle in the densest state. From Eq.1, the values of ϕ_M are given by :

$$\sin\phi_M = \frac{k}{1 + e_{min}} \qquad (7)$$

Thus equi-ϕ_M lines are drawn as straight lines on the k-e_{min} diagram as shown in Fig.3.

7. As mentioned in the paragraph 5, Eq.4 defined the characteristic line, the A-line, on the diagram for natural soils with uniform grading and varying roundness. For soils with similar roundness but varying grading, the following equation can be derived by substituting Eq.6 into Eq.5 :

$$\bar{k} = 0.260 \, e_{min} + (1.114 - 0.544R) \qquad (8)$$

If the soil particles are spherical, the roundness, R, is 1.0. From Eq.8, another characteristic line, named the 'B-line', can be introduced as :

$$\bar{k} = 0.260 \, e_{min} + 0.570 \qquad (9)$$

The b_2-value of 0.570 for natural materials is similar to the value of $b_2 = 0.544$ given by tests on glass beads (Fig.2).

8. The proposed $\bar{k}$-e_{min} diagram with equi-ϕ_M lines and A- and B-lines is shown in Fig.3. The R-values are based on the roundness scale (6 classes) of Powers (ref.5). The position of the granular soils on this diagram depend on the particle shape and grading. In addition, the strength characteristics of these soils can be determined from the contour lines of equi-ϕ_M lines. For example, well graded angular soils have the greatest shear strength because they lie towards the upper left hand side of the diagram. The proposed k-e_{min} diagram can be used to classify cohesionless soils such as sands and gravels.

9. The data used in this study were obtained under the condition of normal confining pressures at which particle breakage seems to be negligible. All natural materials, therefore, must lie to the left of the A- and B-lines, that is the A- and B-lines form a boundary.

10. At higher confining pressure ranges, angular grains tend to break more easily than rounded ones which results in a decreasing k-value. This can become a serious problem for great fill materials thicknesses such as those in high dams. Fig.4 shows the characteristics of k-e_{min} diagram of Marsal's data (ref.6) under high confining pressure. In this figure, as the confining pressure increases, the gradient of the k-e_{min} relationship reduces and finally becomes parallel to the e_{min} axis. That is, at a high confining pressure level, irrespective of the values of e_{min}, the following equation is obtained :

$$k = c \qquad (10)$$

Where, the constant c is dependent on the hardness of the grains. That is, the soils with smaller values of c break more easily.

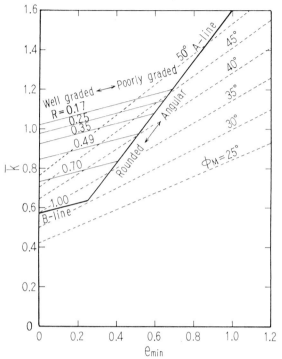

Fig.3. Newly proposed $\bar{k}-e_{min}$ diagram

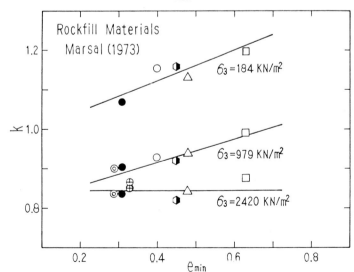

Fig.4. $k-e_{min}$ relationship in high confining pressure ranges

APPLICATION OF $k-e_{min}$ RELATIONSHIP

11. Terzaghi proposed two factors as a measure of compactness for cohesionless materials. These were relative density, D_r, and factor of compactability, F, defined as :

$$D_r = \frac{e_{max} - e}{e_{max} - e_{min}} \tag{11}$$

$$F = \frac{e_{max} - e_{min}}{e_{min}} \tag{12}$$

The factor, F, is a form of classification. The relative density is often specified as the basis for compaction control and strength characteristics.

12. 'Relative compaction C_f' has been used for compaction and quality control of cohesive soils. A basic conception of applying C_f is to reduce a future excessive settlement of embankment. Moroto (ref.7) introduced the following two equations to correspond with the above-mentioned D_r and F :

$$C_f = \frac{1 + e_{min}}{1 + e} \tag{13}$$

$$r = \frac{1 + e_{max}}{1 + e_{min}} \tag{14}$$

Where, e_{max} is the maximum void ratio, and r is a compactability index which represents the ratio of the minimum and maximum volume. The parameter C_f can be expressed in terms of D_r as follows :

$$C_f = \frac{1}{r - D_r(r - 1)} \tag{15}$$

Fig.5 shows the C_f - D_r relationship using the factor r as a parameter in percent. This chart is useful when design is based on D_r and compaction control is based on C_f.

13. The r-values depends on particle characteristics. Its average value is about 1.25(C_f = 90% at D_r = 0%). There is a tendency for the r-value to decrease as the roundness of material increases.

14. It is possible to use the relative compaction, C_f, instead of the relative density, D_r, to predict the angle of internal friction at an arbitrary void ratio e using the k-e_{min} relationship. Substituting Eq.3 into Eq.1 and using Eqs. 7 and 13 give

$$\sin\phi = \sin\phi_M \cdot C_f \tag{16}$$

$$\sin\phi_M = a + \frac{\alpha}{1 + e_{min}} \quad ; \quad \alpha = b - a \tag{17}$$

Eq.16 shows that the strength at an arbitrary density can be directly determined by multiplying the maximum strength by C_f. Therefore, under normal confining pressure and, if particle breakage is negligible, the introduction of a relative compaction C_f factor for cohesionless soils can be more effective for design and compaction control. It has the

210

advantage that: (a) increasing C_f reduces future settlement ;
(b) C_f is related directly to the shear strength and (c) C_f
has been used for cohesive soils and therefore becomes a more
universal parameter.

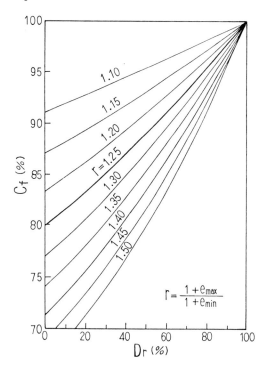

Fig.5.   Relationship between C_f and D_r

15.   For materials with uniform grading and varying
roundness, Eqa.4, 16 and 17 can be combined to give a
relationship between ϕ and C_r :

$$\sin\phi = (1.290 - \frac{0.976}{1 + e_{min}}) \cdot C_f \tag{18}$$

For materials with similar roundness and varying grading,
Eqs.5, 16 and 17 can be combined to give :

$$\sin\phi = (0.260 + \frac{b_2 - 0.260}{1 + e_{min}}) \cdot C_f$$
$$= (0.260 + \frac{0.854 - 0.544R}{1 + e_{min}}) \cdot C_f \tag{19}$$

Fig.6(a) and (b) present the predicted curves of ϕ and C_f
according to Eqs. 18 and 19, respectively.   In Fig.6(b), the
constant b_2 is assumed to be 0.996 for sands D-3 and D-6.   The
agreement between the predicted strength and the actual
strength is very good.

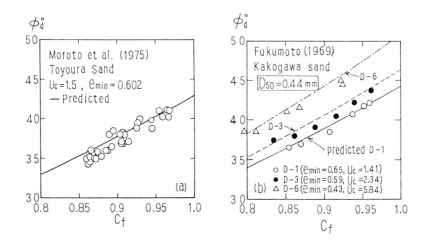

Fig.6. Verification of ϕ_d-C_f relationship on sands

16. For higher pressure ranges, Eqs.7, 13 and 16 can be combined to give

$$\sin\phi = \left(\frac{c}{1 + e_{min}} \right) \cdot C_f = c\rho_d{}^* \qquad (20)$$

Where, $\rho_d{}^* = 1/(1+e)$. From Eq.20, the shear strength is in proportion to the normalized density $\rho_d{}^*$, that is, the material with a smaller initial void ratio has a greater shear strength in higher confining pressure ranges.

CONCLUSIONS

16. The relationship between k and e_{min} for cohesionless soils was examined in detail using data obtained from drained triaxial compression tests. From the results of this investigation, the following main conclusions were drawn :

(1) The proposed k-e_{min} diagram with equi-ϕ_M lines and A- and B-lines can be utilized as a new engineering classification chart for cohesionless soils.
(2) Under normal confining pressure, if the particle breakage is negligible, the introduction of relative compaction C_f for cohesionless soils becomes more effective for compactive control of engineered fills.
(3) The $\sin\phi$-C_f relationship is useful to evaluate the angle of internal friction for cohesionless soils. The relationship between C_f and D_r can be also shown to define a measure of compactability r.
(4) For higher confining pressure ranges, it is very important to notice the density $\rho_d{}^*$ of cohesionless soils.

REFERENCES
1. MOROTO, N. An application of Mogami's strength formula to the classification of granular soils. Soils and Foundations, Vol.22,No.1, pp.82-90, 1982.
2. SHIMOBE, S. and MOROTO, N. Characteristics of $k-e_{min}$ diagram for coarse grained materials. Proc. 46th Annual Meeting of the Japan Society of Civil Engineers, Vol.3, pp.436-437(in Japanese), 1991.
3. FUKUMOTO, T. On the effects of grain size and grading on shearing characteristics of sand. Proc. 4th Annual Meeting of the Japanese Society of Soil Mech. and Found. Eng., pp.475-489(in Japanese), 1969.
4. NORRIS, G.M. The drained shear strength of uniform quartz sand as related to particle size and natural variation in particle shape and surface roughness. Ph.D.Thesis, Univ. of Calif, 1977.
5. POWERS, M.C. A new roundness scale for sedimentary particles. Jour. of Sedimentary Petrology, Vol.23,No.2, pp.117-119, 1953.
6. MARSAL, R.J. Mechanical properties of rockfill. Embankment-Dam Engineering, Casagrande Volume, John Wiley & Sons, pp.193-200, 1973.
7. MOROTO, N. Compactive state and engineering properties of coarse grained soils. Proc., Symposium on the Relative Density of Sand and Engineering Properties, JSSMFE, pp.97-104(in Japanese), 1981.

17. Back-analysis of a pilot scale shear test on coarse granular fill

M. D. BOLTON, Cambridge University, UK, and DA-MANG LEE, Ove Arup and Partners, Hong Kong

SYNOPSIS.   Granular fills often contain large particles which do not fit into conventional test apparatus. Triaxial tests are reported on such a fill, both whole, and after removing larger particles. The peak angle of shearing resistance ϕ_{max} reduces, together with the ultimate angle ϕ_{crit}, as the grading of the material is narrowed. The original soil was also subjected to a pilot test, in which a 2m x 2m concrete base was cast over the compacted fill, loaded with weights, and sheared by jacking. The comparison between these various determinations of ϕ is used to discuss the selection of a design value.

BACKGROUND

1. A wide range of granular materials recovered from excavation, quarrying or demolition may be identified as suitable for compaction as fill. Some of these will contain particles so large that they can not be accommodated in standard soil test apparatus. The original material is then usually "scalped": over-size particles are removed prior to testing the finer soil matrix. The resulting distortion of properties has been investigated.

2. Bolton (ref. 1) recalled that the peak secant angle of shearing in compression tests could be expressed as follows:

$$\phi_{max} = \phi_{crit} + \Delta\phi \tag{1}$$

where ϕ_{crit} is the component for shearing at constant volume which is taken to be a soil constant, and $\Delta\phi$ is the variable component due to dilatancy. $\Delta\phi$ was then shown to be proportional to a relative dilatancy index I_R which was the product of the initial relative density I_D and a newly defined relative crushability I_C, so that

$$\Delta\phi = A\ I_R \tag{2}$$

with A = 5 in plane strain, A = 3 in triaxial strain, and

$$I_R = I_D\ I_C \tag{3}$$

where

Engineered fills. Thomas Telford, London, 1993

$$I_D = \frac{(e_{max} - e)}{(e_{max} - e_{min})} \qquad (4)$$

$$I_C = \ln \, {}^{P_c}/_{p'} \qquad (5)$$

The relative crushability I_C is the logarithm of the ratio of an aggregate crushing stress p_c (to be determined), and the mean effective stress p' at failure.

3. These precepts gave a linear increase of $\Delta\phi$ with relative density at constant p', and a linear decrease of $\Delta\phi$ with log p' at constant density. They fitted the data of 17 sands within about 1 or 2°, with different values of ϕ_{crit} observed in the range 30° to 37° and a single value p_c = 22 000 kPa. The value p_c was selected to give a good fit of triaxial test data to equation 2, and was inferred to be related to particle crushing, on the basis of breakage observed by sieving after tests.

THE PILOT SCALE SHEAR TEST

The trial fill

4. It was decided to select a material that would be suitable for use both in embankment construction and as back-fill to structures, according to the Department of Transport Specification for Highway Works (1986). The selected river gravel was obtained from a site near Crowthorne in Berkshire, convenient to the Transport Research Laboratory at which the pilot test was to take place. A few random samples were taken from the 60 tonnes of fill required for the test, and the averaged grading curve is marked (i) in Fig. 1. The well-graded fill had less than 5% of fines (< 63 μm) and the nominal maximum particle size was 45 mm. This grading was found to be consistent in every compacted layer.

5. The mean limiting densities of the fill, and its derivatives used for triaxial tests, are recorded in Table 1. Since some of the particles were larger than permitted in standard tests, a non-standard brass CBR mould 200 mm in diameter and 215 mm high was used for the dry density. The maximum density was tested by vibrating the dry soil under 14 kPa surcharge, following ASTM D4253a (1983) except that the duration was increased from 8 to 15 minutes. The minimum density was found using ASTM D4254b (1983), filling the larger mould by inverting a thin-walled tube containing the dry fill. Notwithstanding that the coarse gravel was inherently lighter, its removal resulted in successively smaller limiting densities, due to the narrowing of the range of particle sizes.

The compaction of the test fill

6. The test was carried out in a concrete-lined pit, 3.43 m x 6.25 m in plan and dug to 1.3 m depth where natural stiff clay was encountered. Angular aggregates were compacted in a thin layer of cement paste to form a rough and solid base for subsequent compaction of the fill.

	max. particle size d_{max} (mm)	G_s	$\rho_{d,max}$ Mg/m³	$\rho_{d,min}$ Mg/m³
(i) material used in pilot test	45 *	2.62	2.276	1.866
(ii) fill tested in 9" triaxial	38	2.62	2.230	1.820
(iii) fill tested in 4" triaxial	20	2.65	2.049	1.655
(iv) fill tested in 4" triaxial	12	2.65	1.981	1.649

* particles of up to 75 mm are sparsely distributed in the fill

Table 1 Properties of the pilot test fill and its derivatives

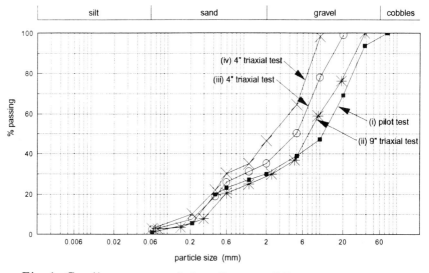

Fig. 1 Grading curves of the pilot test fill and its derivatives

7. The compaction procedure was according to the DTp Specification for Highway Works (Part 2, Series 600, 1986). A Wacker DPU vibrating plate compacter, weighing 605 kg and providing a static loading of approximately 15 kPa (to be doubled due to inertia effects), was used to compact each layer of soil to 125 mm in 6 passes. Fig. 2 shows the data of compaction tests carried out in CBR moulds to BS 1377, for which it was necessary to remove particles larger than 38 mm. Maximum compaction was achieved at about 6% moisture content. The moisture content of the fill as placed varied from 5 to 9%. The compacted dry density of each layer was determined by a Troxler 3411B nuclear moisture-density gauge at 4 to 6

216

locations in each layer, calibrated by conventional sand replacement tests at different densities. It became apparent that some nuclear gauge tests had been affected by the presence of very large particles. When obviously excessive values were discarded, the average dry density of the whole fill was estimated to be 2.23 Mg/m^3, which indicated a relative density $I_D = 90.6\%$. In all, 7 total pressure cells, 20 markers, and 2 piezometers were installed in the fill, as shown in Fig. 3. Further details are given in Lee (ref. 2).

The reinforced concrete pad

8. Shearing in the fill was induced by the horizontal displacement of a lightly reinforced concrete pad cast on top of the fill surface. A low slump of 1 inch was specified to avoid bleed into the fill. Care was taken to prepare the fill surface, using first a stiff brush and then a vacuum cleaner to remove smooth patches of fines left after compaction, without disturbing the body of the fill. A steel plate was cast into the end face of the pad, for jacking. Fig. 4 shows the test arrangement.

9. A vertical stress of 75.4 kPa was provided by a weight of 302 kN – the pad itself, with 48 concrete blocks stacked evenly over it. This magnitude of stress normal to a slip plane was thought typical of earthworks and structures. The horizontal force was provided by a 500 kN hydraulic jack supported on a pair of load cells, acting only 105 mm above the soil surface to minimise the complication of an induced moment. The movement of the pad was picked up by four 50 mm LVDTs and two 300 mm displacement potentiometers. LVDTs and dial gauges were also used to locate the anticipated passive zone of heave.

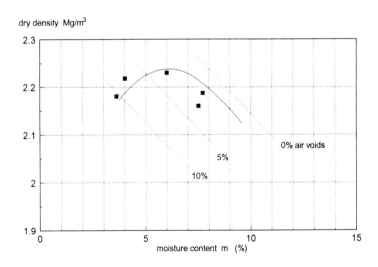

Fig. 2 Compaction tests using CBR mould on pilot material with particles larger than 38mm scalped off

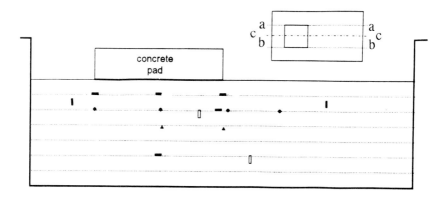

• markers along aa, bb & cc — pressure cells along cc
▯ piezometers along aa ▲ markers along bb & cc

Fig. 3 Positions of buried instruments

Test procedure

10. The test was conducted with the fill fully saturated to eliminate suction. Water was introduced from the base 24 hours prior to testing. Within 0.9 mm of shear displacement it became obvious that the pump for the jack was inadequate. The test was suspended and restarted with a more powerful unit. The shear force had, in that cycle, risen to 62% of the peak value recorded subsequently; its influence on the magnitude of that peak has been disregarded. The shear test proper lasted 90 minutes, the jack being advanced at 1 mm/min for the first 35 mm, increasing gradually thereafter to 5 mm/min up to 138 mm of horizontal displacement. During the drive, the block yawed by up to ± 3° in plan view, but only after achieving peak shear force; this was also disregarded. After the test, the concrete blocks were removed and the reinforced concrete pad was lifted for inspection. The concrete/soil boundary was found to be consistently rough and free from evidence of sliding. The final locations of pressure cells and markers were then measured.

Results

11. Fig. 5 shows the data of (a) shear force, (b) pad heave, (c) soil heave in front of pad – all with respect to the horizontal displacement of the pad. The peak shear strength of 313 kN gives an interface shear stress of 78.2 kPa and a mobilised angle of friction on the base of $\delta = \tan^{-1}(78.2/75.4) = 46.1°$. It is mobilised at a shear displacement of 5 mm. At that stage, the rate of heave of the pad was at its maximum, though the four corners were lifting at different rates. Evidence of soil heave in front of the pad does not show until the displacement is 10 mm, by which time δ has reduced to 43.9°. The rate of softening is a maximum at this point, and two thirds of the drop from peak to

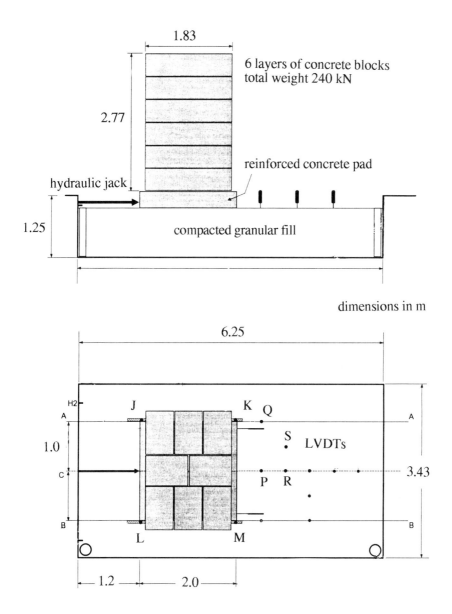

Fig. 4 Arrangement of the pilot test

residual strength has occured after 20mm of displacement. δ
then drops to its ultimate value of 33.7° after a total shear
displacement of only 50 mm, at which stage the mean rate of
heave of the pad had reduced to approximately zero.

horizontal shear force F kN

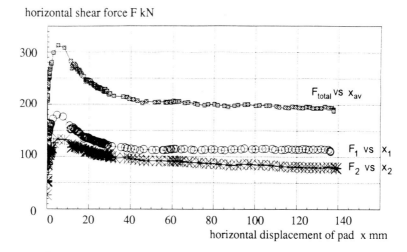

horizontal displacement of pad x mm

vertical displacement z mm

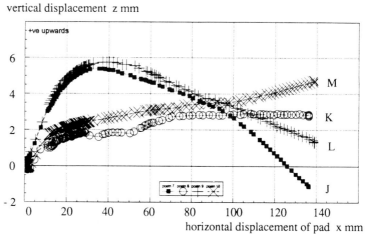

horizontal displacement of pad x mm

heave in soil mm

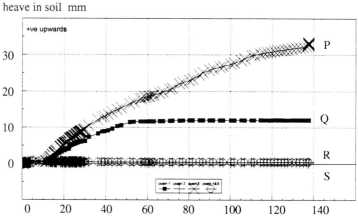

Fig. 5 Pilot scale shearing data   horizontal displacement of pad x mm

12. The pneumatic pressure cells were monitored throughout, but showed evidence of severe local arching effects when the shearing began. No confidence could be placed on their results.

Post-shear investigation and analysis

13. The location of the outcrop of the rupture surface, the magnitude of final surface heave, and the movements of the buried markers, are shown in Fig. 6. A hypothetical zone of shear, consistent with all the evidence, is shaded in. The plastic mechanism is indicative of a bearing capacity failure under a strongly inclined load, as was anticipated. It was clear that the trench was deep enough not to have influenced the results.

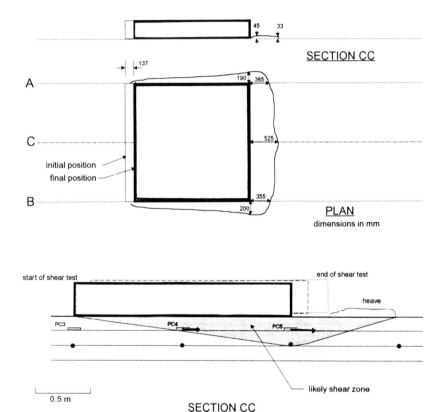

Fig. 6 Approximate shear zone found after the test

14. Sokolovskii's method of characteristics (ref.3) was used to analyse the limiting plastic equilibrium of a base under an inclined load, bearing on heavy soil (the buoyant unit weight was used) with constant angle ϕ. Fig. 7 shows two solutions for ϕ = 47.5° and 50°. The characteristic lines of maximum stress obliquity are inclined, and the angle of shearing ϕ mobilised upon them is now seen to be distinct from the angle of friction δ = 46.1° mobilised on a horizontal plane beneath the pad, at peak strength. As can be seen, the geometry of the plastic zone is very sensitive to the value of ϕ used in the analysis, and 48° was chosen as the appropriate back-analysis. Side effects have been ignored: there is no accepted way of accounting for them.

characteristics drawn for $\phi = 47.5°$

estimated zone of shearing in pilot test

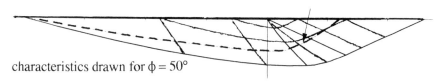

characteristics drawn for $\phi = 50°$

Fig. 7 Shear zones by method of characteristics

15. Sokolovskii's analysis is based on equilibrium, not kinematics, and a perfect match between failure geometries must not be expected. It must be recalled, of course, that evidence of soil heave in the passive zone came *after* peak shear force was developed. This suggests that failure was progressive, and means that no analysis assuming uniform soil properties can be expected to match the evidence exactly, even if it accounted consistently for friction and dilation. In the circumstamces, the error of assuming that the base was a plane of maximum stress obliquity is small. The back-analysis of a pad shear test such as this seems at least as secure as the conventional back-analysis of a direct shear test, which it very strongly resembles.

TRIAXIAL TESTS

16. Triaxial tests were conducted at Cambridge on saturated samples of 100 mm (4 inch) diameter, and one test was carried out at BRE on a sample of 230 mm (9 inch) diameter. The larger particles in the test fill were scalped out to achieve a diameter ratio no smaller than 6, and to test the effect on triaxial ϕ of reducing the spread of grading in this way. Fig. 1 compared the three gradings which were used in the tests with that of the fill in the pilot test on the pad.

17. Fig. 8 shows the results of the tests conducted on samples in the range of relative densities 82% to 92%, and with an effective minor stress of 47 kPa. It will be seen that the 230

mm test with particles scalped at 38 mm offers a much higher peak ϕ_{max} value, and at a very much smaller strain, than the 100 mm tests scalped at 20 mm or 12.5 mm. The increase in initial stiffness can be attributed to the method of compaction. Soil was compacted in 11 layers with a Kango hammer in the 230 mm test, whereas it was vibrated to achieve the test density in the 100 mm configuration. A similar pre-loading effect was observed when compaction was used as a check in further 100 mm tests.

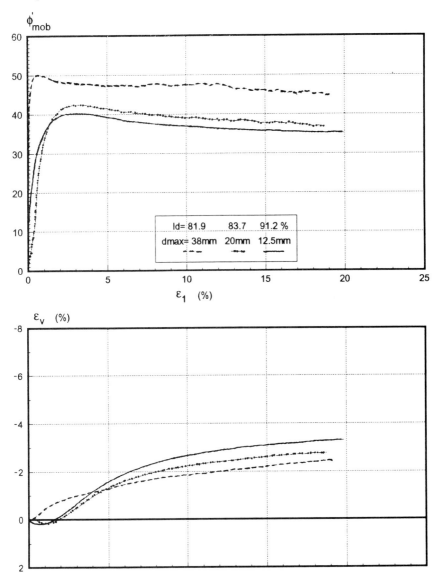

Fig. 8 Triaxial tests on pilot fill with the largest 5, 20 & 40% of the particles removed

18. Fig. 9 compares all the data of ϕ_{max} versus I_D, at a minor effective stress of 47 kPa and a normal effective stress on slip planes of about 75 kPa. The inclusion of large particles is seen to enhance the angles of shearing of all the samples, irrespective of density. It has been shown (ref. 4) that large particles can act as soil reinforcement, due to the tendency of smaller particles to flow around them. The large particles tend to develop compressive stress in the direction of principal compressive strain in the surrounding fine soil matrix, and tensile stress in the direction of principal tensile strain in the matrix. Comparing the large particle with the fine soil it replaces, therefore, it tends both to steal deviatoric stress and to enhance the effective confining pressure of the surrounding matrix. Fig. 9 tends to support the hypothesis that the reinforcement effect of large particles is equally effective at a critical state as at peak strength, though the evidence here could not be regarded as conclusive.

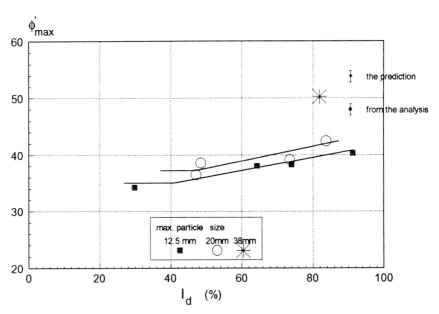

Fig. 9 ϕ versus relative density: analysis of pilot shear test compared with triaxial tests

CONCLUSIONS

19. An angle of shearing resistance can be measured on a test fill by casting a concrete pad on a prepared surface, loading the pad with weights, and jacking it horizontally. The behaviour is brittle, showing evidence of some progressive failure. Peak strength of the test fill reported here was developed at 5 mm displacement, and full softening had occured after 50 mm.

20. The pad shear test can be regarded as an analogue to the laboratory direct shear test and is directly relevant to the

maximum and ultimate sliding resistance of in situ concrete bases. No great error is made by interpreting the point of peak strength by assuming horizontal shear within the soil. However, a thorough investigation revealed that the mechanism conforms to that of a bearing failure with a severely inclined load. Taking into account the combination of vertical bearing and horizontal shearing, the surface mobilising greatest angle of shearing was seen to be slightly dished. This results in an estimate of ϕ_{max} of 48°. A slightly larger error might be involved in assuming that the ultimate mobilised angle of friction δ on the base was a good estimate of ϕ_{crit} since δ fell to 33.7° which is at least 4° smaller than the lowest likely estimate of ϕ_{crit} for the pilot fill.

21.   Triaxial tests gave conservative strengths in the well-graded fill when large particles were scalped prior to testing. The largest test configuration, and the minimum scalping of the test fill, gave a triaxial ϕ_{max} of 50° in a single test. After using equations (1) to (5) to correct for density differences and plane strain, the peak angle of shearing resistance in the pad shear test should apparently have been 58° in the absence of progressive failure. The pilot shear back-analysis agrees more closely with the uncorrected triaxial value. Weakness on bedding planes in the pilot test fill can not be discounted.

22.   It seems likely that some progressive failure did occur in the pilot test. Certainly, the shear strength softened very rapidly post-peak. If a value of ϕ had to be chosen for design, to be used without partial factors covering uncertainty, the critical state value must be used. If this were obtained from the ultimate strength of loose samples of scalped fill, the reinforcing effect of large particles would be lost and a conservative value of ϕ_{crit} would be registered. If designers chose to mobilise ϕ_{crit}, the strains in properly compacted granular fill would be small, less than 0.2% according to the triaxial tests.

REFERENCES
1. Bolton M.D. (1986) The strength and dilatancy of sands. Geotechnique, 36, No. 1, 65-78.
2. Lee D-M. (1993)   The angles of friction of granular fills. PhD dissertation, Cambridge University.
3. Sokolovskii V.V. (1965)   Statics of granular media. Pergamon Press, Oxford.
4. Bolton M.D., Fragaszy R.J., Lee D-M. (1991) Broadening the specification of granular fills, Transport Research Record, No. 1309, 35-41, Washington D.C.

ACKNOWLEDGEMENTS
This work was carried out under a contract placed with Andrew N. Schofield and Associates by the Department of Transport. The authors are grateful for the support of the University of Cambridge and the Transport Research Laboratory. They also thank Building Research Establishment for   assistance with the large triaxial test apparatus.   The opinions expressed in the paper are solely those of the authors as individuals.

18. Stabilisation of embankment clay fills using lime piles

C. D. F. ROGERS and S. GLENDINNING, Loughborough University of Technology, UK

SYNOPSIS. Embankments are commonly constructed from overconsolidated clay fills that exhibit differential, progressively negative, porewater pressures with depth. Recent detailed studies of the long-term performance of clay embankments show many instances of shallow failure, and incipient failure is anticipated in considerably more cases. Current research at Loughborough University examining lime stabilisation of clay soils, particularly via lime piles, has shown that the techniques have application to the better engineering of clay embankments. Lime piles create negative porewater pressures, which reconsolidate the clay along shear planes, and produce enhanced strength by cementation. The potentially adverse effects of pile expansion have been shown not to occur.

INTRODUCTION

1. Overconsolidated clays are commonly, and necessarily, used as clay fills for embankment construction in the UK and recent studies have highlighted some of the engineering challenges that these structures present. The problem of slope failures, particularly shallow planar slope failures, in embankment slopes is being addressed by the owners of transportation routes such as the British Waterways Board, the Department of Transport and British Rail. These failures typically manifest themselves several years after construction and have been shown to be related to parameters such as geology, slope geometry, age of embankment and drainage (ref.1).

2. Studies of failing embankment slopes are particularly valuable in facilitating better design of new embankments and in isolating the causes of failure to be addressed by remedial measures. The choice of remedial measure, however, is controlled by the direct cost of the techniques and both the direct (in terms of lane rentals, etc.) and indirect costs of disruption to traffic. In the case of canals much of the disruption is to the leisure industry, whereas in the case of roads and railways the disruption is to business and freight. It can be argued that these different disruptions should be assigned equal importance, and in any case the best engineering solution would be a remedial measure that avoided traffic disruption.

3. This paper addresses the long-term performance of engineered clay fills by summarising the geotechnical factors that result in embankment failures and describing the potential application of clay stabilisation using lime. This is done both in terms of new embankment construction and, more especially, stabilisation of failed embankment slopes. The most appropriate lime stabilisation technique is that of lime piles, which essentially consist of

Engineered fills. Thomas Telford, London, 1993

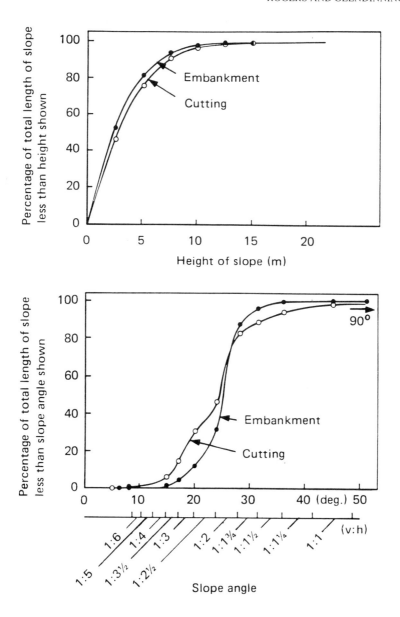

Figure 1. Distribution of height and slope angle of embankments in survey of UK road embankments and cuttings (after Perry[1]).

regularly spaced augured holes filled with compacted quicklime. The results of research into the alteration of the geotechnical properties of remoulded, compacted clay soils by lime piles are then presented to demonstrate the efficacy of this engineering solution.

EMBANKMENT PERFORMANCE POST CONSTRUCTION

4. The most comprehensive study of embankment performance in the UK is that presented by Perry (ref.1), who surveyed 570 km of motorway slopes covering all of the principal geologies. While the study focussed on both cutting and embankment slopes, only the latter will be considered here. The conclusions of this work make recommendations for side slopes to be adopted in new embankment construction in different materials and provide criteria with which to facilitate recognition of slopes that are at risk of failure. The survey revealed significant incidence of shallow slope failure (a total of over 17 km of embankment slope and 5.5 km of cutting slope) and it was estimated that at least three times as many failures will occur in the future in the area surveyed as have occurred over the past 25 years.

5. The survey covered embankments of different heights and slope angles (Figure 1), although 50% of all embankments studied were constructed with a slope of 1:2. The age of the embankments varied up to 25 years, although there were difficulties in determining the age at failure. Three drainage conditions at the bottom of embankment slopes were predominantly encountered: no drainage (26% incidence), open ditch drainage (46%) and French drains (27%). Slope drainage in the form of stone filled trenches, either directly down slope or in a herringbone pattern, was found to be relatively rare in embankments. These data serve to illustrate the types of conditions that might be expected in practice.

6. Embankment failures were generally found to be characteristic of shallow failures, being a combination of translational and circular slippage. 95% of failures were less than 1.5m deep, with the depth to the slip surface ranging from 0.2 to 2.5m overall. Toe erosion, usually created by adjacent excavations, was not a significant cause of failure, although open ditches at the base of embankments were thought to contribute to failure. More importantly most failures were found to occur in overconsolidated clay fills, and particularly the heavily overconsolidated clay fills of the south and south east.

7. The observations on failures reported by Perry (ref.1) provide some indications of how remedial and/or preventative works might be carried out, and even more pertinently cases where they might be inappropriate. For example it was not necessarily the older slopes that failed first, some geologies experiencing most failures in younger slopes. In these cases the implication is that failures, at least at shallow depth, are likely to occur more frequently in the medium rather than the long term. Similarly it was not necessarily the steepest slopes that failed first. These points are discussed in more detail in the next section, although the latter refers to the water shedding capability of steeper slopes. In this respect vegetation is beneficial since it also results in transpiration, and ultimately loss of water from the ground, although tree planting can result in water ingress in the short term and this can affect the porewater pressures within the slope. Shear surfaces induced by compaction, and poorer compaction at the edge of embankments, are also cited as potentially exacerbating the likely ingress of water, although no

Porewater pressure (metres head of water)

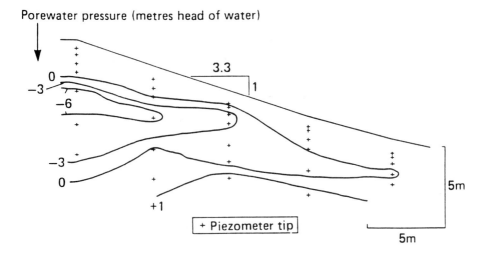

Figure 2.   Typical distribution of porewater pressure with depth in an
embankment monitored by the TRL near Wrotham (after
Crabb and Atkinson[4]).

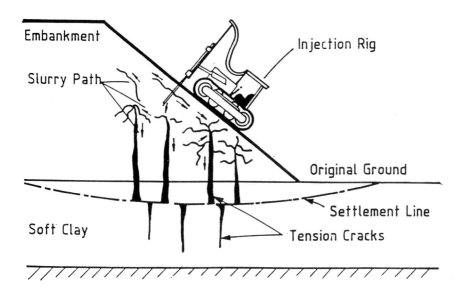

Figure 3.   Slope stabilisation using lime slurry pressure injection (after
US National Lime Association[7]).

229

evidence is available to judge the hypotheses. Cracking of clay in hot weather was also postulated as a means of water ingress.

8. It was reported that remedial works in the case of failed embankment slopes almost invariably took the form of excavation, sometimes in benches, to below the failure surface and replacement of the excavated material with compacted granular, free-draining material such as gravel, crushed rock or brick rubble. Topsoil is sometimes placed over the repaired area to permit vegetation to become re-established. This remedial process will typically require closure of part of the transport route, even if only the hard shoulder on a motorway for example, and requires excavation plant and significant lorry movements to both dispose of excavated material and deliver the granular fill. It is apparent that remedial methods that require small plant and minimal associated traffic would be of advantage.

GEOTECHNICAL CONSIDERATIONS OF EMBANKMENT SLOPE STABILITY

9. The problem of time-dependent failure of embankment slopes in overconsolidated clays occurs as a result of equilibration of porewater pressures. Excavation of overconsolidated clays and recompaction in embankments results in stress relief and considerable negative porewater pressure. These high suctions cause the embankment slopes to achieve a high degree of stability immediately after construction. It is therefore only after water ingress, and thus an increase in porewater pressure from a negative base, that problems of instability arise.

10. To investigate this phenomenon the Transport Research Laboratory has undertaken a programme of monitoring of embankment slopes (ref 2,3). Three failing embankments were instrumented to measure porewater pressure with the typical distribution (ref.4) shown in Figure 2. It can be seen that here, as in the other two cases, the porewater pressures at depths of up to 2m were at or close to 0 kPa, whereas further into the slope the porewater pressures became negative and proved to remain approximately constant. It is thus apparent that the depths at which slip planes typically occur (< 1.5m reported in paragraph 6) lie within the zone that is affected by seasonal porewater pressure variations (ref.4). It is equally apparent, therefore, that the prevalence of shallow slips within 25 years of construction of motorway embankments is a result of the porewater pressure distributions and that deep-seated slips are only likely to occur after a greater period of time, depending on the local water regime. While this can be considered to be a generalisation only, it is true also of the experience of the British Waterways Board.

11. The porewater pressure regime can be considered to have stabilised, and thus the ultimate stability of a slope reached, when the water contents of the soil are in equilibrium and the ingress of rainwater and groundwater is balanced by the egress by drainage and vegetation (ref.4). The controlling factor in this equilibration process is the facility for rainwater to penetrate the slope. This perhaps explains the greater apparent stability of steeper slopes of the same material in certain cases reported by Perry (ref.1), since steeper slopes will tend to shed rainwater better. Hence although the disturbing forces in a steeper slope are greater, it is the balance between these and the (restoring) negative porewater pressures that controls stability. It is apparent, therefore, that these steeper slopes are likely to fail, but only after a greater period of time than the flatter slopes, unless the effects of drainage and vegetation are sufficient to maintain the negative porewater pressure regime.

230

12. The shallow depth to the failure plane in the slips being considered here results in normal total stresses of up to 30 kPa on the shear plane. If the porewater pressure at these depths is zero, then the effective stresses will similarly be a maximum of only 30 kPa. The strength measurements required for design should thus be based on laboratory tests at these small stresses, a subject addressed by Crabb and Atkinson (ref.4) in the context of the curvature of the peak strength failure envelope over such a stress range. Their conclusions, however, were that the strength parameters required for design were those of the fully softened, critical state rather than those of peak strength. Similar considerations are likely to apply to residual strength measurement, since, under such low normal stresses, well defined shear planes are unlikely to form and would in any case require movements of several metres to become fully aligned. At much higher normal stresses the movement required to achieve residual strength conditions is likely to be hundreds of millimetres.

13. The important conclusion from this discussion is that if an effective apparent cohesion can be achieved, perhaps by the cementing actions of lime stabilised clay, then this will dominate the stability of shallow slip planes on which the effective normal stress is low. Thus a small effective cohesion equivalent to 1-2 kPa might be sufficient to raise the factor of safety of a shallow slope to a sufficiently high level for long-term stability.

USE OF LIME IN EMBANKMENT STABILISATION

14. The technique of using lime to improve the engineering properties of clay soils is reasonably well established, if not yet fully understood. The basis of lime stabilisation is the chemical reaction between the lime and the clay mineral constituents of clay. This reaction can be divided into two parts: modification and stabilisation. Modification takes place within approximately 24-72 hours of mixing lime and clay together. It describes the rapid flocculation of clay particles, which changes the cohesive nature of the material. The clay becomes more friable and granular in nature and its strength is improved. Stabilisation occurs more slowly, although there is disagreement about whether this set of reactions only occurs subsequent to modification. Stabilisation concerns the long-term strength development brought about by the generation and subsequent crystallisation of cementicious compounds from the reaction of lime with the siliceous clay components.

15. The strength gains that can be achieved by full stabilisation are potentially very considerable indeed, and in general are far higher than necessary for many civil engineering applications. The gain in strength from a geotechnical viewpoint occurs in two distinct ways as a result of the two processes (ref.5). Modification results in an increased frictional capacity of the material, or an increase in its effective angle of internal friction. The material in this state is far less malleable than prior to lime addition, mixing and compaction, and its tensile capacity is assumed to be high, although perhaps erroneously high due to interpretation of test data as though the strength is equivalent to undrained shear strength. Stabilisation yields a strong brittle, material in which the high tensile capacity results from cementation bonds.

16. It is thus possible that lime modified, rather than lime stabilised, clay might have considerable benefits to embankment construction since the material could be constructed with significantly steeper sideslopes. The

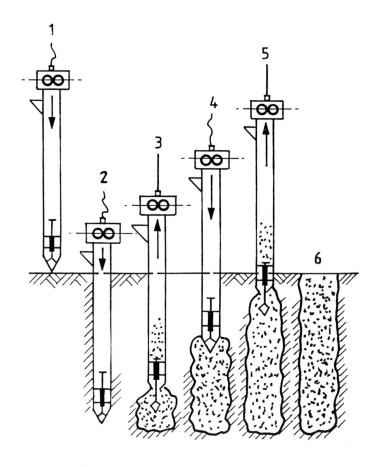

Figure 4.   Procedure for construction of lime piles in soft clay soils
(after Ingles and Metcalf[9]).

savings made by reduced landtake and material handling would be offset against the extra costs of importing lime. If full stabilisation were not required, however, the costs might not prove to be excessive. Repair of a cutting slope using lime-clay mixing techniques has been successfully carried out (ref.6), and embankment construction would provide an extension to this principle.

17. The use of lime as a deep stabilising technique has been pioneered overseas and takes several forms. Lime slurry pressure injection has been established as a stabilising technique in the US (ref.7). As the name suggests, lime in the form of a slurry is injected into the ground under controlled high pressure using specialist equipment (Figure 3). It has been used to treat expansive clay soils, silts and other soft ground, as well as to stabilise failed embankment slopes. There are potential problems with positive porewater pressure generation using this technique however. Lime columns (ref.8) are similarly of questionable value to slope stabilisation.

18. Lime piles, however, have considerable potential for slope stability. Ingles and Metcalf (ref.9) show one method of construction, primarily for soft soils, as illustrated in Figure 4. A hollow tube is pushed into the soil to the required depth of pile. Quicklime is then forced into the tube under pressure as it is withdrawn. The pressure forces open the end of the tube allowing the lime to fill the cavity below. After each metre of hole is filled the end of the tube is closed and used to compact the lime. The procedure for embankment slopes would be based on augured holes, rather than driven tubes, but the principle remains the same. A substantial clay plug should be inserted into the top of the holes to seal the quicklime from contamination, by carbonation, from the atmosphere.

19. In terms of slope stability, the important benefits of lime pile stabilisation are that the quicklime has a great affinity for water and will produce high negative suctions within the slope until the lime has become slaked, or hydrated. These negative porewater pressures will remain in the slope until removed by an external source of water. This suction has the additional benefit that the shear plane, which consists of disturbed, sheared material, will reconsolidate in the same way that it would if subjected to an external positive effective stress. In addition the lime will migrate to some extent into the surrounding clay and create a stabilised zone around the pile. This effect is not, however, as extensive as reported in the (variable) literature on the subject, nor as extensive as reported in an earlier paper by Rogers and Bruce (ref.8). Part of the reason for this is the difficulty of measurement of ion migration (free calcium ion contents being a misleading indicator) and the facility for edge contamination in tests in laboratory boxes. Without water transportation of calcium ions, migration can be expected to be of the order of tens of millimetres. Nevertheless the strength of the annular treated zone around each pile, together with that of the pile itself where applicable, results in increased strength along the shear plane.

20. Several American authors describe the use of lime piles in slope stabilisation, including Handy and Williams (ref.10), Anon (ref.11) and Lutteneger and Dickson (ref.12). The first authors illustrate their paper with a case study. Quicklime piles 150mm in diameter on a 1.5m grid were successfully used to stabilise a failing slope where more conventional treatments had failed. Design methods were entirely based on predicted migration of the lime and the soil type in relation to the subsequent lime clay reaction. However it should be noted that, although the slope stabilisation

proved successful, the designed mechanism of stabilisation by migration appears suspect from the results of current research at Loughborough and some other mechanism might have accounted for the stability.

LABORATORY INVESTIGATIONS

21. Laboratory investigations at Loughborough have studied a number of aspects of the problem of lime pile improvement of clay soils. These include work on the rate and distance of migration of calcium ions through the clay surrounding a lime pile, which has shown that migration occurs to distances of typically less than 50mm and is complete within one month. Reactions following ion migration will occur over longer periods, however. The expansion of lime piles on hydration has been studied with the preliminary conclusions that the (significant) expansion does not result in lateral consolidation of the clay. This finding is to be expected from considerations of stoichiometry, but is in contrast with certain reports in the literature. Clay dehydration has been monitored, again with the conclusion that what is claimed in the literature is exaggerated unless large quantities of water boil off as a result of the heat generated. Studies of porewater pressure reduction have also been carried out, and these have been used to plan the experiments described hereafter. The above work is the subject of parallel papers currently under preparation.

22. The experimental work outlined above provides information on the changes in certain properties of the clay, but little information on how the material forming the shear plane will react. The idea of porewater pressure reduction was first raised by Rogers and Bruce (ref.8). However, its effects were then only considered to be temporary: an immediate effect which would be beneficial before the longer-term clay stabilisation reactions took over. Its consequences were only considered over the time in which it was present, i.e. while the reaction was taking place. It is now thought that the longer-term effect on any shear failures present within a slope could be more crucial. If the reduction in porewater pressure is considered as an increase in effective stress on the shear plane, it will result in consolidation of the remoulded zone associated with a shear failure. This consolidation might be sufficient to raise the shear strength in this area above the fully-softened critical state or residual value, whichever is applicable. This rise in shear strength might in turn be sufficient to prevent further failure. The rise in shear strength due to over-consolidation of the shear plane has also been studied, using a 100mm shear box.

23. The aim of this work was thus to observe the effect of over-consolidation on the shear strength of a pre-formed shear plane. The direct shear box apparatus was used for its ability to create samples with shear planes within them at known, and variable, vertical effective stresses. An initial assessment was made by consolidating the samples to a high normal effective stress (150kPa) and creating the shear plane using uni-directional shear. This allowed peak and residual shear strengths at this stress to be found. Samples were then subjected to a further vertical stress of 20kPa (i.e. 170kPa in total), the incremental value being shown to reflect the pore water pressure reduction in the field. After the period of consolidation samples were allowed to swell under a vertical stress of 150kPa once more and the new value of strength was determined by reshearing the sample. Little change was observed under these high levels of stress, as would be experienced in deep-seated slip planes.

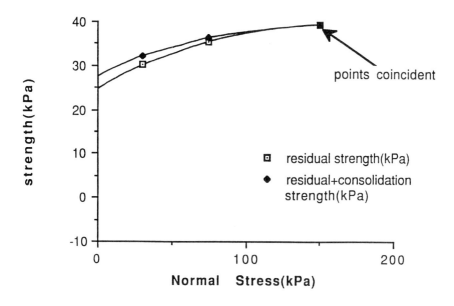

Figure 5. Influence of overconsolidation of the shear plane by a 20 kPa increment of effective normal stress.

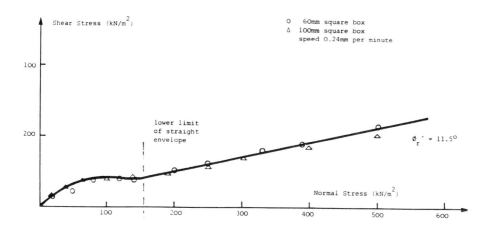

Figure 6. Residual strength envelope for Lias clay determined by reversal shear box tests (after Anayi et al [13]).

235

24. It was considered that at such a high normal effective stress, a very well defined shear plane would be developed with a high degree of particle orientation. In such a case slight over consolidation would have little beneficial effect. In the case of shallow slips the shear plane definition will be poorer, the degree of particle orientation will be lower, and the effect of a 20kPa reduction in porewater pressure will be proportionately greater. In order to achieve realistic results for shallow slips in overconsolidated clays, two approaches were considered:

1. Consolidate to a low normal stress, shear, over-consolidate and shear again; or
2. Consolidate to a high normal stress, allow to swell under a low normal stress, shear under a low normal stress, over-consolidate and shear again.

25. The latter approach has been attempted, with the anticipation that an ill-defined shear plane would result and would be more susceptible to re-alignment of particles by over-consolidation. The initial consolidation pressure and increment of over-consolidation pressure used were again 150 and 20kPa respectively, while swelling and shearing took place at normal effective stresses of 30 and 75kPa.

26. The results using china clay (liquid limit = 60%, plastic limit = 36%) at a moisture content (37%) close to its plastic limit were peak and residual strengths of 74 and 40kPa respectively when sheared under an effective normal stress of 30kPa. After over-consolidation (i.e. at a total effective normal stress of 50kPa) an immediate increase in shear stress of 2.06kPa was observed on resumption of shearing. At such a normal stress, assuming c' to be zero, this would be equivalent to an increase in effective angle of internal friction (ϕ') of 2.7°. On repetition of the same experiment the observed increase was 3.9°. A control test was performed, stopping and resuming shear with no over-consolidation and this gave no observable increase in ϕ'.

27. A further sample was tested using the same experimental procedure but with the shear plane created at an effective normal stress of 75kPa and over-consolidation at a total effective stress of 95kPa. An increase in shear strength of 0.85kPa was observed after over-consolidation. This would be equivalent to an increase in ϕ' of 0.65° if c' is assumed to be equal to zero.

28. Presentation of these results is problematic due to the changing value of effective normal stress. Also, percentage change is difficult to estimate due to the low value of initial normal stress, the points thus lying in the non-linear region of the Mohr-Coulomb failure envelope and so distorting initial values. Bearing these points in mind, the sequence of results has been illustrated in Figure 5.

29. The results suggest that modest over-consolidation of a relatively shallow shear plane will give rise to a noticeable increase in shear strength, the effect being most pronounced when an ill-defined shear plane has been developed, i.e. a low normal stress. The trend of both the residual strength envelope and the envelope of enhanced shear strength due to over-consolidation by an increment of 20kPa appears to be one of curvature at low normal stresses. This is in agreement with previous work at Loughborough (ref.13,14), for example with tests on Lias clay shown in Figure 6. It is also apparent that the curves diverge more as the normal effective stresses acting

on the shear plane reduce. Thus the greatest difference was found in the tests that represent perhaps the upper limiting normal effective stress (30kPa) for the shallow land slips being considered here (i.e. assuming zero porewater pressure and a depth to shear plane of < 1.5m).

30. It should be noted that the improvements in strength reported here are small, even at effective normal stresses of 30kPa, but that the results are nevertheless particularly encouraging. Firstly at lower normal stresses still, the shear plane definition will become progressively poorer and the effects might be expected to be proportionately greater. In addition the required

improvements, in terms of ϕ' or (especially) apparent effective cohesion, are also very small to increase the factor of safety to a value greater than unity. Finally the effect of over-consolidation of the shear plane is only one means of stabilisation, and the improvements in strength due to the cementing reactions in the annular zones surrounding the piles, for example, will also considerably improve the stability of the slope.

CONCLUSIONS

31. Clay fill embankments have been shown to develop shallow, typically less than 1.5m deep, failures planes several years after construction as the magnitude of the negative porewater pressures reduce on equilibration. Current repair techniques, though generally successful, require much plant activity and are relatively expensive to carry out. Such failures could be potentially avoided if lime-modified clay were used as fill in embankments, the additional costs involved in chemical stabilisation being offset against reduced landtake and reduced volumes of material to handle.

32. Stabilisation of shallow failures can be effected by the use of lime piles, which can be constructed with small plant and a minimum of associated lorry movements. This will remove the need for traffic restriction that would otherwise apply in some cases. The improvement in stability is brought about, in part, by the generation of negative porewater pressures and consolidation of the shear plane. It has been shown that such consolidation is most effective at low normal effective stresses, such as exist in the cases under consideration. The technique, which is being further researched at Loughborough University, could thus provide for better engineering maintenance of clay fill embankments.

ACKNOWLEDGEMENTS

33. The work reported herein forms part of a research programme funded under the SERC LINK scheme. The programme is supported in equal measure by SERC and industrial contributions ftom Geotechnics Ltd, Cementation Piling and Foundations Ltd, The British Waterways Board and Buxton Lime Industries. Their support is gratefully acknowledged.

REFERENCES

1. Perry J. *A survey of slope condition on motorway slopes in England and Wales.* Research Report RR199, Transport Research Laboratory, Crowthorne, Berks, 1989.

2. Crabb G I and West G. *Monitoring porewater pressures in an embankment slope.* Proceedings of the international symposium on failures in earthworks, Institution of Civil Engineers, London, March 1985, pp 406-410, Thomas Telford Ltd of London.

3. Crabb G I, West G and O'Reilly M P. *Groundwater conditions in three highway embankment slopes*. Proceedings of the ninth European conference on soil mechanics and foundation engineering, Dublin, August 1987, pp 401-406, AA Balkema of Rotterdam.

4. Crabb G I and Atkinson J H. *Determination of soil strength parameters for the analysis of highway slope failures*. Proceedings of international ICE conference on slope stability engineering developments and applications, Isle of Wight, April 1991, pp 13-18, Thomas Telford Ltd of London.

5. Rogers C D F and Bruce C J. *The strength of lime-stabilised British clays*. Proceedings of Lime Stabilisation 90 Symposium, Sutton Coldfield, March 1990, pp 58-72, BACMI, London.

6. Cobbe M I and Wrench P N. *Saddington cutting slip repair - research into the use of lime stabilised clay as an embankment fill*. Proceedings of Lime Stabilisation 90 Symposium, Sutton Coldfield, March 1990, pp 73-83, BACMI, London.

7. National Lime Association. *Lime slurry pressure injection*. Bulletin 331, NLA, Arlington, Virginia, USA, 1985.

8. Rogers C D F and Bruce C J. *Slope stabilisation using lime*. Proceedings of international ICE conference on slope stability engineering developments and applications, Isle of Wight, April 1991, pp 395-402, Thomas Telford Ltd of London.

9. Ingles O G and Metcalf J B. *Soil stabilisation, principles and practice*. Butterworths, 1972, (out of print).

10. Handy R L and Williams W W. *Chemical Stabilisation of an Active Landslide*. Civil Engineering (37), 8, pp62-65, 1967.

11. Anon. *Subgrade Improved with Drill - Lime Stabilisation*. Rural and Urban Roads, October 1963.

12. Lutteneger A J and Dickson J R. *Experiences with Drilled Lime Stabilisation in the Midwest USA*. Proceedings of Fourth International Symposium on Landslides, 1984.

13. Anayi J T, Boyce J R and Rogers C D F. *Comparison of alternative methods of measuring the residual strength of a clay*. Transportation Research Record 1192, Transportation Research Board, National Research Council, Washington DC, USA, p 16-26, 1988.

14. Boyce J R, Anayi J T and Rogers C D F. *Residual Strength of soil at low normal stresses*. Proceedings of Fifth International Symposium on Landslides, Lausanne, Switzerland, July 1988, pp 85-88.

19. Some observations on the behaviour of an engineered contaminated cohesive fill

A. AL-TABBAA, The University of Birmingham, UK

SYNOPSIS.   Some of the geotechnical properties of contaminated cohesive fills using combinations of kaolin clay and Leighton Buzzard sand were investigated. The behaviour of the fills as the sand content increased was as expected. The contaminants used were sodium chloride and polyethylene glycol. The results show that both contaminants affect the behaviour of the fill: the liquid and plastic limits increased, the maximum dry density and unconfined compression strength decreased. The permeability to sodium chloride increased and that to polyethylene glycol depended on its concentration. The permeability was also found to be dependent on how the contaminant solution was introduced into the soil.

INTRODUCTION

1.   Engineered cohesive fills are extensively used to form the lining of landfill sites to act as a barrier for the containment of contaminants. An understanding of the behaviour of materials used for these liners is essential in the design of such barriers. The paper considers particular cohesive fills formed by mixing speswhite kaolin clay and different percentages of Leighton Buzzard sand. The paper will looks initially at the change in behaviour of the fill as the percentage of the sand in it is increased. The soil behaviour is measured in terms of tests such as the liquid and plastic limits, compaction, unconfined compression strength and permeability.

2.   Fill materials used for landfill liners depending on their origin may be contaminated with non-hazardous contaminants which might be assumed to be harmless. Some contaminants, however, change the engineering properties of their host soil and this behaviour might change as the concentration of the contaminant in the soil changes. Hence, in this paper the behaviour of the fill discussed in paragraph 2 is compared with that when the fill is subjected to contamination. Two contaminants are considered in this paper. A salt in the form of sodium chloride and a soluble organic polymer in the form of polyethylene glycol.   The research work presented here forms an initial part of a research programme at Birmingham University investigating the effect of contamination on clay chemistry and the consequences relating to the clay's engineering behaviour.

THE FILL MATERIAL

3.   The fill material used in all the tests reported here was made up from a mixture of speswhite kaolin clay and Leighton Buzzard sand. These two materials were chosen because their physical and mechanical properties are well documented. Speswhite kaolin clay has 80% of particles smaller than 2μm and the Leighton Buzzard sand used was a uniform medium sand with 100% passing through 300μm sieve and retained by 225μm sieve.

4.   Two contaminants are used. The first is a salt in the form of sodium chloride (NaCl). Salts occur naturally in many soils and are a common constituent in most contaminant streams. Salt input to the soil may occur from the addition of waste (e.g. from the petroleum industry), fertiliser applications, precipitation and irrigation. The second is a soluble organic polymer: polyethylene glycol $[HO(C_2H_4O)_nH]$ with an average molecular weight of 2000 and will be referred to as PEG in the paper. This contaminant is chosen as a contaminant which might be found in household waste and is also used as a soil conditioner in agriculture. PEG is a neutral polar organic chemical which does not exhibit a net charge but has an asymmetric distribution of electron density resulting in an appreciable dipole moment, which is an indication of its polar characteristics.

5.   The various uncontaminated and contaminated cohesive fill materials formed are shown in Table 1 below. Each of the twelve different combinations of fills is mixed at a number of different moisture or solution contents ranging between 7.5% and 40%. The clay and sand were first mixed dry and then the required amount of solution was added and mixed. The soil was then left for 24 hours to stabilise. Both contaminant solutions were made up at a concentration of 35g/100ml of water. This concentration is around the saturation level for sodium chloride and the molarity concentration is 5M. PEG can be dissolved in water at very high concentrations, hence a similar mass ratio to that of the sodium chloride solution was chosen which gave a molarity concentration of 0.14M. The concentration of the contaminant has a large effect on the behaviour of the soil and a detailed investigation of this effect is beyond the scope of this paper. Therefore the above concentrations were kept constant throughout the testing programme. The water used in all the tests was deaired water.

Table 1: A table showing the symbols used for the different combinations of fills and contaminants.

	Percentage of Leighton Buzzard sand in fill			
	0%	**15%**	**30%**	**45%**
water	A	B	C	D
NaCl	E	F	G	H
PEG	I	J	K	L

6.   Because of the simplicity of the preparation method, the samples tested were prepared by mixing the dry soil with the contaminant solution. It would be expected that the behaviour of these fills is different from that of the fills prepared by mixing the dry soil with water first and then permeating the contaminant solution through it. The samples which are water permeated first would be closer to the situation which might be encountered in practice. A preliminary investigation was carried out to look at the extent of this effect and the results of these tests are reported in the appropriate sections below.

TESTING PROGRAMME, RESULTS AND DISCUSSION

7.   Samples of the twelve different fills were tested for their geotechnical properties using the liquid and plastic limits, compaction, unconfined compression strength and permeability. The results are discussed separately below.

The Liquid and Plastic Limits
8.   The values of the liquid and plastic limits were measured for each of the twelve soils. The cone penetration test was used to obtain the liquid limit and the 3mm thread method for the plastic limit. Fig. 1 shows a plot of the liquid and plastic limits against the sand content in the soil. The sand content is defined as the mass of sand over the

total dry mass of the soil. It can be seen, as expected, that as the sand content increases, the liquid and plastic limits of the soil decrease and as the sand content approached 50% and hence the clay content approached 50%, the liquid and plastic limits reduced to around half their values for the pure clay soil (ref. 1). Fig. 1 also shows the effect of the two contaminants on the liquid and plastic limits of the fills where it can be seen that the presence of both sodium chloride and PEG in the pore water increases both the liquid and plastic limits. The increase in the plastic limit was larger than that in the liquid limit showing that the presence of the contaminant reduced the plasticity index.

9. Since the ionic concentration in the sodium chloride solution used is very high, the adsorbed layer becomes thinner and this produces an open flocculating clay structure. This results in higher liquid and plastic limits than the originally more dispersed structure when the clay was mixed with deaired water which agrees with the experimental observations. However, very few references were found on the effect of contaminants on the liquid and plastic limits. Ref. 2 summarises the effect of different concentrations of sodium chloride solutions on the liquid and plastic limits of some illite and montmorillonite clays reported by a number of researchers. It showed contradicting results where in some cases the liquid and plastic limits increased and in others they decreased. This was attributed mainly to the use of different solution concentrations and to the test method used.

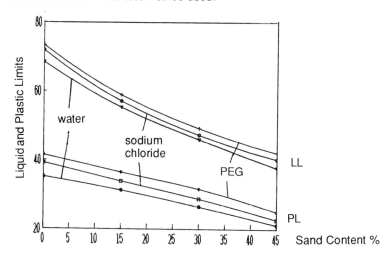

Fig. 1. Liquid and plastic limits for all twelve fills considered.

10. The bonding between PEG and the clay particles is by electrostatic bonding i.e. the positively charged carbon in it is attracted to the negatively charged clay particles. It has been reported that when clays are permeated with an organic chemical, changes in interlayer spacing occurs. This has been related to the dielectric constant of the organic chemical. The dielectric constant of a medium is defined as ε in the following equation: $F = (Q\,Q')\,/\,(\varepsilon\,r^2)$, where F is the force of attraction between two charges Q and Q' separated by a distance r in a uniform medium. The interlayer spacing between the clay particles is proportional to the dielectric constant of the pore fluid. If ε decreases the interlayer spacing decreases and the double layer around the clay particles shrinks. For organic chemicals with dielectric constant lower than that of water, the individual clay particles will contract as a result of thinner interlayer spacing and this provides the opportunity for the clay particles to orientate themselves into a flocculating structure. The dielectric constants of water and PEG are 79.5 and 37.7 respectively.

241

11. Samples which were water compacted first and then permeated with the contaminant solutions were tested for their liquid and plastic limits. These tests showed a further similar increase in both the liquid and plastic limits compared to the results in Fig. 1. This can be explained by the fact that when the water is mixed with the dry fill first, it separates the clay particles and creates a better environment for the contaminant, which are introduced later, to bond with the clay particles.

Compaction tests

12. Samples from each of the twelve fills in Table 1 were tested for their compaction behaviour using the Dietert apparatus. The soils were compacted in a 38mm diameter mould using 10 blows on one side of the sample only and the resultant dry density calculated. Three samples from each soil were tested for repeatability and the average value of the dry density reported. The results were found to be within 3% of the average value. A plot of dry density against moisture content for the soils mixed with the three different solutions of deaired water, sodium chloride and PEG are shown on Figs 2(a, c and e) respectively. The moisture content in this case is plotted using its standard definition of the ratio of the mass of water or solution to the mass of dry solid, denoted by m_{ct} as in total moisture content.

13. The results for the water compacted samples shown on Fig. 2(a) show, as expected, that as the sand content increases in the cohesive fill, the maximum dry density increases i.e. the compaction behaviour improves. The maximum dry density increased from $1.403Mg/m^3$ for the pure clay fill to $1.68Mg/m^3$ for the 45% sand fill. This figure also shows that as the sand content increases and hence the clay content decreases, the optimum moisture content reduces, the optimum moisture content reduced from around 30% for the pure clay samples to around 17.5% for the 45% sand samples i.e. the optimum moisture content approached half its value as the clay content approached 50% of the dry soil mass. This is obviously due the fact that the water when mixed with the dry soil is absorbed mainly by the clay, which as its content reduces the moisture content increases. Hence the results in parts (a, c and e) in Fig. 2 are plotted in parts (b, d and f) where the moisture content is defined there as the ratio of the mass of water to the mass of clay only, denoted by m_{cc} as in clay moisture content. The results now show that the optimum moisture content defined in those terms is the same for all four different sand content fills and equal to around 30%.

14. The compaction graphs for the sodium chloride solution samples shown on Figures 2(c&d) and for the PEG solution samples on Figs 2(e&f) show a general compaction behaviour similar to that observed for the water compacted samples. The optimum moisture content defined in terms of the clay content m_{cc} was found to be around 30% for the sodium chloride samples but increased to 37.5% for the PEG solution. It is more convenient to compare the effect of the three different solutions on a plot of maximum dry density against sand content as shown on Fig. 3. Fig. 3 shows that both contaminant solutions have caused a reduction in the maximum dry density of the fills. The reduction was up to 2% and 4% for the sodium chloride and PEG respectively with no specific pattern.

15. Compaction tests were carried out on water compacted samples which were subsequently permeated with each of the two contaminant solutions. One sample from each of the four fills (A to D in Table 1) was tested with each permeant. The samples chosen were those at the optimum moisture content. At the end of the permeation test the moisture content increased and it was the compaction results at this moisture content which were compared. The results show that these samples gave up to twice the reduction in maximum dry density compared to those samples which were mixed dry with the contaminant solutions. These results support the argument in paragraph 11 above relating to the effect of the initial mixing the dry soil with the water.

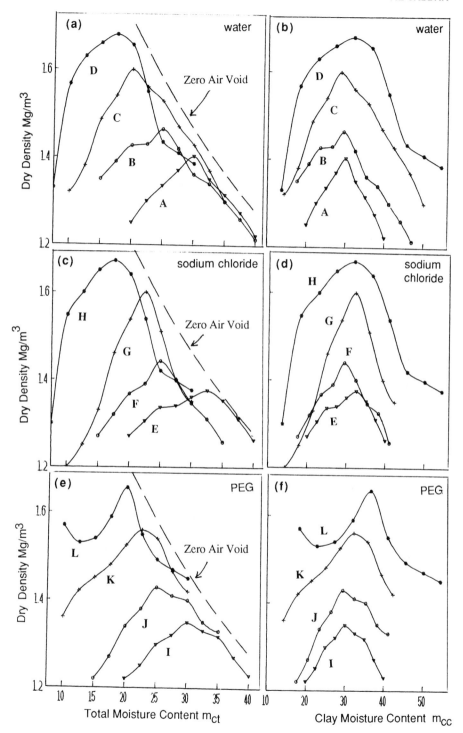

Fig. 2. The compaction curves.

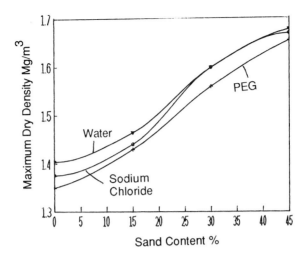

Fig. 3. Maximum dry density versus sand content for all twelve fills.

Unconfined Compression Strength (UCS)

16.   The height of the 38mm diameter soil samples compacted in the Dietert apparatus was trimmed to 76mm and then tested in the unconfined compression apparatus for their UCS. Three samples from each soil batch were tested and an average value taken. The results were found to be within 30% of the average. The UCS values from these tests are plotted against total moisture content m_{ct} on Figs 4 (a), (b) and (c) for the water, sodium chloride and PEG solution samples respectively. The results in the figures show a similar pattern of behaviour to one another and the shape of the graphs is similar to those of the compaction behaviour. Fig. 4(a) shows that the maximum UCS for the water compacted fills remained more or less the same, at an average value of 152kPa, as the percentage of sand in the fill increased. This behaviour implies that the percentages of sand added were too small to change the behaviour of the cohesive fill considered. The moisture contents at which this maximum strength occurred were, in most cases, slightly lower than the optimum moisture content for compaction. This result agrees with that reported by (ref. 3) where a decrease in the moisture content of about 2% was observed. In the tests reported here where the moisture content was reported to the nearest 2.5% showed a difference in most tests of 2.5%.

17.   The UCS of the contaminated samples shown on Figs 4(b) and (c) for sodium chloride and PEG respectively show a decrease in the strength of the fill. As the sand content increased, the strength values seemed to slightly increases for the sodium chloride samples and decrease for the PEG samples. A constant average value for the maximum UCS of the sodium chloride and PEG samples can be reported as 130kPa and 120kPa respectively which is a decrease to 85% and 79% compared to the water compacted samples. Samples which were water compacted first and then permeated with the contaminant solution gave lower values for the UCS for both contaminant solutions compared to those reported in Figs 4(b) and (c). The reduction was in the region of 10-30% and was less as the sand content increased. This observation is consistent with the compaction behaviour discussed in paragraph 15.

18.   The correlation observed in the tests between the UCS and the index properties agrees with that reported by (ref. 4) i.e. as the liquid limit of the soil increased the UCS decreased. Also the effect of the contaminant on the UCS agrees with the arguments discussed in paragraphs 9 & 10 above which suggest that a flocculating structure has formed in both cases which gives lower strength for the soil.

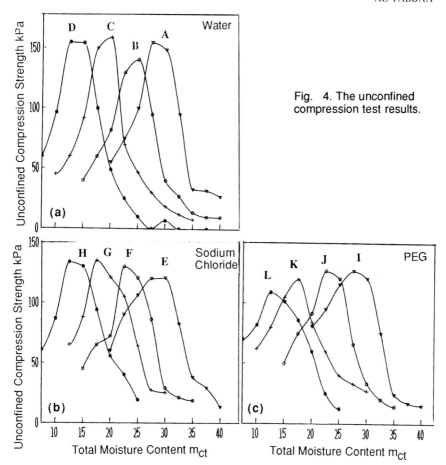

Fig. 4. The unconfined compression test results.

Permeability

19. Samples with moisture contents around the optimum value were chosen as this condition is likely to be similar to the soil used in practice for the landfill liner. The fill was compacted in a rigid wall cell and a flow pump test was conducted on the sample. The samples were compacted to dry densities similar to those achieved in the compaction tests. The void ratio at the beginning of the tests ranged between 0.55 and 0.8 and the moisture content between 15-30%. At the end of the tests the void ratio and moisture content increased to 0.8-1.0 and 25-35% respectively as the permeant solution replaced the air in the pores. The permeant solution was forced to flow from the top of the sample at a constant flow rate (Δq) vertically down through it. At the inflow, a connection was made to a pore pressure transducer to measure the pore fluid pressure (Δu) developed at the inflow position. The pore fluid pressure at the lower end of the sample was taken as zero. The flow was allowed to continue for 24 hours first to enable the flow rate to come to equilibrium. The thickness of the sample was chosen to be between 1 and 2 cm so that after the 24 hours the solution in the whole sample would have been replaced by the inflow solution. After that the flow was monitored for 1-4 days. The vertical permeability (k_v) was then calculated using Darcy's law: $k_v = (\Delta q \ \gamma_w) \ / \ (\pi R^2 \ \Delta u)$, where R is the radius of the sample and γ_w is the unit weight of water. Three different flow rates where used on each sample and an average value of the permeability reported, the variation being within 10% of the average. The tests were conducted under no applied vertical stress.

20.   A number of different tests were carried out on pure kaolin and 45% sand samples only and all the results are shown on Fig 5 where the permeability is plotted on a logarithmic scale. To differentiate between the two sets of results the percentage of the sand is indicated next to the letter symbols at each point. In all the tests the height of the sample was maintained constant and hence swelling and softening due to the presence of the permeant solution was prevented. It was, however, not possible to measure the extent of any seepage induced consolidation in the equipment used but the pore pressure difference across the height of the sample was not permitted to exceed 50kPa in any of the tests and it is assumed that any seepage induced consolidation under these conditions would be minimal.

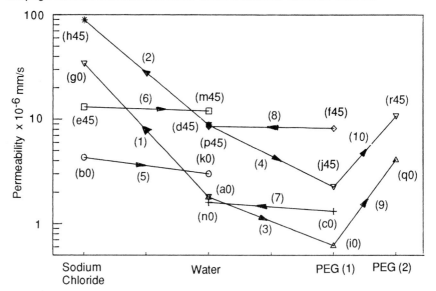

Figure 5. Permeability test results.

21.   In the basic series of tests the sample was permeated with the same solution as that used originally in forming the sample when the soil was mixed dry with the permeant solution. The following comments can be made on these tests:

(i)   The permeability of the water compacted pure kaolin clay samples was measured at 1.8×10^{-6} mm/s as shown by point (a0) on Fig. 5. This value is higher than the permeability of saturated kaolin consolidated from a slurry to the same void ratio of around 0.5×10^{-6} mm/s (ref. 5). The increase is obviously due to the clay being compacted which tends to produce a non-uniform sample in the sense that the sample tends to be less compacted at the edges and tends to have large flow channels compared to the clay sample consolidated from a slurry .

(ii)   The permeability of the sodium chloride pure kaolin sample permeated with sodium chloride solution is shown by point (b0) of 4.3×10^{-6} mm/s. This shows that the presence of sodium chloride increases the permeability as expected from the effect discussed in paragraph 9 above where the resulting flocculating structure due to the presence of the sodium chloride solution increases the permeability.

(iii)   The permeability of the PEG pure kaolin sample permeated with PEG solution was 1.32×10^{-6} mm/s as shown by point (c0) on Fig. 5. Hence this concentration of PEG causes a reduction in the permeability of the fill. This is expected since at this low concentration, the dielectric constant of the solution will be similar to that of water and so its effect on shrinking the double layer is negligible. In this situation it is the viscosity of the solution which has the dominating effect on the changes in the

246

permeability (ref. 7). Since the viscosity of the PEG solution is higher than that of water, the permeability will decrease at low concentrations.

(iv)   Fig. 5 shows that the fill sample containing 45% sand denoted by point (d45) showed a permeability of 8.8×10^{-6} mm/s which is an increase of about five times that of the pure kaolin sample. The corresponding permeability values for the sodium chloride is 13.1×10^{-6} mm/s denoted by point (e45) and for PEG of 8.2×10^{-6} mm/s denoted by point (f45) which shows a similar trend to the results obtained for the pure kaolin samples.

22.   The second series of tests involved permeating one solution compacted samples with a different solution permeant. Hence the water compacted samples were permeated with sodium chloride and PEG separately and vice versa. The result are also presented on Fig 5. The test paths are shown by numbered arrowed lines. The following remarks can be made on these tests:

(i) The water compacted pure kaolin sample (a0) was permeated with sodium chloride solution and gave a permeability value as shown by point (g0) of 34.2×10^{6} mm/s, and test path (1). The same procedure carried out on the 45% sand samples (d45) gave permeability values of 89.2×10^{-6} mm/s denoted by point (h45) and test path (2). This indicates that the presence of the particular concentration used of sodium chloride solution increases the permeability of the pure kaolin samples by about 19 times. Increase in the permeability of clays due to sodium chloride solutions has been reported (refs 2,7). It can also be seen that this permeability is higher than the permeability of the sodium chloride compacted samples permeated with sodium chloride solution shown by point (b0). As discussed previously, mixing the clay with water first leads to separation of the clay particles and it then becomes easier for the Na^+ cations to attach themselves to the surface of the clay particles.

(ii) When PEG solution was permeated through the water compacted pure kaolin sample (a0), the resultant permeability value was 0.62×10^{-6} mm/s denoted by point (i0) and test path (3). A similar observations is seen for the 45% sand sample (d45) giving permeability values of 2.26×10^{-6} mm/s shown by point (j45) and test path (4). This shows that the introduction of the PEG solution gives a lower permeability of the fill to the PEG solution that to water. The permeability reduced to about 0.33-0.25 of that for water and this again is attributed to the effect of the high viscosity of the PEG solution. As the permeant solution passed through the sample it was initially diluted by the water already present in the soil and hence at the beginning of the test the permeability was similar to that of water. It then gradually decreased with time as the concentration of the permeant solution in the sample approached the inflow concentration.

(iii) The two sodium chloride solution compacted samples, (b0) & (e45) were permeated with deaired water. The resulting permeabilities are given by points (k0) of 3×10^{-6} mm/s and (m45) of 12×10^{-6} mm/s and paths (5) & (6). The corresponding values for the two PEG solution samples are given by points (n0) of 1.6×10^{-6} mm/s and (p45) of 8.5×10^{-6} mm/s and paths (7) & (8). These results suggest a slight change in the permeability values which implies that passing water through the sample has little effect on extracting the contaminants from the soil.

23.   An additional test was carried out in which the concentration of the PEG solution permeated through the water compacted, (i0) & (j45), sample was doubled, i.e. 0.28M, for both the pure kaolin and the 45% sand samples. This solution gave a permeability of 4.1×10^{-6} mm/s as shown by point (q0) and test path (9) and 10.7×10^{6} mm/s shown by point (r45) and test path (10). So, it seems that this increased concentration of PEG is starting to show the effect of the dielectric constant on increasing the permeability of the fill to this organic chemical. It would therefore be expected that as the concentration of the solution is increased that the permeability would increase. Similar observations have been reported in the literature with regard to other organic liquids where the high viscosity controls the flow of the liquid at low

concentrations and hence reduces the permeability, and above a threshold concentration the permeability increases due to the shrinking of the double layer around the clay particles due to the low dielectric constant of the organic chemical (e.g. refs 6,8). Ref. 8 reported that a concentrated solution of ethylene glycol would increase the permeability of a smectite clay by up to 180 times with an initial decrease in permeability with diluted solutions. The effect of a concentrated solution of PEG, which has a much higher molecular weight than ethylene glycol, is expected to produce a higher effect (ref. 9). The effect of different concentrations of the PEG solution is being investigated at present.

CONCLUSIONS

24. This paper presented the results of a preliminary investigation into the effect of two contaminants on the geotechnical properties of a cohesive fill as the percentage of sand in it is increased. This work forms an initial part of a research programme into the effects of contaminants on the engineering behaviour of clays. The conclusions reached here relate to the concentration of the two contaminant solutions used in the testing programme. The tests carried our so far indicate that the liquid and plastic limits of the fills increased and the unconfined compression strength decreased for both contaminant solutions. The permeability to the sodium chloride solution increased. The permeability to the PEG decreased initially and increased when the concentration was doubled. It was also concluded that the permeability is affected by the way the contaminant solution is introduced into the soil. These tests show that the presence of such untoxic contaminants could change the geotechnical properties of the soil in a way which detrimental to the design of landfill liners using such materials.

ACKNOWLEDGEMENTS

The author wishes to express her thanks to David Lynock and Barry Simkiss for assisting with the experimental work and to Joanne Farnworth for her comments on the chemistry aspects of the work.

REFERENCES

1.   SEED, H.B., WOODWARD, R.J. and LINDGRESS R. Fundamental aspects of the Atterberg limits. Journal of the Soil Mechanics and Foundation Engineering Division, Proceedings of the ASCE, 1964, Vol. 90, No. SM6, pp 75-105.
2.   HO, Y.A. and PUFAHL, D.E.   The effects of brine contamination on the properties of fine grained soils. Geotechnical practice for waste disposal '87, Geotechnical special publication No. 13, ASCE, 1987, pp547-561.
3.   Road Research Laboratory. Soil Mechanics for Road Engineers. HMSO, 1961.
4.   W ROTH C.P. and WOOD D.M.   The correlation of index properties with some basic engineering properties of soils. Canadian Geotechnical Journal, 1987, Vol. 15, No. 2, pp137-145.
5.   AL-TABBAA A. and WOOD D.M. Some measurement of the permeability of kaolin. Geotechnique 37, 1987, No. 4, pp499-503
6.   YONG R.N., MOHAMED A.M.O. and WARKENTIN B.P.   Principles of contaminant transport in soils. Development in Geotechnical Engineering, Volume 73, Elsevier, Amsterdam, 1992.
7.   FERNANDEZ F. and QUIGLEY R.M.   Viscosity and dielectric constant control on the hydraulic conductivity of clayey soils permeated with water-soluble organics. Canadian Geotechnical Journal, 1988, Vol. 25, pp582-589.
8.   ANDERSON D.C., BROWN K.W. and THOMAS J.C.   Conductivity of compacted clay soils to water and organic liquids.   Waste Management and Research, 1985, Vol. 3, pp 339-349.
9.   DE BOODT M.F., HAYES M.H.B. and HERBILLON A.   Soil colloids and their associations in aggregates. Plenum Press, London, 1990.

20. Description, classification and specification of chalk for use in earthworks

J. R. GREENWOOD, Travers Morgan Consulting Group, UK

SYNOPSIS. The current Department of Transport classification and specification for chalk is critically reviewed. A revised classification scheme is proposed based on intact dry density and moisture content. A specification for acceptability is proposed based primarily on the natural moisture content of the chalk. The study revealed the shortcomings of the existing descriptive scheme to relate to the use of chalk as an engineering fill. Amendments to the descriptive scheme are proposed to assist with identification of relevant engineering properties at the investigation stage.

INTRODUCTION
1. Most of the major transport routes radiating from London encounter the chalk deposits of South and East of England. Chalk is usually a white, porous jointed limestone having greatly varying strength. It has a natural moisture content varying typically from 8 to 36%. During earthworks operations the excavation and compaction processes may break down the natural structure of the chalk releasing some of this water and generating fine material. If the amount of fines and released water, added to any free water found in the joints, is high enough then the result is a temporarily unstable fill material which may remain unstable for a period varying from days to weeks (ref 1, DTp advice note HA44/91).
2. Classifications and specifications for use of chalk in earthworks have been developed as experience in construction has been gained.

EXISTING APPROACHES AND PROBLEMS
3. The historical development of chalk classifications and specification was discussed by Greenwood, 1989 (ref 2). Whilst other design factors may be relevant, the particular nature of chalk is such that it is generally the trafficability and response to compaction which governs its acceptability for reuse in embankment fill.
4. The Department of Transport's 6th edition Specification, 1986 (ref 3) and the current 7th edition, 1991 (ref 4) permit the Engineer to include chalk as a granular (class 1) or cohesive (class 2) material or to treat it as a special material (class 3) under Clause 605. The options are listed in Table 1. The Notes for Guidance accompanying the Specification recommend designation based on ground investigation results and the proportions of chalk material falling into classes A, B, C or D in accordance with TRRL report LR806 (ref 5, Ingoldby and Parsons, 1977).
5. A modification of the TRRL classification chart was subsequently presented by the Department of Transport in

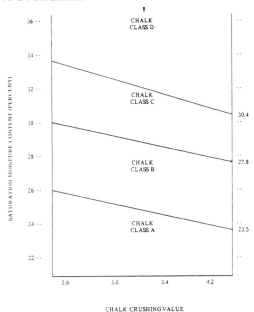

Figure 1   Modified Chalk Classification Chart (Departmental Advice Note HA/44/91 DTp 1991).

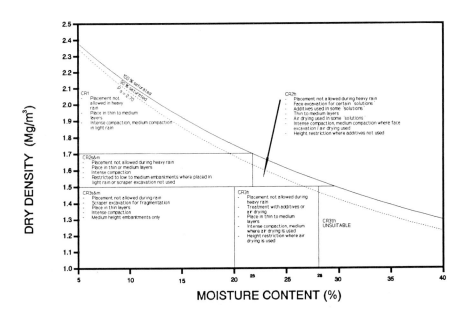

Figure 2   French Classification Scheme for chalk and Conditions of Use (1983).

250

Advice Note HA44/91 (ref 1) as shown in Figure 1. A flow
chart was given in HA44/91 to assist the designer in
classifying the chalk for use with the Specification.
    6. Various points of concern were noted by Greenwood, 1989
(ref 2) in the application of the LR806 chart and the use of
the Chalk Crushing Value (CCV) test:- The CCV test procedure
was cumbersome, the chalk grade was not particularly dependent
on the CCV, and the use of saturation moisture content alone
did not reflect the degree of saturation. This point was made
by Guy, 1989 (ref 6), and the classification was insensitive
to highly weathered chalk grades (class D) as only intact
lumps can be selected for CCV testing.
    7. The publication of advice note HA44/91 did clarify the
initial confusion regarding use of the LR806 classification
and CCV for site control. They should not be used. The
Clause 605 requirements have proved valuable and appropriate
but although the designer can select SMC and/or natural
moisture content (mc) to determine acceptability of chalk
there is only limited guidance on the application and choice
of limits. In practice use of SMC alone can be rather
misleading as it does not represent the degree of saturation
unless it is related to natural moisture content. SMC is in
reality a measure of the dry density (which must be measured
to determine SMC) but is less direct because the particle
density of the soil grains is required to be measured or
assumed. The LR806 classification could therefore benefit
from review together with the site control specification and
compaction levels. It is considered that a simple system of
classification and designation of chalk in the Specification
would help to avoid contractual misunderstandings.

PROPOSED CLASSIFICATION
    8. Research and case studies presented at Brighton, 1989 by
various authors confirm that the most easily measured property
which is indicative of the likely behaviour of the chalk is
the intact dry density (γd). This parameter gives a good
indication of the susceptibility of the chalk to crushing
although it has been recognised that other factors such as
textural features can contribute (ref 7, Mortimore and
Fielding, 1989).
    9. The current French classification system was introduced
by Rat and Schaeffner 1983 (ref 8) and is based on dry density
(γd) and natural moisture content (mc). This is shown in
Figure 2. The system is currently being updated (ref 9
Schaeffner, 1991) to include rather more categories of
material. The divisions between categories may not however be
appropriate for conditions encountered in the UK.
    10. It is proposed that dry density is used as the basic
parameter for classification of chalk in the UK with the
addition of natural moisture content to define the state of a
particular specimen. The use of these two parameters on the
standard γd - mc chart (as used for compaction testing) should
adequately reflect the important and relevant engineering
characteristics of the chalk. It is recommended that the 100%
and 90% saturation lines are included on the chart, based on
the typical particle density of 2.70 for chalk. These lines
may be revised if a significant local variation in particle
density is measured. Use of the γd–mc chart in the UK will
mean that a similar basic classification system is in use
throughout the UK and France.
    11. The current LR806 classes A, B, C and D are shown in
Figure 3 plotted on the γd–mc chart. The hardness
classification proposed by Mortimore et al 1989 (ref 10) is
also shown for reference.

251

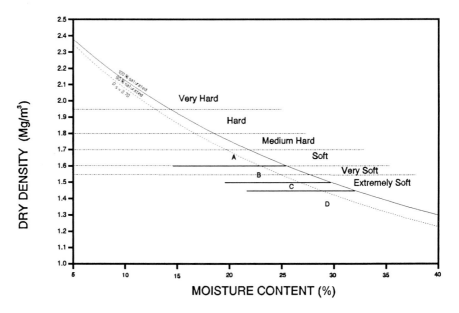

Figure 3   Hardness/strength classification proposed by
Mortimore et al (1989) and current TRRL classes
approximately represented on γd - mc chart.

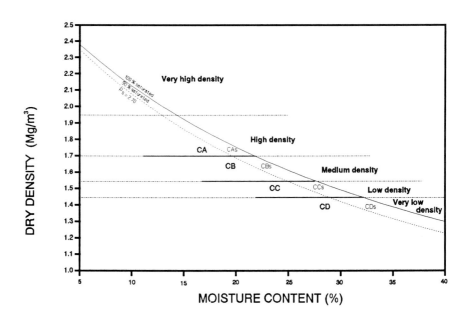

Figure 4   Proposed classification chart with revised chalk
classes and revised terms for Mortimore's hardness scale.

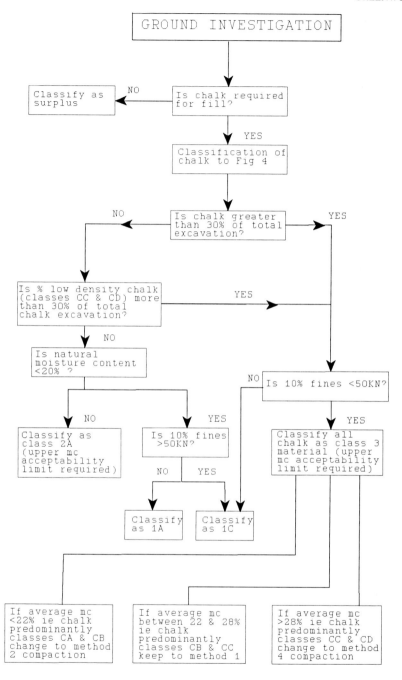

Figure 5 Proposed flow diagram for classifying
chalk for use as general fill

253

Table 1 Current Chalk Classes and Compaction Methods (Department of Transport Specification, 1991)

Class	Requirement	Compaction Method
1A	SMC <20%	Method 2
1C	10% fines >50kN	Method 5
2A	SMC >20%	Method 1
3	Degradable chalk. Optional use Special require-ment (CL605 apply)	Method 4 if SMC >26% Method 1 if SMC 20-26% Method 2 if SMC <20%

Table 2 Guidance on construction use for revised classification chart

Chalk Class	Preferred Compaction Method	Comments/Notes		Workable in light rain
		Likely Plant	Instability	
CA	2	Any	Not expected	Yes
CA_s	2	Any	Not expected	Probably
CB	1^+	Any	Not expected	Yes
CB_s	1	Face shovel	Unlikely	No
CC	1^+	Face shovel	Unlikely	Possibly
CC_s	4	Face shovel	Probable	No
CD	4	Face shovel	Possible	No
CD_s	$(4)^*$	Face shovel	Expected	No

Notes:
- * use may not be economical in embankment fill
- + amend to method 2 if 'dry' ie mc <22%
- CA_s indicates chalk material between 90% and 100% saturation

Table 3 Proposed chalk classes and compaction methods (relating to Department of Transport Specification 1991)

Class	Requirement	Compaction Method
1A	mc <22%	Method 2
1C	10% fines >50kN	Method 5
2A	mc ≥22%	Method 1
3	Optional use (CL605 to apply)	Method 4 if mc >28% (CC/CD) Method 1 if mc 22-28% (CB/CC) Method 2 if mc <22% (CA/CB)

12. It is noted that the LR806 classes represent a comparatively narrow range of chalk densities ie $\gamma d = 1.45$ to $1.60 Mg/m^3$ and it is considered that the classes should be extended over the range 1.45 to $1.70 Mg/m^3$ as shown in the proposed classification chart (Figure 4). The revised classes are referred to as CA, CB, CC and CD. Mortimore's hardness classification is renamed in terms of density, as discussed later.

13. The classification boundaries are deliberately stopped well short of the zero moisture content axis as a reminder that most chalks occur naturally at around 70 - 100% saturation and chalk in a very dry state may require special compaction if reused in earthworks.

14. A further subdivision of the chalk classes CA to CD is proposed (ie classes CA_s to CD_s) for chalks having a degree of saturation (S) between 90% and 100%. These are the chalks which are most likely to be problematic during earthworks.

15. Guidance on the use of each class of chalk in construction, under the proposed classification is presented in Table 2.

INVESTIGATION AND DESIGN

16. It is anticipated that the γd - mc chart will be used for classification of chalk at the investigation stage. This will enable quantities of the various classes of chalk to be estimated and its likely behaviour predicted. Testing at investigation stage will include moisture content, dry density determinations on the chalk lumps (or chalk mass of highly weathered chalks) and checks on the particle density and 10% fines value. Modifications to the chalk descriptive scheme at investigation stage are presented in the following sections.

ACCEPTABILITY AND CONSTRUCTION CONTROL

17. It is proposed that acceptability of chalk for general earthworks be controlled by moisture content alone. Whilst on first sight this would seem to permit chalk of lower classes to be used, they would be restricted to drier chalk of that class which should cause no particular problems in earthworks. For example, an upper moisture content limit of 28% would exclude most of the wetter class CC chalks and permit only class CD chalk that is well on the dry side of the saturation line. A revised flow diagram for classifying chalk for general fill is given in Figure 5. The DTp chalk classes and compaction methods may be revised as given in Table 3.

18. If, for engineering design purposes, a restriction on the class of chalk is required, then dry density criteria may be specified in addition to the natural moisture content.

19. The main site control test during construction will be the natural moisture content, with periodic measurement of the dry density of the intact chalk as a check of the classification. Frequency of testing should be compatible with other acceptability testing.

PROPOSED DESCRIPTIVE SCHEME

20. A detailed description of the chalk is required at the investigation stage to identify features which are likely to affect the earthworks operations. The Brighton conference (1989) highlighted a number of shortcomings in the descriptive scheme with the use of the Mundford grades and SPT correlations being particularly criticised. Various suggestions were put forward for improvement. The introduction of a 'hardness' term into the description as proposed by Mortimore et al (ref 10) would be of particular benefit in classifying chalk. However as Mortimore's hardness

Table 4 Field Description of Intact Chalk Density (after Mortimor
Roberts and Jones, 1989) and proposed density banding for earthwor
application

PROPOSED DENSITY BANDS	HARDNESS SCALE (Mortimore et al)	FIELD IDENTIFICATION
Very Low Density		
1.45 Low Density	Extremely Soft <1.55 Mg/m^3	Crushed to putty when struck with the hammer head; steel pin pushed in by hand; geological hammer pick penetrates with very light blows yielding considerable 'splashes of putty'; crushed easily by hand and remoulds to putty easily.
1.55	Very Soft 1.55 – 1.60 Mg/m^3	Crushed to putty when struck with hammer head; steel pin hammered in by hand; when struck with the pick this chalk generates 'splashes of putty'; broken with ease by hand.
Medium Density	Soft 1.60 – 1.70 Mg/m^3	Part fractured, part crushed and sometimes has a flinty ring when struck with hammer head; steel pin and pick head hammered in fairly easily; broken by hand with slight difficulty.
1.70	Medium Hard 1.70 – 1.80 Mg/m^3	Fractured with a flinty ring when struck with hammer head (clean, conchoidal fractures in the purer varieties); some resistance to hammering in steel pin or pick end; broken with difficulty by hand.
High Density	Hard 1.80 – 1.95 Mg/m^3	Fractured when struck with hammer head, chips and fragments fly-off after heavy blows; great difficulty in hammering in steel pin or pick; broken with great difficulty by hand.
1.95	Very Hard 1.95 – 2.40 Mg/m^3	Slight ring when struck with hammer head and chips fly-off when struck hard; slight dent only made by steel pin and pick end of hammer; impossible to break with hand pressure.
Very High Density	Extremely Hard >2.40 Mg/m^3	Rings when struck with hammer head, splinters after many heavy blows; impossible to hammer in steel pin or penetrate with a geological pick.

scale relates to intact dry density it is proposed that the name 'density' is used to avoid confusion with 'soft' and 'hard' applied as a general soil descriptive terms (Table 4). It would be desirable to reduce the categories of intact dry density to provide boundaries which are meaningful for earthworks, foundations, tunnelling etc. The following are suggested as relevant to earthworks as indicated on Table 4.

Very High Density	> 1.95 Mg/m³
High Density	1.7 – 1.95 Mg/m³
Medium Density	1.55 – 1.7Mg/m³
Low Density	1.45 – 1.55Mg/m³
Very Low Density	<1.45Mg/m³

21. Table 5 presents a descriptive scheme for chalk which attempts to incorporate the best features of BS5930, (ref 12) Spink & Norbury, 1989, (ref 13) and Mortimore et al, (ref 10) to result in a more widely applicable chalk description of greater value in earthworks design. The scheme is largely based on that of Spink and Norbury, with the principal differences being an amendment to the rock material weathering definitions (Table 6) and the inclusion of the intact chalk density terms and other rock mass descriptions derived from Mortimore et al (Table 4). The remaining terms are as defined in BS5930. A shorthand code is suggested by Lord (CIRIA RP458) to describe the fracture aperture and spacing as an alternative to the Mundford grade. Example descriptions using the suggested scheme are given in Table 7. Further details of the scheme are given by Greenwood, 1993, (ref 11).

FURTHER DEVELOPMENT
22. A Steering Group has been established to direct the current CIRIA/TRL Research Project RP458 on Engineering Properties of chalk. It is intended to merge the proposals presented in this Paper with the CIRIA work to prepare a descriptive scheme which can be applied to chalk in all aspects of construction.

CONCLUSIONS
23. A revised classification scheme for chalk earthworks has been developed based on the fundamental parameters of dry density and moisture content. Control in the field may be by moisture content alone once the characteristics of the chalk are established.
24. The descriptive scheme for chalk has been updated to be more applicable to earthworks and now includes a 'density' term. A CIRIA/TRL steering group will further consider this scheme for general application to all chalk usage.

Table 5   Suggested descriptive scheme for chalk

ROCK MATERIAL			ROCK MASS						WORD ORDER FOR DESCRIPTIONS
WEATHERING STATE	STRENGTH (MPa)	DRY DENSITY (Mg/m³)	BEDDING STRUCTURES	Discontinuity Spacing (1D)	% Comminuted Chalk Matrix	% Coarser Fragments	Discontinuity Aperture	(Weathered State) GRADE	
	Very Weak	Very Low	Structureless Melange						Rock material colour : rock material weathering : "CHALK fragments" : rock material strength and intact dry density. Structure : "composed of" : percentage of coarse fragments : fragment roundness : fragment size : fragments : "set in" : percentage of matrix: soil strength of matrix: matrix colour : grain size and nature of matrix : presence and nature of flints. (Grade)
	1.25	1.45	or						
Completely Weathered			Remoulded Melange		>35%	<65% (D_f)		VI	
	Weak	Low	or						
	5.00	1.55	Poorly Structured		<35%	>65% (D_c)		V	
Highly Weathered	Moderately Weak		Thinly Laminated	Discontinuity Spacing (1D)			Discontinuity Aperture	NOT RECOMMENDED	Rock material colour : rock material weathering : "CHALK": rock material strength and intact dry density. Bedding structure : discontinuity type : discontinuity spacing : discontinuity width and nature of infill if appropriate : discontinuity orientations (in situ observations only): presence and nature of flints:other features. (Grade)
	12.50		6mm	Extremely Close (5)			(C)		
Moderately Weathered	Moderately Strong	Medium	Thickly Laminated 20mm	20mm			>3mm open or infilled		
	50	1.70	Very Thin 60mm	Very Close (4) 60mm					
Slightly Weathered	Strong	High	Thin 200mm	Close (3) 200mm			<3mm open or infilled (B)	III	
	100	1.95	Medium 600mm	Medium (2) 600mm				II (where blocky)	
Fresh	Very Strong	Very High	Thick 2m	Wide (1) 2m			Closed or Tight or Very Tight or Annealed (A)	I (where massive and high to very high density)	
	200		Very Thick	Very Wide					Note (A) (3) etc is a shorthand representation of discontinuity aperture and spacing suggested by Lord et al (CIRIA RP456) to replace the Mundford Scale
	Extremely Strong								

Table 6 Chalk Material Weathering Definitions

WEATHERING STATE	FEATURES
Completely Weathered	The chalk material is discoloured and completely altered to an engineering soil. The material fabric is largely preserved.
Highly Weathered	The chalk material is noticeably weaker than the fresh material. Material is discoloured with alteration penetrating deeply from margins.
Moderately Weathered	The chalk material is noticeably weaker than the fresh material (as determined in the field). Material is discoloured with alteration starting to penetrate in from the margins.
Slightly Weathered	The chalk material shows no loss of strength. Some discolouration or staining.
Fresh	The chalk material shows no loss of strength. No discolouration.

Note: definitions are broadly based on ANON (1972) (ref 14) although similar to Hawkins (1986) (ref 15)

Table 7 Example chalk descriptions using the
suggested descriptive scheme

GRADE	DESCRIPTION
D_f (Mundford VI)	White, highly weathered, CHALK fragments, very weak, very low density. Structureless melange composed of 20% rounded gravel to occasionally cobble size fragments set in 80% soft, cream, sandy silt size chalk matrix, with occasional rinded flint cobbles (Grade D_f) *
	White, fresh, CHALK fragments, very weak, low density. Poorly structured melange composed of 40% (?) subrounded to subangular gravel to cobble size fragments set in 60% (?) soft, pale grey, clayey silt size chalk matrix, with seams of black nodular and sheet flint. (Grade D_f) +
D_c (Mundford V)	White, moderately to highly weathered, CHALK, very weak, medium density. Structureless melange composed of 70% subangular to rounded medium to coarse gravel size fragments set in 30% soft, light brown, sandy silt size chalk matrix. (Grade D_c) *
C4 (Mundford IV)	White, slightly weathered, CHALK, very weak, low to medium density. Bedding fractures very closely spaced, open up to 20mm, infilled with powdery chalk; joints widely spaced; open >3mm, infilled with sheet flint, inclined at 60 - 70° (Grade C4) +
B3 (Mundford III)	White, slightly weathered, CHALK, weak, medium density. Bedding fractures closely spaced, subhorizontal, tight; joints closely spaced, subvertical, open <3mm, infilled with silt size comminuted chalk; with medium spaced subhorizontal bands of closely spaced rinded flint cobbles (Grade B3) *
A2 (Mundford II)	White, fresh, CHALK, moderately weak, high density. Bedding fractures medium spaced, tight; joints medium to widely spaced, tight, subvertical; with subhorizontal stylolites at 8.40m and 9.60m (Grade A2) *
A1 (Mundford I)	White, fresh, CHALK, moderately strong to strong, high density. Joints widely spaced, tight, subvertical (Grade A1) *

Descriptions based on examples given in:
*   Spink & Norbury (1989)
+   Mortimore et al (1989)
Grades based on proposals by Lord et al (CIRIA RP458, 1993)

ACKNOWLEDGEMENTS

This Paper is based on the research project carried out by
Travers Morgan Limited for the Transport Research Laboratory.
The comments and support of Mr Tony Toombs and other staff at
TRL were greatly appreciated. Thanks are due to Travers
Morgan staff and in particularly Tom Skailes for their
assistance.

The Author is grateful for the willing cooperation of the
individuals and organisation contacted in France and the UK
who responded with valuable discussion and comment.

The Paper is published by permission of the Highways
Engineering Division of the Department of Transport. The
views expressed are those of the Author and should not be
attributed to the Department of Transport.

REFERENCES
1   **DEPARTMENT OF TRANSPORT (1991)** Earthworks – design and
preparation of contract documents. HA44/91, HMSO London.
2   **GREENWOOD, JR (1989)** Specification of Chalk for Highway
Works. pp 421–423. Proc Int Conf on chalk, Brighton 1989,
Published by Thomas Telford 1990
3   **DEPARTMENT OF TRANSPORT (1986)** Specification of Highways
Works, Part 2, Series 600; Earthworks and Notes for Guidance
6th Edn, HMSO, London.
4   **DEPARTMENT OF TRANSPORT (1991)** Manual of Contract
Documents for Highway Works. Volume 1 : Specification for
Highway Works (Dec 1991): HMSO (MCHW 1).
5   **INGOLDBY, HC & PARSONS, AW (1977)** The classification of
chalk for use as a fill material, TRRL, LR806.
6   **GUY, GD (1989)** Classification and assessment of chalk on
the M25 around the Gade Valley, Kings Langley, Hertfordshire,
pp 441–448. Proc Int Conf on chalk, Brighton 1989, Published
by Thomas Telford 1990
7   **MORTIMORE, RN & FIELDING, PM (1989)** The relationship
between texture, density and strength of chalk. pp 109–132.
Proc Int Conf on chalk, Brighton 1989, Published by Thomas
Telford 1990
8   **RAT, M & SCHAEFFNER, M (1983)** Classification des craies et
conditions re reutilisation en remblai. Bull. Liaison Lab. P
et Ch. 123, pp 65–74.
9   **SCHAEFFNER, M (1991)** Les materiaux evolutifs dans la
nouvelle classification RTR. Les Remblais en Materiaux
Evolutifs, Paris 1991
10 **MORTIMORE, RN, ROBERTS, LD & JONES, DL (1989)** Logging of
chalk for engineering purposes. pp 132–152.
11 **GREENWOOD J R (1993)** Classification and Specification of
chalk for Earthworks, TRL Report in preparation.
12 **BS5930 :** 1981 Code of Practice for site investigations.
13 **SPINK, TW & NORBURY, DR (1989)** The engineering geological
description of chalk. pp 153–159.
14 **ANON (1972)** The preparation of Maps and Plans in Terms of
Engineering Geology. Q.J. Eng. Geol. Lond. 5, pp 293–382.
15 **HAWKINS AB (1986)** Rock Descriptions. Site Investigation
practice : Assessing BS5930, Eng. Geol. Spec Pub 2, Geological
Society, London

21. Specifications for and performance of compacted opencast backfills

S. J. HODGETTS, J. M. W. HOLDEN and C. S. MORGAN, Scott
Wilson Kirkpatrick, UK, and J. N. ADAMS, British Coal Opencast, UK

SYNOPSIS A major Review of information relating to the
controlled restoration of British Coal Opencast coal sites
has been completed recently. A description is given of
specifications which were used to control the placement and
compaction of backfill at the studied sites. Results are
presented and discussed in respect of the standard of
compaction achieved, the characteristics and magnitude of
the backfill settlements and the rate and degree of
groundwater recovery. Recommendations are made for future
restoration specifications which vary according to the
timing and type of development.

INTRODUCTION

1 Opencast coal sites are being restored increasingly by
means of schemes of controlled compaction in order to allow
development of the sites for a variety of after-uses.
Restoration of the void by placing and compacting suitable
backfill materials in thin layers ensures that subsequent
settlements will be of a relatively small and predictable
magnitude. Development of the site with roads or housing or
for commercial, retail and industrial use, can proceed
safely within a relatively short period of time after
completion of backfilling. Typical periods have been
between one and four years. Where development may be
delayed, relaxations in the standards of deposition and
compaction in the lower parts of the backfill may be
adopted, thereby effecting economies in the reinstatement
process.

2 Monitoring the standard of compaction achieved during
restoration and the magnitude of early backfill settlements
and groundwater recovery allows the likely long-term
performance of a site to be assessed. It also provides a
better understanding of the behaviour of backfills and can
be used to generate more confident predictions of the
distribution, duration and magnitude of settlements in fills
restored to differing compaction standards.

3 Until recently the considerable amount of information
obtained from restored opencast coal sites had not been
compiled and analysed as a whole, but an initiative by

262

British Coal Opencast has resulted in a major review being undertaken by Scott Wilson Kirkpatrick (ref.1). The review provides analyses of monitoring data, includes a study of past specifications for compaction operations, examines compaction trial results and literature relevant to the performance of both compacted and uncompacted sites, and provides recommendations for future restoration specifications with standards varying according to the timing and type of development envisaged. This paper briefly describes the reviewed sites and typical backfilling operations and summarises some of the main findings of the study.

TYPICAL OPENCAST BACKFILLS IN THE UNITED KINGDOM
4   In the United Kingdom coal is worked from strata belonging to the Carboniferous Coal Measures. These strata comprise cyclical deposits of mudstones, siltstones and sandstones with subordinate amounts of coal and ironstone. In opencast workings mudstones are generally predominant and vary from fresh and slightly weathered rock to highly weathered material comprising mostly clay-sized particles. Glacial drift deposits of boulder clay and sands and gravels and, in certain areas, deposits of Triassic age may overlie the Coal Measures strata. Opencast workings are associated increasingly with land reclamation and improvement and the incorporation of made ground, old foundations and reworked backfill in the reinstatement is often required.
5 Traditionally, backfill placement was simply carried out by the machinery used for the excavation of the overburden; draglines casting the material directly into the dump area, truck-transported spoil which was end-tipped over a loosewall or scraper-transported spoil being placed in layers at various levels. The result of these methods are large volumes of backfill having greatly varying degrees of compaction dependent upon the machinery and technique used. Where controlled and compacted restoration of the site is required, the backfill material must be placed in relatively thin, uniform layers. The drift deposits and weathered overburden are often worked by scrapers, whilst the more competent strata are excavated and deposited by means of draglines or large face shovels and dump trucks. Hard sandstones are loosened and fractured by blasting and the resulting material often comprises blocks in excess of one cubic metre in size which, nevertheless, may need to be deposited in the zone of controlled restoration. At the studied sites the backfill placed by controlled restoration was typically up to 50 m to 60 m in depth and occasionally 100 m or greater.

REVIEWED SITES
6   Table 1 indicates the British Coal Opencast (BCO) sites which have been backfilled for a specific after-use and for which generally both compaction and settlement monitoring

data were available for review. The majority of these sites have been restored throughout the full depth by means of controlled compaction in accordance with a specification. Figure 1 shows the location of these sites together with those of a further twelve sites which either are being restored currently or have been completed recently, and for which less information is available. There are an additional eighteen sites at which development has taken place but for which little or no monitoring data are available.

Table 1   Details of Studied Sites

British Coal Region and Site (Period of Backfilling)	Max depth (m)	After -use	Specification for Compaction Method	Performance
Central West				
Bilston (87-89)	40	Housing industrial	'Blue book' method - layer thickness and no of passes increased by 50% below 20 m	-
Flagstaff (85-86)	25	A42 trunk road, industrial	'Blue book' method - materials with mc >omc - 2%	materials with mc in range omc - 2% to omc - 4% dry density>1.85 Mg/m³
High Lane (90-91)	120	Minor road	'Blue book' method upper 30 m/1 m layers below	-
Ketley Brook (89-90)	50	Housing	'Brown book' method	-
Lounge (87-90)	40	A42 trunk road	'Blue book' method - materials with mc > omc	materials with mc omc to omc - 4%: dry density >2.0 Mg/m³
Newdale (88-89)	28	Campus development	-	'Brown book' performance: generally dry density > 95% mdd or air voids < 5%
Patent Shaft (88-90)	40	Retail	'Blue book' method	-
Central North				
Barnabas (86-88)	25	Industrial	'Blue book' method revised to achieve dry density > 90% mdd	-
Dixon (87-92)	50	Major road industrial	'Blue book' method revised following trials	-
Pithouse West (90-91)	57	Road, industrial	'Blue book' method - layer thickness and no. of passes increased by 50% below 30 m	-
Tinsley Park (91-93)	95	Airport, industrial	'Brown book' method upper 5 m/thick layer method below	-
Scottish				
Blindwells (78-82)	60	A1 trunk road	'Blue book' method upper 15 m/layer tipping below	-
Cadgerhall (-81)	19	A76 trunk road	'Blue book' method upper 15 m	-

Note: mc = moisture content (%)
        mdd = maximum dry density (Mg/m³))determined in BS 1377 (1975)
        omc = optimum moisture content (%))

        Test 12 (2.5 kg rammer) Barnabas, Flagstaff, Lounge Sites
        Test 13 (4.5 kg rammer) Newdale Site

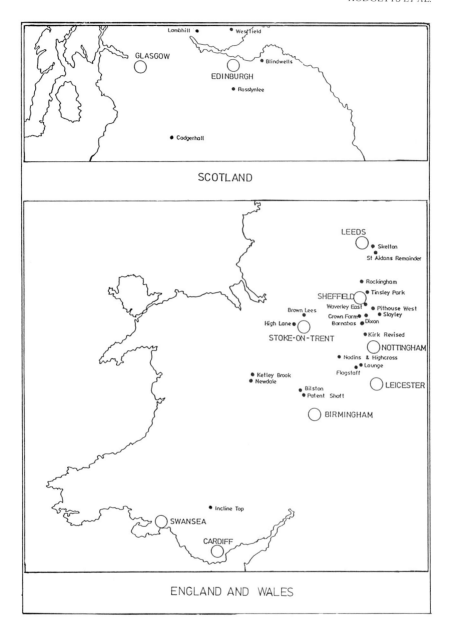

Fig 1.   Location of British Coal Opencast Sites restored by
controlled compaction

SPECIFICATIONS FOR COMPACTION OPERATIONS
7   The construction of engineered fills is generally
controlled by means of a specification to ensure that the
requisite standard is achieved. The specification dictates,
amongst other things, the type of material which is
acceptable for incorporation in the works and the degree of

265

compaction which the material is to receive. In recent years, highway earthworks compaction has been carried out either in accordance with the Department of Transport (DTp) Specification for Road and Bridgeworks 1976 (ref.2), known as the 'blue book', or the DTp Specification for Highway Works 1986 (ref.3), known as the 'brown book'. In compacted opencast backfills, similar post-reclamation fill properties and performance are required to those of soil and rock embankments and hence specifications for controlled backfilling have been based generally on these readily available specifications. However, the DTp specifications have been developed for materials excavated from relatively shallow cuttings with scrapers, face shovels and other plant which are smaller and less powerful than those used in opencast operations. These factors affect the nature of the fill available for re-use and render the DTp specifications not entirely appropriate for opencast operations.

8   Table 1 shows that the studied sites have been restored in accordance with either a method specification or a performance specification. The former defines the compaction method in terms of plant type, number of passes and layer thickness. Compaction to a performance specification allows the Contractor to choose the compaction method provided it achieves the specified 'end product' which can be a minimum value of dry density, a relative compaction value or sometimes a maximum value of air voids. Performance specifications are generally used in order to obtain a higher and more consistent standard of compaction than can be achieved by means of the method specification.

9   At the majority of the studied sites in the BCO Central West and Central North Regions the full depth of the backfill has been restored by placing and compacting the fill in relatively thin layers, generally between 200 mm and 400 mm thick. These sites have been planned for development within a period of 3 years after backfilling. At a number of the studied sites the backfill has been placed in thicker layers often without specific compaction being applied. Within the proposed road corridor at the High Lane site, for example, the majority of the backfill was placed in approximately 1 m thick layers without compaction except for trafficking by dump trucks. The upper 30 m of fill were placed by 'thin layer' method compaction and a minor road was built shortly after completion of backfilling. At the Tinsley Park site the majority of the fill, which is up to 100 m deep, has been placed in approximately 800 mm thick layers to which compaction was imparted by means of heavy vibratory rollers; development of an airport runway, a business park and access roads is proposed. The Blindwells site in the Scottish Region was restored largely by dragline placement, and/or shovel and truck, in layers of largely uncontrolled thickness, although the upper 15 m were placed and compacted in layers of less than 1 m. The diverted A1

266

trunk road was constructed across the site some three years after backfilling. The performance of the road in the seven years since it was opened has been generally satisfactory and little maintenance has been required so far. The Cadgerhall site was similarly restored in the upper 15 m; however, this site was a maximum of 19 m deep and consequently there was little uncompacted fill. A major road is likely to be built some thirteen years after backfilling.

10   At a number of the sites where the full depth of backfill was restored by controlled compaction a performance specification was adopted and Table 1 indicates the minimum standards of compaction which were applied. At Lounge and Flagstaff sites it was intended that materials with moisture contents higher than optimum moisture content minus 2% and optimum moisture content, respectively, would be compacted by method compaction. However, in practice the great majority of the materials had moisture contents below these limits and was placed in accordance with the performance criteria.

11 Specifications based on both the 'Blue Book' (ref.2) and the 'Brown Book' (ref 3) define the properties of materials suitable for use in the compaction zones in a similar way. Materials are classified as either granular or cohesive according to grading; the maximum grain size must be less than two thirds of the compacted layer thickness 'Brown Book' (ref.3) acceptability limits are set in terms of material moisture content for all fills other than rockfill and the limits are generally related to the optimum moisture content (omc) as measured in laboratory compaction tests and are set in order to ensure that an adequate state of compaction can be achieved. Typical allowable moisture content ranges for mudstones and siltstones are omc -2% to omc + 2% where omc is measured using the 4.5 kg rammer compaction test. In the case of wet cohesive material (which has a lower moisture content limit of Plastic Limit -4%) the upper moisture content limit is set in order to ensure that the requirements of trafficability and slope stability are met. Compliance with these requirements should ensure that subsequent settlements are tolerable. In highway earthworks moisture content limits are commonly set in terms of Moisture Condition Value. However the MCV has not been used on any of the studied opencast sites mainly because the test is only applied to wet superficial materials which were generally present in minor proportions.

12 Where method compaction is carried out in accordance with specifications based on either the 'Brown Book' (ref.3) or 'Blue Book' (ref.2) the majority of material excavated at opencast sites, other than rockfill, is placed and compacted by the same method in layers up to 275 mm thickness. However, at a number of the sites studied a 50% increase in this thickness was permitted provided that the number of passes of compaction plant was increased proportionally.

Materials compacted to a performance specification are typically deposited in layers not exceeding 350 mm uncompacted thickness. This thickness is the maximum which can be tested by the standard nuclear density gauge and in both types of specification the in situ testing of compacted fills is conducted typically by means of this type of instrument with periodic check testing by means of sand replacement density tests. Rockfill material is placed and compacted in thicker layers, 800 mm in the case of specifications based on the 'Brown Book' (ref.3). Specifications have typically required that rockfill has a minimum 10% fines value of 50 kN when tested following a 5 day period of soaking in water. This ensures that argillaceous materials, which may be durable in the short term but which soften once inundated, are deposited and compacted in thin layers. If they were classified as rockfill and deposited in thicker layers the post compaction air voids percentage would be high and large settlements could occur. Coarse material such as sandstone may also be incorporated as isolated boulders and individual large compacted.

13   Where a performance specification is adopted, compaction trials are conducted in order to confirm that the method of compaction proposed by the Contractor is appropriate. In the case of method compaction, compaction trials are also carried out generally to confirm that the required state of compaction can be achieved. Additional passes can be instructed in order to increase the compaction if necessary.

14   Sometimes, the overall programme and restoration requirements may demand the incorporation of materials which would normally be considered to have unacceptable moisture content, ie too wet or too dry. The former may be, for example, soft glacial clays or washery discard and are usually placed in thin layers between layers of drier and coarser materials. Materials which are too dry will require additional compactive effort and/or the addition of water in order to ensure than an adequate state of compaction in terms of percentage air voids is achieved; however, the scale of opencast operations is such that attempts to increase significantly the fill moisture content are mostly unsuccessful.

15   Compaction specifications also describe the geometry of the compaction zone at its interface with uncompacted fill and stipulate the slope of excavated faces at the boundary of the compacted fill. Adoption of appropriate slope gradients is important where roads or other structural developments are to be constructed over these zones and consideration of differential settlement are paramount.

16 The Review (ref.1) presents recommendations for future controlled backfilling projects and provides three model specifications for restoration to different standards. These specifications are:-

1) a performance specification for compaction in thin
layers;
2) a method specification for compaction in thin layers;
3) a 'thick layer' method specification for placement of the
lower portion of fill in 1 m layers and the upper 20 m by
controlled compaction in thinner layers.
The review of monitoring data has provided a better
indication of the type of specification required for
different forms of development, with particular regard to
the timing of development after backfilling, and guidance is
provided on the selection of an appropriate specification
for differing circumstances. For example, when development
with building structures or a major road is required within
2 years after backfilling, the performance specification (No
1) will be necessary. When development is required within 2
to 5 years after backfilling, the method specification (No
2) is likely to be appropriate, although for minor roads the
thick layer method specification (No.3) also may be
suitable. For development in the longer term, either the
method specification or the thick layer specification will
be appropriate depending upon the site conditions.
Correlations between standard of restoration and backfill
settlement are described further below.

MECHANISMS OF BACKFILL MOVEMENT
17 Figure 2 illustrates the characteristics of backfill
movements. Earthfills, rockfills and opencast backfills may
be subject most commonly to movements as a result of creep
settlement, settlement on inundation or heave. These
mechanisms often act in combination depending upon the site
circumstances. Further movement can be generated as a result
of the imposition (and possibly subsequent removal) of loads
such as overburden mounds, highway embankments or

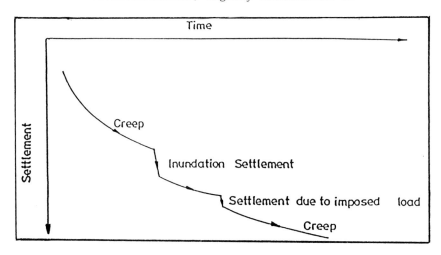

Fig 2. Idealised Settlement Behaviour of Opencast Backfill

structures, although, for deep opencast backfills, settlements resulting from building loads are insignificant compared with the 'natural' ground movements. The movements are not restricted to the vertical plane and lateral displacements can occur, generally in locations where backfill overlies buried excavation walls.

18 Creep settlement is the steady long-term movement which occurs under conditions of constant stress and moisture content. The mechanism for creep settlement is one of gradual rearrangement of the material fragments due to failure of the contact points between fragments. Relatively rapid settlements take place on inundation of a body of fill by ground or surface water. This can occur from depth as groundwater rises into the fill on cessation of any local pumping operations. Alternatively, the fill may become saturated as surface water percolates downwards. Small heave movements typically occur within the upper few metres of well compacted backfill as a result of ingress of surface water. The rate at which these latter movements take place depends on the rate of surface water saturation, but they are mostly complete within a few months after backfilling. In addition, elastic expansion of the backfill material occurs on removal of overburden mounds from the surface of the backfill. The rate at which the resulting heave movements take place generally decreases with time.

19 Examples of movements which have taken place are shown in Figure 3. The Figure displays early heave movements followed by settlements on inundation and creep; the movements are larger than those typically measured at well compacted sites. The magnitude of strains on inundation and creep settlements recorded at the studied sites are described in detail below.

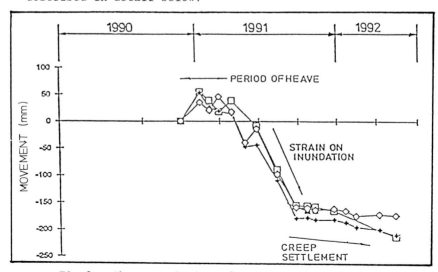

Fig 3.   Characteristics of Movements at Opencast Site

SETTLEMENT ON INUNDATION AT THE STUDIED SITES

20 Settlements of backfill due to inundation by ground or
surface water often form a significant proportion of the
overall settlement which occurs after completion of
backfilling, particularly in backfills which have not been
well compacted. Consequently, prediction of the
likely time scale during which such movements will occur
is an important consideration for safe building and
infrastructure development. The rate and timing of
settlements resulting from groundwater inundation depend
largely on the characteristics of the groundwater recovery
and, in most cases, settlements occur rapidly following
inundation. Table 2 shows that, following completion of
backfilling, at the majority of the studied sites the
groundwater rose quickly to a level similar to that prior to
working. The recovery often took place at a rate of up to
about 2 m per month and was generally complete within about
12 months after completion of backfilling and cessation of
any pumping operations associated with opencast mining. At
those sites a significant proportion, often greater than
50%, of the full backfill depth was inundated. However, at
the Barnabas, Flagstaff and Lounge sites only small rises in
groundwater level were observed because of the effects of
pumping from local deep mines. The small rises which
occurred were fed apparently by perched water entering via
the pit boundaries. Groundwater recovery is not yet complete
at the Blindwells site because of the influence of adjacent
opencast workings. At the sites where groundwater recovery
has occurred it appears that the rate of rise was not
controlled by the composition and hence permeability of the
backfill. The permeability appears to have been
sufficiently high such that external hydrogeological factors
were more critical.

Table 2. Groundwater Recovery at Studied Sites

Site	Compaction area (ha)	Rate of groundwater rise (m/month)	Time complete after backfilling (months)	Final height of groundwater level above base of fill (m)	Proportion of fill saturated (%)
Barnabas	25	0.5-1.0	12	0-6	0-30
Bilston	21	>1.0	<12	23-29	40-50
Dixon	12	≤2.0	12	10	40-70
Flagstaff	25	≤1.0	>18	0-9	0-40
Ketley Brook	12	≤2.5	8	3-22	20-70
Lounge	15	-	3-12	<5	0-20
Newdale	16	≤2.5	10-21	6-11	30-60
Patent Shaft	26	≤7.5	8-15	12-30	60-90
Pithouse West	7	≤4.0	1	38-40	80-85

21 Settlements which occurred due to inundation by ground and surface water were monitored at a number of sites by means of both surface markers and extensometers. The upper half of Figure 4 shows settlements (expressed as strains) which have occurred at various depths within the backfill at the studied sites. The strains are due to inundation by both surface water (open symbols) and groundwater (filled symbols) and were extracted from the extensometer data. The

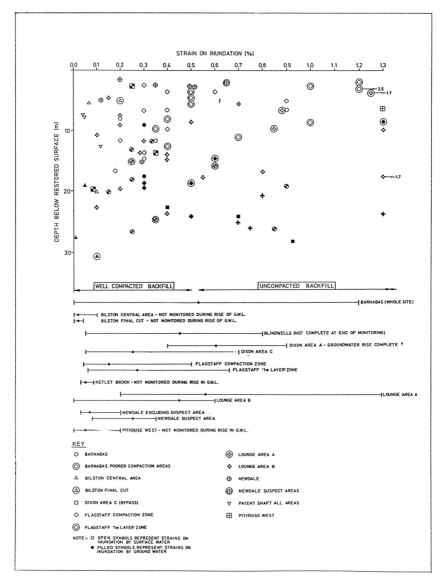

Fig 4. Strain on Inundation (%) versus Depth Below Restored Level and Overall Strain on Inundation through Depth

lower half of the illustration shows the average and range of strains on inundation recorded at surface level by both surface markers and extensometers. The magnitude of total strains measured at restored surface level depends greatly on the stage at which observation commenced, ie whether or not monitoring was carried out during groundwater, restoration and the degree to which inundation occurred. At the sites at which only a small proportion of the fill was saturated by rising groundwater there was greater potential for settlements due to surface water inundation over a prolonged period of time. At the Barnabas and Lounge sites settlements due to saturation continued for up to four years after backfilling. At the Blindwells site settlements due to ingress of surface water into the uncompacted backfill have continued for over eight years. At the well compacted sites where groundwater rose quickly to saturate a large proportion of the fill, settlements due to ingress of surface water in the upper unsaturated layers have been small and were generally complete within less than 2 years after backfilling. Figure 4 shows that strains due to saturation in well compacted backfill were generally less than 0.4% whilst strains in less well compacted backfill or uncompacted backfill were often greater than 1.0%.

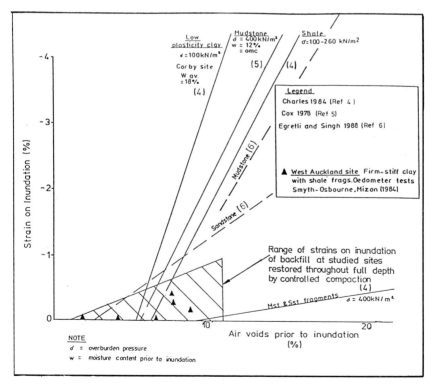

Fig 5. Strains on Inundation In Situ and Laboratory Tests

273

22   In an attempt to predict the magnitude of settlement
occurring on inundation, a number of researchers have
carried out laboratory investigations (ref.4, 5, 6).
The results of these are summarised in Figure 5.   The
experiments generally involved inundation of samples
confined within large-diameter compression cells.   For
comparison, Figures indicates the range of values for
strains on inundation measured within well compacted
backfill at the studied sites and also shows published
results from the West Auckland site (ref.9). For such
fill with an air voids content of less than 10% the strains
on inundation will be small, often less than 0.4%.   The only
laboratory tests which relate directly to one of the studied
British Coal sites are those by Egretli and Singh (ref.6).
However, Figure 5 appears to suggest that the laboratory
experiments may over-estimate the magnitude of strains which
are likely to take place. Nevertheless, they do indicate the
significant increase in magnitude of strains which occurs in

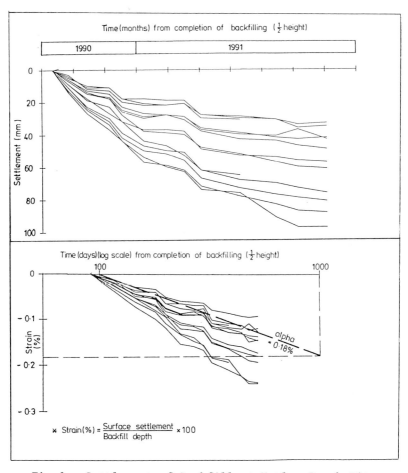

Fig 6.   Settlement of Backfill at Ketley Brook Site.

uncompacted backfills with percentage air voids greater than about 12%.

CREEP SETTLEMENT AT THE STUDIED SITES

23 Sufficient settlement monitoring data were available from the majority of the reviewed sites to allow separation of the creep and inundation settlement components. Most long-term creep movements took place at a rate which appears constant with log time. Figure 6 shows movements recorded at the Ketley Brook site. The movements largely comprise creep settlements although some inundation settlement occurred in the early stages. The creep compression rate parameter (alpha) is defined as the amount of creep strain occurring within a log cycle of time, usually measured from the time at which half the full height has been placed. Accurate estimation of the creep compression parameter ideally requires monitoring to have taken place over a period of not less than about two to three years, depending on the duration of inundation settlements. The Review (ref. 1) indicates that for a homogeneous body of fill compacted to a give standard the magnitude of creep strains does not vary significantly with depth through the fill, and the surface settlement at any point is proportional to the underlying fill depth. Small seasonal movements of a few millimetres generally occur within the upper 5 m of backfill; however, for areas of deep fill these movements form an insignificant part of the overall movement.

24 Figure 7 summarises the rates at which creep settlements have taken place at the studied sites. It shows the average and range of creep compression parameter for sites restored to different dry densities. The best correlation for use in prediction of creep settlement is that between alpha and percentage air voids because the value of percentage air voids is a unique measure of the packing state of the material, whereas a value of dry density is not. However, values of percentage air voids were only calculated at a small number of the studied sites and Figure 7 shows that overall there is a good correlation between alpha and dry density. The majority of the data relates to the fully compacted sites but some data are available for uncompacted or partly compacted fill. At the compacted sites restored by method compaction average dry densities were typically in the range 1.85 to 2.0 Mg/m^3, representing a relative density of 85%-90% compared with that obtained in the 4.5 kg rammer compaction test. At the Lounge and Newdale sites which were restored by means of a performance specification average dry densities were greater than 2.0 Mg/m^3, equivalent to a relative density of 95% with 4.5 kg rammer. The creep settlement performance of the compacted sites is very consistent. The average value of alpha typically lies in the range 0.1% to 0.2%. These figures, within backfill of 30 m depth, equate to an average creep settlement of between 30 to 60 mm in a log cycle of

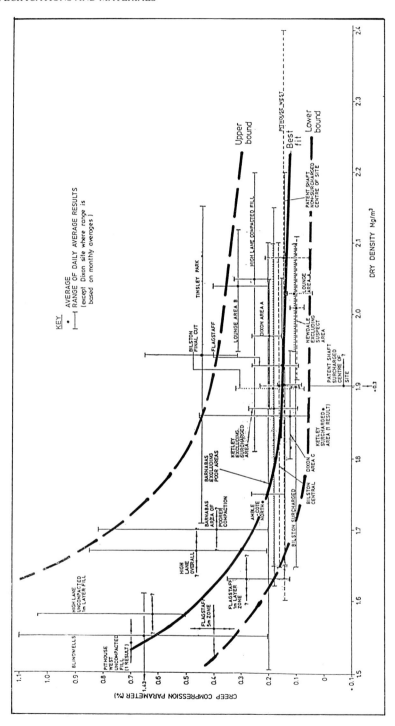

Fig 7 Creep Compression Parameter versus Dry Density for Studied Sites

time, eg between 1 year and 10 years after backfilling. Settlements of this magnitude can be accommodated by most forms of development provided differential settlements are commensurately small.

25 Creep settlements in the main excavation Area B of the Lounge Site have occurred at a slightly higher rate than expected and a possible explanation for this is given by Thompson et al(ref.8). The compaction zone in this area lies immediately adjacent to the side wall of the excavation over the majority of its length and only very small creep settlements were measured in the early stages of monitoring. Shear stresses which would have developed between the backfill and the excavation wall may have initially slowed down the rate at which creep settlements occurred. Consequently, the creep settlements have effectively started at a point in time significantly later than the time of placement of half of the backfill at any location in the site and as a result the alpha value measured from the monitoring data appears larger than expected.

26 At Newdale site an area of 'suspect' backfill was placed which comprised material in layers which were thicker than the maximum specified and which included some large boulders. There is little evidence in the in situ test results of a lower standard of compaction. However, creep settlements in this area have been slightly larger than for other areas of the site.

27 At Bilston site the backfilling of the final cut was subject to various modifications including the accommodation of large lumps of material at less than 30 m depth and the incorporation of a larger proportion of fine washery discard than elsewhere. However, Figure 7 shows that these factors have had little influence on the settlement performance of this area when compared with the central area of the site.

28 At Barnabas site, the target minimum dry density was 90% of maximum dry density in a standard (2.5 kg rammer) compaction test, which is somewhat lower than the minimum densities specified on other sites. Markers and extensometers in a number of areas of the site showed significantly larger creep movements which suggest that the fill is less well compacted. At the Bilston, Ketley Brook and Patent Shaft sites a portion of the completed backfill was surcharged by overburden mounds. Figure 7 shows that this had the effect of reducing the rate at which creep settlements occurred. At Patent Shaft surcharging of relatively shallow depths of backfill has resulted in heave movements.

29 There is less information relating to creep movements in uncompacted or partly compacted backfills. Figure 7 summarises information from the Blindwells, Flagstaff, High Lane, Pithouse West and Tinsley Park sites. The majority of the movement which has occurred at the Blindwells site has taken place in the lower layers of uncompacted backfill. Data from 36 surface markers and 4 extensometers show that

277

creep settlements have taken place at an average rate of
0.65% per log cycle of time; however, there is a much
greater variation in settlement rate when compared with the
fully compacted sites. At Flagstaff site, an area where it
was intended that the fill should be deposited in 1 m layers
without compaction was monitored by means of 3 extensometers
and 6 surface markers. In practice, this zone included some
compacted fill placed in layers of less than 1 m thickness.
Consequently, the area has performed better than might have
been expected. An area of the same site where it was
intended that the fill should be tipped in 5 m layers was
also monitored for a short period. It is likely that this
fill was generally placed in thinner layers. Figure 7 also
shows the overall performance of the High Lane site and the
performance of the uncompacted fill only. At the Pithouse
West site one marker was positioned over uncompacted layer
tipped fill. At the Tinsley Park site a relatively high
average dry density has been achieved by placing and
compacting dry mudstones and some siltstones in 800 mm
layers. However, much of the backfill is larger in size than
generally permitted by compaction specification and the
percentage air voids are typically in the range 10% to 20%.
Consequently, settlements are taking place at a higher rate
than at sites where fill has been compacted in thinner
layers (ref.10).

30 Figure 7 indicates that for fill placed for the greater
part in 1 m thick layers without specific compaction the
average creep compression parameter is about 0.45% or more.
For fill placed largely in thick layers by end tipping or
dragline the average alpha value is at least 0.65%. These
figures equate to average creep settlements, within a 30 m
depth of backfill, of at least 135 mm and 195 mm
respectively in one log cycle of time, eg a period 1 to 10
years after backfilling. Figure 7 also shows that the range
of alpha values measured at an uncompacted site is greater
than that for a compacted site. For well compacted sites the
typical variation from the average alpha value for the site
is +/- 40% of the average. For uncompacted sites the
variation increases to between 70% and 100%. Knowledge of
likely localised variations in settlement is important when
assessing whether or not a site is suitable for structural
development.

CONCLUDING STATEMENT

31 The Review (ref.1) has consolidated available
information relating to the controlled restoration of
British Coal Opencast sites. Some of the main findings
relating to the specifications adopted in controlled
compaction operations and subsequent site performance have
been described.

32 Backfill settlements at the studied sites commonly
comprise both rapid settlements due to inundation by ground
and surface water and steady long-term creep movements. The

278

magnitude of movements following backfilling depends largely on the state of compaction, the backfill material type and the degree of inundation. Placement and compaction of suitable backfill material in thin layers generally ensure that percentage air voids are less than about 10% and that subsequent settlements are small. Uncompacted and less well compacted backfills may undergo total settlements of the order of 1.0% to 2.0% of the backfill thickness following backfilling and may not be suitable for structural development for many years.

33 The specification chosen for use in restoration of a site will depend upon the magnitude of post-development settlement that can be tolerated and hence the type and timing of development. The Review has indicated the likely magnitude of settlement that can be expected for differing standards of restoration.

REFERENCES

1   Scott Wilson Kirkpatrick (1993) A State-of-the-Art Review of the Compaction of Opencast Backfill. British Coal Opencast, Mansfield, Nottinghamshire

2   Department of Transport (1976) Specification for Road and Bridgeworks.

3   Department of Transport (1986) Specification for Highway Works

4   Charles J A (1984) Some geotechnical problems associated with the use of coal mining wastes as fill materials. Proc Symp on the Reclamation, Treatment and Utilisation of Coal Mining Wastes. NCB Minestone Executive, Durham. Paper No 50

5   Cox D W (1978) Volume change in compacted clay fills, pp 79-86. Proc Conf on Clay Fills. Institution of Civil Engineers, London

6   Egretli I and Singh R N (1988) A laboratory investigation into the effects of air void and water saturation on the collapse settlement of opencast mine backfill. Mining Science and Technology $\underline{7}$, pp 87-97

7   Smyth-Osbourne K R, Mizon D H (1984) Settlement of a factory on opencast backfill. Proc 3rd Int Conf on Ground Movements and Structures, session v

8   Thompson J, Holden J M W, Yilmaz (1990) Compaction
     of opencast backfill beneath highways and associated
     developments. Proc 3rd Int Symp on the Reclamation,
     Treatment and Utilisation of Coal Mining Wastes,
     British Coal Corporation, Minestone Services, Glasgow
     pp 293-312

9   Charles J (1993) Building on fill: geotechnical
     aspects. Building Research Establishment, pp 104

10   Morgan C S, Holden J M W, O'Brien P S (1993)
     Engineered fill for Sheffield and Rotherham City
     Airport. International Conference on Engineered Fills,
     Newcastle upon Tyne

22. Polystyrene as an ultra-lightweight engineered fill

R. L. SANDERS, Babtie Geotechnical Ltd, UK, and R. A. SNOWDON, Transport Research Laboratory, UK

SYNOPSIS. The paper reviews the development of polystyrene as a fill material. The manufacturing process to form polystyrene is briefly outlined and related to the properties of the various grades and forms of polystyrene. Design principles currently recommended in the United Kingdom and Europe for polystyrene fills are discussed, in particular, the means of protecting polystyrene from damage in service. Guidance is given on the detailing of polystyrene fill in design and on practical construction aspects. The paper concludes with an assessment of the potential uses for polystyrene fill in the design of engineered fills, highlighting the advantages over other lightweight fills.

INTRODUCTION

1.   Experience in several countries has indicated that expanded and extruded polystyrene may be used successfully as lightweight fill in situations where conventional construction would be impractical or uneconomic. This paper reviews current practice in the use of polystyrene as fill material, and the properties of the available materials.   The paper presents recommendations for the design, specification and construction of embankments incorporating polystyrene fill material.

HISTORY OF POLYSTYRENE FILL USE

2.   The use of polystyrene as an engineering fill commenced in 1965 in Norway as an insulating layer in road pavement construction. From these trials it was evident that polystyrene could be used in greater thicknesses as a load-bearing bulk fill in embankment construction. Expanded polystyrene was first used for embankment construction in 1972 at Flom, near Oslo, Norway, for the reconstruction of a 1.5 m high embankment which had been affected by substantial settlements over 30 years.

3. Up to 1980, around 25 embankments incorporating expanded polystyrene fill were constructed in Norway with a total 35,000 m³ of polystyrene materials being placed. Polystyrene continues to be used extensively in Norway and Sweden. Norwegian experience with the use of expanded polystyrene fill is reviewed in ref 1 and 2.

4. Expanded polystyrene fill has been used in France since 1983, when a trial embankment was constructed for a motorway widening project in Provence (ref 3) and the approach embankment to a bridge near Montpellier was reconstructed using this material (ref 4, 5 and 6). Polystyrene was first used in the United Kingdom in 1985 on the A47 Great Yarmouth Western Bypass in Norfolk, where the material was used to form the approaches to a bridge over a railway line, on a site underlain by soft clays and peat. The design and performance of this embankment is reviewed in ref 7 and 8. Subsequently, polystyrene fill has been used at Thanet Way, Kent, the A120 Dovercourt Bypass, Essex, the Havengore Bridge, Foulness, Essex, and on other schemes. A total of over 50,000 m³ of polystyrene fill has been placed in the United Kingdom.

5. Polystyrene has also been used for fill in Canada, U.S.A., the Netherlands, Germany, Yugoslavia, Japan and Ireland.

MATERIAL PROPERTIES
Manufacture of Polystyrene

6. Polystyrene products are produced in two forms:
- expanded polystyrene   -   extruded polystyrene

To produce polystyrene, styrene monomer beads containing an expanding agent are expanded by dry steam up to 40 times their original volume. The beads are then allowed to mature. To form expanded polystyrene the beads are then placed in a mould and further steam is added to induce final expansion and fusion to the required shape. The amount of expansion allowed prior to drawing off the beads from the first stage of expansion directly affects the final density of the polystyrene block. Extruded polystyrene is formed by continuous extrusion of the foamed material through a mould, a process which 'builds-in' a certain amount of compression into the material. Because of the difference in the production processes, there are some inherent differences in the other physical and mechanical properties of the two materials. Polystyrene products do not develop their full strength until approximately 2 days after final moulding.

7. Polystyrene fill is supplied as factory produced blocks or boards. It is not technically feasible to foam the material in position on site. Mobile production plants have been considered for overseas projects in remote areas where large quantities of material are required. This is unlikely to be an economic option in the United Kingdom as a mobile production plant would cost in the order of £500,000 to £1,000,000. The standard commercially available size for expanded polystyrene blocks for fill use is 2440 x 1220 x 610 mm, but cut blocks of other reduced thicknesses can be produced. Extruded polystyrene is generally produced in the form of boards, up to a maximum section size of 610 x 110 mm.

CLASSIFICATION OF POLYSTYRENE

8. In accordance with BS 3837: Part 1: 1986 (ref 9) and Part 2: 1990, (ref 10) polystyrene is classified into grades according to compressive stress/strength as

determined by BS 4370: Part 1: 1988 (ref 11) Method 3. Details are given in the table below.

Grade	Minimum Compressive Stress/Compressive Strength* kN/m^2	
Expanded Polystyrene		
Impact Sound Duty (ISD)	25	
Standard Duty (SD)	70	
High Duty (HD)	110)	
Extra High Duty (EHD)	150)	generally used as fill
Ultra High Duty (UHD)	190)	
Extruded Polystyrene		
E1	100)	
E2	150)	
E3	200)	
E4	300)	'skin' retained
E5	350)	
E6	100)	'skin' removed
E7	250)	

* Where the maximum stress in the test occurs prior to 10% strain the result is reported as compressive strength.

9. Expanded polystyrene is produced in SD, HD, EHD and occasionally UHD grades for use as fill. Extruded polystyrene grades E1 to E5 retain skins formed during manufacturing. Grades E6 and E7 have the skins removed by planing. Extruded polystyrene grades E1 and E2 are not used as the minimum strengths are well below what can be currently manufactured. It should be noted that extruded polystyrene can be manufactured to far exceed the minimum strength of grade E5.

10. Expanded polystyrene beads formed prior to final moulding cannot be used as 'loose fill' as there would be insufficient interbead friction to give adequate strength. Placement and containment of the beads and compaction of any overlying pavement would be extremely difficult.

PHYSICAL PROPERTIES
Density
11. The most important characteristic of polystyrene is its extremely low density. Typical dry densities for expanded polystyrene materials are as follows:

Expanded Polystyrene Grade BS 3837: Part 1: 1986	Dry Density (Typical Density Value) (Mg/m³)
SD	0.015
HD	0.02
EHD	0.026
UHD	0.03

This compares to a minimum of 12 Mg/m³ for PFA fill and 18 Mg/m³ for sand fill.

12. The dry density of the commonly available extruded polystyrene products generally varies between 0.028 and 0.055 Mg/m³, although densities of up to 0.10 Mg/m³ have been produced.

13. The density of polystyrene materials has a direct influence on the other physical properties, such as compressive strength. It should be noted the density of the material in service is significantly greater than the dry density as a result of the absorption of water.

Behaviour in Compression

14. The behaviour of expanded and extruded polystyrene under uniaxial compression shows marked differences as shown below:

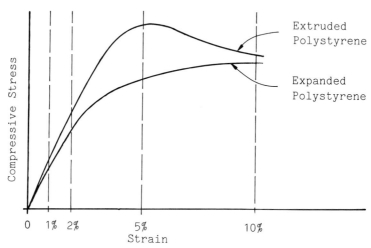

Figure 1: Typical Stress-Strain Curves for Compression

15. Expanded polystyrene behaves as a linear elastic material up to strains of around 1 to 2% where yield occurs. Thereafter the material shows typical strain hardening behaviour in strain-controlled tests, with the compressive stress continuing to increase, but at a decreasing rate, with increasing strain. After a strain of around 10%, very little further increase in compressive stress occurs.

16. Extruded polystyrene shows linear elastic behaviour over a similar range of strain to expanded polystyrene. After yield occurs, the material shows strain

softening behaviour. In strain controlled tests, a peak compressive strength is reached at a strain of around 5%. This peak stress is generally quoted as the compressive strength. Extruded polystyrene behaves in a brittle manner at failure.

17.   Tests in Norway (ref 2) indicate that the compressive strength of polystyrene is maintained even after repeated loadings of up to 80% of its compressive strength. For engineering purposes, the stress at which yield occurs is more significant that the maximum compressive strength as, if the yield point is exceeded, non-recoverable plastic strains will accumulate under repeated loading. In general, the yield stress may safely be taken to be the compressive stress at 1% strain.

18.   Creep tests at constant loads indicate that the tendency for long-term creep deformations increases with the applied stress. Where the stress is such that the initial compressive strain is around 1%, the rate of creep is small. Creep appears to be negligible if the initial strain does not exceed 0.5%. Laboratory tests have been undertaken to determine creep under cyclic loading (ref 12). The findings indicate that as long as cyclic loads are below the yield stress no creep deformation occurs. This is confirmed by the results of long term monitoring of polystyrene embankments (ref 1).

19.   Typical compressive strengths for polystyrene fill materials are as follows:

Grade (BS 3837: Part 1: 1986 Part 2: 1990)	Compressive Stress at 1% Strain (Manufacturer's Data) (kN/m²)	Compressive Stress/ Strength at 10% Strain (BS 4370: Part 1: 1988 Method 3) (kN/m²)
Expanded Polystyrene:		
SD	21	70
HD	45	110
EHD	70	150
UHD	100	190
Extruded Polystyrene:		
E1 - E7	135 - 370	250 - 670

(Values for extruded polystyrene are indicative of the range of materials currently available in the United Kingdom).

20.   Polystyrene undergoes very little lateral deflection under vertical load, with Poissons ratio being between 0 and 0.02.

Shear Strength and Frictional Resistance

21.   Polystyrene has a high shear strength in relation to its compressive strength. Typical values for various grades of expanded polystyrene (ref 13) are as follows:

Moulded Density (kg/m³)	Equivalent Grade (BS 3837 Part 1: 1986) (BS 3837 Part 2: 1990)	Shear Strength (kN/m²)
15	SD	90 - 120
20	HD	120 - 150
25	EHD	150 - 190
30	UHD	190 - 220

22. The frictional resistance between blocks is of greater engineering significance in embankment design than the shear strength of the intact material. A coefficient of friction of 0.5 is usually taken for the moulded faces of the blocks, equivalent to an angle of friction of 27°. The angle of friction of cut or broken surfaces is likely to be somewhat higher.

Water Absorption

23. Polystyrene materials have a low potential for water absorption despite their extremely low density. The individual beads that make up the materials have a closed cellular structure and therefore will not absorb water. The void space available for water uptake is, therefore, limited to the interconnected pore spaces between the beads, which form a small proportion of the total volume. The permeability of the material is low, but water may be drawn into the materials by capillary action to a limited extent.

24. Typical water absorption values for various grades of expanded polystyrene for short term exposure are as follows (ref 13):

Grade	Water Absorption by Volume	
	After 7 Days	After 1 Year
SD	3.0%	5.0%
HD	2.3%	4.0%
EHD	2.2%	3.8%
UHD	2.0%	3.5%

25. Extruded polystyrene absorbs water to a much lesser extent with water absorption by volume being generally in the range of 0.05% to 0.2%.

26. Long term monitoring in Norway (ref 1) indicates that for expanded polystyrene fills permanently below the water table, the water absorption approaches 9%. For fills that are periodically submerged in water, water contents of up to 4% have been observed. For fills permanently above the water table, the capillary rise appears to be small, around 200 mm, and above this the moisture content rarely exceeds 1% by volume.

27. Experience has shown that the absorption of water has little influence on the strength or compressibility of the fill.

Chemical Properties

28. Polystyrene is chemically resistant to a wide range of substances but attack of the surface of the polystyrene, causing shrinkage on prolonged exposure, can occur with the following:

- some vegetable oils   - diesel fuel
- most animal fats and oils   - phenol

-       paraffin oils

29. There are some substances which, in large quantities or high concentrations, can attack polystyrene foam to a significant degree, generally by dissolving the material. These include:

-       most hydrocarbons, including petrol and tar oils
-       chlorinated hydrocarbons and chlorofluorocarbons (CFCs)
-       other organic solvents such as ketones, ethers and esters
- -   concentrated acids.

30. Many paints, adhesives, cleaning fluids and construction products (eg. curing compounds, bituminous sprays or paints) contain organic solvents.

31. Norwegian experience of polystyrene fills constructed over peat indicated no problems have resulted from generation of methane. Deleterious substances may, however, be present in significant quantities in contaminated ground and may also be introduced by accidental spillage.

32. Detailed information on the chemical resistance of expanded polystyrene has been published by (ref 13).

Biological Properties

32. Polystyrene is not susceptible to bacterial or fungal attack and therefore will not undergo biological degradation. Polystyrene has no nutritional value for animals.

Combustibility

33. Polystyrene is a combustible material and will burn readily when ignited in the presence of large volumes of oxygen. To sustain a fire, an oxygen supply of 150 times the bulk volume of burning material is required. The rate at which heat is released by combustion of polystyrene is rapid and a fire will propagate rapidly once started.

34. Polystyrene can be supplied with a flame retardant additive which reduces the risk of ignition. The additive restricts the early stages of development of the fire such that, if the source of ignition is removed, the fire will be extinguished.

Ultra-Violet Light

35. Exposure to ultra-violet light causes discolouration and embrittlement of the surface of the polystyrene. This is of no significance to finished engineered fills.

Testing

36. Testing of the polystyrene should be undertaken at random by taking samples cut from a large block of polystyrene. The majority of testing for polystyrene is covered by BS 4370 (ref 11 and 14).

37. Of all the test methods listed under BS 4370, there are essentially three main properties to test prior to the installation of the polystyrene as an engineered fill.

38.   Measurement of Dimensions
        During construction, stability of a polystyrene fill depends to a certain extent upon the consistency of the block dimensions.

39. <u>Compressive Strength</u>
To ensure that the material has sufficient strength to withstand the imposed loads.

40. <u>Density</u>
The value of density for polystyrene blocks is relatively simple to determine and assists in assuring the consistency of the supplied material.

DESIGN CRITERIA
<u>Design Strength</u>
41. The design strength should be taken to be the compressive stress at 1% strain for all normal loading conditions.

42. The compressive stresses induced by the weight of normal fill and any pavement cover, vehicle loads and any other imposed loads must be determined and appropriate grades of polystyrene selected such that the compressive stress does not exceed the design strength of the polystyrene. Compressive stresses induced by abnormal loads may exceed the design strength but should not exceed 80% of the minimum compressive stress/compressive strength at 10% strain. Where frequent abnormal loads are expected, consideration should be given to the design strength being defined by the compressive stress imposed by the abnormal loading.

43. For concentrated loads, such as wheel loads, it is usually sufficient to assume that the loads are distributed uniformly over a finite area. For polystyrene, a cover of natural fills or pavement materials, it is usual to assume an 'angle of spread' of the load of 1 (horizontal) to 2 (vertical) as shown below:

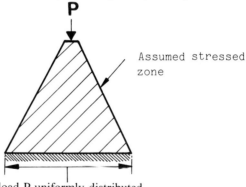

load P uniformly distributed
Figure 2: Load Distribution

44. Where a concrete slab is introduced above the polystyrene, a greater degree of 'load spreading' may be assumed through the slab. For lightly reinforced concrete, an 'angle of spread' of 1 to 1 is usually assumed.

45. The compressive stress induced by the weight of fill and pavement cover should be assumed to be constant with depth through the polystyrene.

46. For highway embankments, higher grade (EHD and above) polystyrene materials will be required for the first few layers below the pavement. For lower layers, fill below verges and side slopes or other non load bearing engineering

fills, lower grade materials are usually sufficient.

Design Density

47. The value of density used in settlement calculations must allow for the absorption of water under service conditions.   The following densities are recommended for all grades of material for settlement analysis:

Expanded Polystyrene:

Fills permanently below water	=	100 kg/m^3
Fills submerged for short periods only	=	50 kg/m^3
Fills permanently above water	=	50 kg/m^3
Density for buoyancy calculation	=	20 kg/m^3 (minimum)

Extruded Polystyrene:

Dry density should be used for all conditions.

48. The potential for flotation of the completed embankment should be checked for the maximum design flood level. For these calculations the dry density values of polystyrene should be used. The weight of the polystyrene and its normal fill and pavement covers must exceed the uplift force resulting from water pressure on the base of the polystyrene.

49. Where water levels permit, polystyrene embankments may be designed with compensated foundations such that there is no net loading of the foundation strata. This is achieved by excavating the ground beneath the embankment and back filling the excavation with polystyrene fill. The potential for long term foundation settlements and foundation failure is thus completely avoided provided equilibrium is maintained, e.g. construction on adjacent and land could result in serious problems.

Coefficient of Friction

50. It is recommended that the coefficient of friction between polystyrene blocks is taken to be 0.5, equivalent to an angle of friction of 27°.

Shear Strength

51. In stability calculations the shear strength for the polystyrene blocks should be used except for horizontal slip surfaces where the shear strength along boundaries should be used. Any structure should be constructed so as to ensure there are no deep continuous joints, thus maintaining material contact of the blocks.

Weight of Fill Cover and Pavement

52.   The minimum thickness of fill cover over polystyrene fill will be determined by the requirements for:
-       protection of the polystyrene
-       imposed loads
-       requirements for accommodating services etc.
-       landscaping.

53. In the design of embankments incorporating a mixture of polystyrene and normal density fills, it is necessary to consider the effects of settlement on the

polystyrene fill. Differential settlement of the polystyrene should be kept to a practical minimum, as large differential movements may lead to problems with the placement of the blocks, tilting of the structure and surface drainage. In extreme cases the stability of the fill may be affected. It is desirable to keep the thickness of fill above polystyrene to the practical minimum such that the majority of the denser fill materials are placed prior to placement of the polystyrene. Where surcharge loading to accelerate consolidation is required, this should be undertaken as a preloading exercise prior to placement of the polystyrene.

54. On sloping sites, the potential for sliding of the embankment along its base and the overall effect of the embankment weight on slope stability should be considered. On steeply sloping sites, it may be necessary to anchor the fill to the slope to resist sliding. This is best achieved by introducing flat horizontal members between the polystyrene layers.

55. Stability under horizontal loading may be a problem where large loads act across the width of the embankment, e.g. impact forces on embankment sides or parapet walls. The overall stability may be simply checked by consideration of sliding on an appropriate basal plane. Internal stability (i.e. internal shearing or punching of the fill) may also need to be considered. Significant variations in level as a result of, for example, poor construction or wide dimensional tolerance, could result in internal voiding and the consequent loss of shear strength between blocks. The surface of the basal layer, usually comprising sand, should therefore remain within strict tolerances to ensure overall stability after completion of the embankment.

Lateral Pressures on Structures

56. Expanded and extruded polystyrene materials have a very low Poisson's Ratio and therefore will impose low lateral forces on adjoining structures. Monitoring of polystyrene structures undertaken by the Norwegian Road Research Laboratory (ref 1) indicates that provided a suitable air gap is allowed between polystyrene fill and an adjoining wall at the time of construction, no lateral pressures will be applied to the wall by the polystyrene in the long term. Tests on polystyrene fills placed in contact with a bridge abutment (ref 2) indicate that, for restrained fills, the lateral load applied to the adjoining wall is between 0.1 and 0.2 times the vertical surcharge load to the polystyrene. In Norway it is normal practice (ref 1) to adopt a very conservative assumption and apply a uniform horizontal pressure of 10 kN/m^2 against adjoining abutments and walls.

57. Polystyrene fills, by themselves, generally have a very limited capacity to resist horizontal loads, which derives solely from friction between the blocks. Consequently, any lateral pressures applied to the polystyrene may be transmitted to adjoining structures. Therefore, any fills or slopes adjoining the polystyrene should be designed to be self-supporting.

58. Retaining walls need to be designed to resist lateral pressures exerted by the normal fill and pavement cover over the polystyrene. The structures may also need to resist horizontal forces applied to the fill surface.

Pavement Design

59. Where an engineering fill supports a highway the CBR of the polystyrene needs to be assessed. During the design of the A47 Great Yarmouth Western

Bypass, laboratory CBRs of 2 to 5% for extruded polystyrene were measured. Expanded polystyrene gives values <2% and pavement capping layer design should be based on these values unless particularly high grades of expanded material are tested to see if the CBR values are >2%. For trunk roads this results in a minimum thickness of 1 m from finished road level to the upper surface of the polystyrene. A thinner construction may be feasible on less heavily loaded routes.

60. In Norway, some problems have been encountered with differential icing between pavements constructed on polystyrene fill and normal construction, particularly in late autumn. This appears to be the result of the low thermal capacity of polystyrene compared with earth fills of similar volume, and the insulating properties of the material which restrict heat flow from the underlying ground. This problem may be overcome by providing a sufficient thickness of pavement materials. Based on Norwegian practice, (ref 15), it is suggested that a minimum pavement thickness of 800 mm should be adopted.

61. In Norwegian and French design guidelines (ref 16 and 17), a lightly reinforced concrete slab between 100 and 150 mm is provided beneath the pavement to assist load transfer to the polystyrene. This enables reduction of the total thickness of cover over the polystyrene or, for the same cover thickness, a lower grade of polystyrene may be used below the pavement. The overall construction costs are not greatly affected by inclusion of a concrete slab. Therefore in the United Kingdom, except for two schemes, the concrete slab has been omitted, and experience suggests that this does not have any detrimental effect on the performance of the pavement.

PROTECTION OF POLYSTYRENE
Potential Hazards
62. Expanded and extruded polystyrene are vulnerable to damage resulting from a number of potential causes. The various potential hazards are listed below.

Malicious damage	Ultra violet light
Petrol and chemical attack	Accidental impact
Heat and flame	

Malicious Damage
63. Polystyrene fills are potentially vulnerable to malicious damage but a soil cover or hard facing will conceal and discourage casual vandalism. Malicious damage is only likely to be a significant hazard during the construction stage. Nevertheless, in areas of high vandalism risk, the designer should consider whether there is a risk of damage to other elements of the scheme which may expose the polystyrene fill or result in consequential damage to it.

Protective Measures
64. Suitable protective measures should be incorporated in the design. The protection is required to perform the following functions:
- act as an inert insulation layer to protect against heat and flame (and ideally to restrict air supply to the polystyrene)
- exclude ultra violet light

-   absorb or spread impact forces
-   reduce vulnerability to malicious damage.
-   provide an impermeable barrier to exclude petrol, organic solvents and other deleterious substances

Heat, Flame and Ultraviolet Light Protection

65. Except for providing an impermeable barrier, protection can be fulfilled by covering the polystyrene with a layer of inert fill material, pavement materials or concrete slab. On vertical fill faces, a reinforced concrete or masonry facing will provide the same functions. Where embankments are particularly vulnerable to vehicle impact (e.g. approaches to bridges over highways) additional protection from impact in the form of safety barriers may be advisable.

Chemical Attack

66. Most fills, pavement materials and concretes are permeable to petrol and other fluids/vapours and thus there needs to be additional protection against deleterious fluids/vapours. On landfilled or contaminated sites it may also be necessary to provide an impermeable layer below the polystyrene. This may also need to incorporate a vapour barrier to prevent migration of methane or other deleterious gases into the fill.

67. The French design guidelines recommend the use of a clay fill cover to side slopes and this could also be used above polystyrene. However, the effectiveness of this method is dubious as, during dry weather, shrinkage of the clay may occur leading to formation of cracks, allowing ingress of fluids. In addition, on side slopes a clay cover is inherently less stable in the long term than one formed from granular materials.

68. It has been the practice in some countries to provide an impermeable membrane on highway schemes below the pavement and verges only. However, in a large spill it is unlikely that the discharge fluids will be confined to these areas of the highway embankment. It is therefore recommended that impermeable membranes are provided below the cover to side slopes, to vertical faces behind walls, and any other surfaces that may come into contact with fluids as they drain away from the point of discharge.

Impermeable Membranes

69. A variety of impermeable polymer membranes are available for use in engineered fills. Many are not resistant to organic solvents, i.e. petrol, diesel, and are therefore unsuitable for use as protection to polystyrene. There are, nevertheless, a limited range of materials that are impermeable and chemically resistant to hydrocarbons, organic solvents and acids, the most important of these being high density polyethylene and polyester. Polyethylene materials are covered by BS 3412: 1976 (ref 15) in which HDPE is defined as material with a minimum density of 0.940 Mg/m^3 and ASTM D883 (ref 16) where HDPE is defined as material with a minimum density of 0.941 Mg/m^3.

70. HDPE membranes are a type of thermoplastic semicrystalline geomembrane. The advantages of this type of membrane include excellent chemical resistance, the ability to form effective seams by thermal and/or

extrusion methods and a large range in thickness. The disadvantages include a sensitivity to seam cracking, the need for high quality seam workmanship and high thermal expansion/contraction. HDPE membranes demonstrate a high resistance to many chemicals including hydrocarbons and organic solvents which may attack polystyrene.

71. HDPE membranes are available with either smooth or roughened texture. Measurements have shown the friction angle between a smooth HDPE membrane and sand is 16° compared to at least 30° for roughened HDPE membranes. Increased protection against accidental damage during construction is provided from the adoption of thicker HDPE membranes (2 - 3 mm thickness) but thinner membranes are required to be of sufficient flexibility to be used on stepped surfaces (e.g. below the cover to embankment side slopes).

72. A 2.5 mm thick, preferably, roughened membrane is recommended for use below the pavement and verges, where the risk of mechanical damage and chemical attack is greatest. On vertical faces, and below side slopes, a 0.5 mm thick membrane may be used without significantly compromising the degree of protection.

73. Polyester membranes have been used to form impermeable membranes over polystyrene, e.g. A47 Great Yarmouth Western Bypass (ref 8). These materials have good chemical resistance to hydrocarbons and organic solvents, but are not resistant to concentrated alkalis. The light polyester membranes have also proved difficult to lay and join in practice. Consequently the use of polyester membranes is not recommended.

74. The specification for membranes should contain requirements for the following properties: Thickness, Density, Strength (tensile properties and elongation), Toughness (resistance to tear and puncture), Durability (chemical resistance) and Seamability (Homogeneous fusion of liner sheets). No suitable British Standard test methods exist and reference should be made to ASTM standards D374, D751, D638-M, D882, D883, D1004 and D1505 and DIN standards 53370, 53479 and 53455 (ref 16-25 inclusive).

Impermeable Coatings

75. Polyurethane, sprayed in situ, was initially used for protection in Norway and has subsequently been used in France (ref 3). However, successful spraying of the polyurethane proved difficult and is susceptible to adverse weather conditions. Polyurethane also produces toxic gases if set alight. Consequently, the use of this material is not recommended.

DETAILING OF POLYSTYRENE EMBANKMENTS
Abutments and Retained Fill

76. Polystyrene fill material may be placed directly against bridge abutments and vertical facing walls, although it is advisable to allow an air gap on vertical facing wall to avoid lateral pressures being applied to the wall. The air gap should be closed off at the top (e.g. by a flexible membrane) to prevent ingress of loose fill material. An air gap width between 20 and 30 mm has been used on two projects in Norway (Fryenlund pers.comm.). The protective impermeable membrane should be carried down the vertical face of the polystyrene to guard against infiltration of fluids between the structure and the fill. Drainage outlets

will also be required at the base of the wall.

77. Polystyrene blocks may be trimmed to fit the exact shape of the adjoining structure, although this may be difficult for complex or sharply curved surfaces and therefore the form of the adjoining structure should be made as simple as possible. Alternatively, the polystyrene may be placed to the approximate shape of the structure and any void infilled with sand. This option is less advantageous as it will increase the fill weight and may lead to lateral pressures being applied to the wall.

Transition Zones

78. Where polystyrene fill construction adjoins a normal fill embankment, transitional arrangements will be required to accommodate differential settlements between the two types of construction. For simple embankment designs, a transition may be achieved simply by progressively reducing the thickness of the polystyrene. The length of transition zone is determined by the anticipated magnitude of differential settlement and the maximum angular deflection that may be tolerated for the finished surface of the embankment. Where ground improvements or other measures (e.g. surcharging) are required, the design of the transition zone is potentially more complex. From a practical viewpoint it is advisable to adopt uniform ground treatment for both the normal fill and polystyrene fill sections so that the simple approach described above may be adopted.

79. In designing the transition zone, consideration should be given to the stability of the normal fill slope at the interface with the polystyrene. In order to reduce stresses and deflections in the polystyrene fill, it is recommended that the normal density fill element of the transition embankment is placed at the base of the embankment (apart from the minimum cover required for protection, etc.) - see Figure 3.

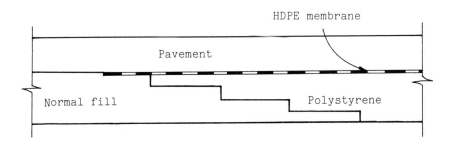

Figure 3: Typical detail of transition zone (long section)

Polystyrene Fill Layers

80. Polystyrene is supplied in the form of blocks or boards for economy of production. It is possible to cut the blocks and boards to any shape but the inevitable wastage of material adds significantly to the construction costs and therefore it is desirable to minimise the amount of site cutting. It is usual to restrict trimming of blocks to straight cuts perpendicular to the largest face of the block as this is easiest and involves the least wastage of material.

81. Polystyrene fills should be formed of successive layers of uniform thickness, with each layer having edges perpendicular to the plane of the previous layer. Where the thickness of fill varies, as is the case below side slopes, this is achieved by stepping of successive layers (Figure 4). This stepped arrangement has a practical advantage in that it assists retention of soil cover on side slopes.

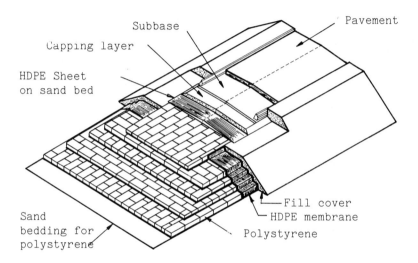

Figure 4: Typical Construction of a Polystyrene Fill Embankment

82. Polystyrene fill layers should be planar, but may be laid to a single gradient, to follow approximately the finished level of the embankment, if desired. Curved surfaces, crowns and valleys, and complex arrangements of falls of the engineered fill structure should be avoided as these will make it difficult to place successive layers so as to form a tightly packed and stable mass.

83. In plan, polystyrene fill layers may be trimmed to any desired outline, but it is recommended that simple straight edges separated by the minimum number of angles are adopted to facilitate site cutting and placement of the protective membranes.

84. For detailed design drawings it is usually sufficient to indicate the limits of each layer, referenced to a convenient setting out line and set of base levels. Detailed block layout plans are not normally necessary but may assist construction.

Drainage of Polystyrene Fills

85. Adequate drainage of the upper surface of polystyrene fills is essential as accumulation of water will prevent adequate compaction of the overlying fill and pavement materials. Good drainage will also assist protection from petrol and chemical spillage by assisting dispersal of fluids away from the polystyrene and its protective membrane. It is recommended that on large horizontal surfaces the protective membranes are laid on a sand bed prepared with falls towards the edges of the embankment. The sand bed has the further advantage of increasing friction between the protective membrane and the polystyrene fill.

86. Where polystyrene fill is placed against slopes (e.g. in landslide repairs), drainage may also be required along the interface between the polystyrene and the ground to relieve water pressures resulting from seepage from the ground.

87. Temporary dewatering may be required in some situations to enable the placement of polystyrene below normal ground water levels. Such dewatering should be maintained until sufficient weight of cover has been placed to resist flotation.

Accommodation of Services, Street Furniture, etc., for Highway Embankments

88. Careful detailing is required to ensure that services, sewerage and the foundations of lighting columns, safety fences and road signs can be accommodated within the embankment without damaging the polystyrene fill or compromising the protection to it.

89. It is usual practice to increase the depth of fill cover below highway verges to about 1.5 - 1.7 m and to place drainage and services within this fill cover. Provided that reasonable care is taken, trenches can be excavated in the fill cover without undue risk of damage to the polystyrene fill. The thicker polyethylene membranes used for protection to the polystyrene have sufficient strength and toughness to resist impact damage from an excavator bucket.

90. Concrete bases are recommended for safety fences, lighting columns and signs. Driven posts should not be used as there is a risk of puncturing the protective membranes.

91. It is recommended that sumps for drainage gullies are accommodated in the verge rather than below the carriageway edges as is usual practice.

92. Service ducts passing beneath the carriageway should, as far as practicable, be accommodated within the capping layer above the polystyrene. Alternatively trenches may be formed within the surface of the polystyrene fill and lined with the protective membrane. In such cases, the services should be installed in this trench prior to placement of the capping layer. The services may be sealed over with a protective membrane with resealing being relatively simple should further access be necessary.

CONSTRUCTION OF POLYSTYRENE FILLS

Storage and Handling

93. The handling of expanded and extruded polystyrene materials on site presents few difficulties. The commonly supplied sizes of blocks can be lifted easily by hand and large volumes may be moved rapidly by, for example, a small mobile crane. This ease of handling can be a considerable advantage where access to the site for vehicles is difficult (e.g. steeply sloping sites). Some care is,

however, necessary to avoid damage to the blocks by impact or abrasion.

94. Polystyrene materials are vulnerable to damage from a number of causes if left unprotected and it is therefore desirable that the material is placed in its final position in the works as soon as practicable to reduce the potential for damage. Where site storage of materials is necessary, care is required to prevent deterioration. The blocks must be carefully stacked in a regular and stable manner and should be covered by sheeting to exclude ultra violet light. It is also necessary to anchor stockpiles against wind. Good site security is also important as the stockpiled material will be vulnerable to the theft and malicious damage.

Placement

95. Successful construction of polystyrene fills is dependent to a large degree on the preparation of the surface on which the polystyrene blocks are to be laid. It is essential that this surface is smooth and planar, and this is best achieved by placing a bedding layer of sand and levelling this by hand tools. The surface of this bedding layer should be left loose, such that the first layer of blocks may be "bedded in" to the sand to give a firm stable base to the fill. It may be necessary to re-level the sand to achieve a level upper surface to the first layer of polystyrene. It should be stressed that additional care taken in preparation of the sand bed and laying of the first layer of polystyrene will be worthwhile as this will greatly facilitate placement of the subsequent layers of blocks.

96. In each layer, the polystyrene blocks should be laid so as to overlap the joints in the layer below and, where practicable, the orientation of the long axes of the blocks should be at right angles to that in the layer below (Figure 4). Blocks may be cut to size to suit the required fill outline, using a hot wire, straight saw blade or a chainsaw. Fixing or joining of polystyrene blocks is not normally required for long term stability. The Norwegian and French design guides recommend the use of double sided steel fasteners to fix blocks together. These are considered to be of dubious value. If fixing of blocks is required to ensure temporary stability during construction, it is recommended that this be achieved by driving lengths of mild steel reinforcing bar through the blocks into the layer beneath (the length of bar should be approximately one and one half times the layer thickness).

97. Placement of polystyrene should be programmed such that the material is covered as soon as possible after placement, and the area of polystyrene exposed should be kept as small as practicable. Where it is necessary to leave areas of polystyrene exposed for long periods, the polystyrene should be covered with sheeting to exclude ultra violet light and blocks should be pinned or weighted as necessary to prevent accidental disturbance of the fill.

Plant Restrictions

98. Restrictions on operation of plant over polystyrene are necessary to prevent damage to the material. In general no plant should be operated over polystyrene fill until at least 150 mm thickness of normal fill cover has been placed. Thereafter it is recommended that the maximum weight of compaction plant is limited to 6000 kg/m width of roll until filling is completed. Vibratory compaction plant should not be used within 500 mm vertically and 2 m laterally from polystyrene fill. It is recommended that other construction plant shall be

limited to a maximum applied pressure of 20 kN/m².

99. Where the operation of plant directly on polystyrene fill is unavoidable, temporary mats should be provided to distribute plant loads such that the design strength of the fill is not exceeded.

Installation of Protective Membranes

100. The protective membranes should be laid as soon as practicable after laying polystyrene. Individual sheets should be jointed to form a continuous impermeable membrane. For the high density polyethylene sheets this should be carried out by extrusion welding. Welding should be carried out clear of the polystyrene surface and the joint should be allowed to cool sufficiently before it is placed in contact with the polystyrene. Care should be taken to ensure that a complete seal is formed along each joint. At the joints, the sheets should be lapped, with the lap being positioned so that fluids tend to drain away from the joint.

101. Minor damage to the membranes may be repaired without the need to replace whole sheets by placing a patch over the damaged area, provided that there is no significant damage to the polystyrene fill beneath. The patch should extend at least 200 mm outside the damaged area and should be joined to the membrane by extrusion welding.

102. A minimum of 50 mm sand bedding is recommended between the polystyrene and the protective membrane under the pavement to provide an even surface for laying the membrane and a passage of escape for any trapped water vapour.

Potential Benefits and Uses of Polystyrene Fill

103. Polystyrene is an expensive material in comparison with conventional soil fills, but its extremely low density can result in considerable cost savings in situations where substantial structures, foundations or ground improvements would otherwise be required. In situations where differential settlements or long term secondary compression of organic soils are anticipated, the use of polystyrene may also result in a substantial reduction in maintenance costs. The light weight of polystyrene also enables construction to occur in highly constricted areas or where heavy vehicles or construction traffic is restricted. In the United Kingdom polystyrene has been beneficially used as an engineered fill in the following embankment construction situations:

- Over very weak and/or compressible soils to avoid stability and settlement constraints.
- Over services and structures sensitive to settlement.
- Over weak structures to avoid structural failure.
- Adjacent to potentially unstable slopes to avoid loading the slope.
- As a transition between bridges and normal fill embankments.
- To avoid negative skin friction and lateral loading on adjacent piles.
- As backfill to retaining structures/bridge abutments to enable lighter retaining walls to be used and/or to avoid pile foundations.
- In widening existing embankments.
- As infilling to viaduct arches to support the arches without causing additional foundation loads.

104. Polystyrene has also been used in the United States to backfill around underground structures to reduce earth pressures and permit lighter structural wall design (ref 24). In this situation it also provides thermal insulation to reduce the heating requirements for the building.

105. In Japan it has been successfully used to reduce small amplitude ground vibrations caused by road traffic (ref 30). In Norway and Japan polystyrene has been used to stabilise unstable slopes by replacement of unstable material and in the United States to reconstruct a failed highway embankment (ref 31).

106. Development of the use of low strength compressible polystyrene will in the future permit controlled yielding of soils and a brief resume of this is given in (ref 32).

CONCLUSION

107. Polystyrene can be used as an engineered fill in a wide variety of situations. Design of such fills requires both a knowledge and understanding of the physical properties of the material and consideration of protective measures to avoid damage by physical and chemical processes. Successful construction relies upon careful preparation of the site and programming.

ACKNOWLEDGEMENT

108. The work described in this paper was carried out under a contract placed on Babtie Geotechnical by the Transport Research Laboratory, and the paper is published by permission of Highways Engineering Division, DOT and the Chief Executive of TRL.

109. Copyright Controller of HMSO 1993. The views expressed in this paper are not necessarily those of the Department of Transport. Extracts from the text may be reproduced, except for commercial purposes, provided the source is acknowledged.

REFERENCES
1. **AABOE, R. (1987).** 13 years of experience with expanded polystyrene as a lightweight fill material in road embankments. In - Plastic Foam in Road Embankments. Norwegian Road Research Laboratory Publication No. 61.

2. **FRYDENLUND, T.E. & AABOE, R. (1988).** Expanded Polystyrene - A Superlight Fill Material. International Geotechnical Symposium on Theory and Practice of Earth Reinforcement, Fukuoka, Japan, 5th - 7th October 1988. pp 383-388. Balkema, Rotterdam.

3. **MAGNAN, J P, BAILLY, J C, BONDIL R (1990).** Les Remblais en Polystyrene Expanse de l'autoroute A8 a Mandelieu. Bulletin de Liaison des Laboratoires des Ponts et Chausees. No. 165, Jan - Feb 1990. pp 17-32.

4. **LASSAUCE, P & ANTOINE, R (1985).** Le Remblai en Polystyrene Expanse du Pont des Quatre Canaux a Palavas-les-Flots : 1. Point de vue du maitre d'oeuvre. Bulletin de Liaison des Laboratoires des Ponts et Chausees. No. 136 Mar - Apr 1985. pp 21-29.

5. **MIEUSSENS, C (1985).** Le Remblai en Polystyrene Expanse du Pont des Quatre Canaux a Palavas-les-Flots: 2. Aspects Geotechniques. Bulletin de liaison des Laboratoires des Ponts et Chausees No. 136 Mar-Apr 1985. pp 30-36.

6. **CHAZAL, P. & TESSONNEAU, D. (1985).** Le Remblai en Polystyrene Expanse du pont des Quatre Canaux a Palavas-les-Flots: 3. Point de vue de l'Enterprise. Bulletin de Liaison Laboratoires des Ponts et Chausees No. 137. May-June 1985, pp 25-27.

7. **WILLIAMS, D & SNOWDON, RA (1990).** A47 Great Yarmouth Western Bypass: Performance During the First Three Years. Transport and Road Research Laboratory Contractor Report 211.

8. **McELHINNEY, A H & SANDERS, R L (1991).** A47 Great Yarmouth Western Bypass. The Use and Performance of Polystyrene as a Lightweight Fill. Transport and Road Research Laboratory Contractor Report 296.

9. **BS 3837: Part 1: 1986.** Specification for Boards Manufactured from Expandable Beads. British Standards Institution, London.

10. **BS 3837: Part 2: 1990.** Specification for Extruded Boards. British Standards Institution, London.

11. **BS 4370: Part 1: 1988.** Methods of Test for Rigid Cellular Materials. Part 1. Methods 1 to 5. British Standards Institution, London.

12. **MAGNAN, J P AND SCRATRICE, J F (1989).** Proprietes Meanique du Polystyrene Expanse pour les Applications en Remblui Routier. Bulletin de Liaison des Laboratories des Ponts and #Chausees No. 164, pp 25-31.

13. **SHELL PLASTICS (1988).** General Properties of Expanded Polystyrene. Styrocell Technical Manual, Technical Bulletin, May 1988. Shell Plastics, London.

14. **BS 4370: Part 2: 1973.** Methods of Test for Rigid Cellular Materials. Part 2. Methods 6 to 10. British Standards Institution, London.

15. **REFSDAL, G (1987).** EPS Design Considerations. In - Plastic Foam in Road Embankments, Norwegian Road Research Laboratory. Publication No. 61.

16. **PUBLIC ROADS ADMINISTRATION, NORWAY (Road Research Laboratory) (1980).** Guidelines on the use of Plastic Foam in Road Embankment. Norwegian Road Research Laboratory, Oslo.

17. **LABORATOIRE DES PONTS ET CHAUSSEES (1990).** Utilisation de Polystyrene Expanse en Remblai Routier - Guide Technique. Laboratoire Central des Points et Chausees, Paris.

18. **BS 3412: 1976.** Polystyrene Materials for Moulding and Extrusion. British Standards Institution, London.

19. **ASTM D883 (1990).** Standard Definitions of Terms Relating to Plastics. American Society for Testing and Materials, Philadelphia, USA, Volume 08.01.

20. **ASTM D374 (1988).** Standard Test Methods for Thickness of Solid Electrical Insulation - Method C. American Society for Testing and Materials, Philadelphia, USA, Volume 08.01.

21. **ASTM D751 (1989).** Standard Test Methods for Testing Coated Fabrics. American Society for Testing and Materials, Philadelphia, USA, Volume 08.03.

22. **ASTM D638-M (1990).** Standard Test Method for Tensile Properties of Plastics (Metric), American Society for Testing and Materials, Philadelphia, USA, Volume 08.01.

23. **ASTM D882 (1990).** Standard Test Methods for Tensile Properties of Thin Plastic Sheeting. American Society for Testing and Materials, Philadelphia, USA, Volume 08.01.

24. ASTM D1004 (1990). Standard Test Method for Initial Tear Resistance of Plastic Film and Sheeting. American Society for Testing and Materials, Philadelphia, USA, Volume 08.01.

25. ASTM D1505 (1990). Standard Test Method for Density of Plastics by the Density Gradient Technique. American Society for Testing and Materials, Philadelphia, USA, Volume 08.01.

26. DIN 53370: 1976. Testing of Plastic Films: Determination of Thickness by Mechanised Feeling. Beuth Verlag GmbH, Berlin, Germany.

27. DIN 53479: 1976. Testing of Plastics and Elastomers: Determination of Density. Beuth Verlag GmbH, Berlin, Germany.

28. DIN 53455: 1981. Testing of Plastics: Tensile Test. Beuth Verlag GmbH, Berlin, Germany.

29. ANON (1991). Polystyrene Foam Eases the Burden at Syracuse Mall. Civil Engineering. American Society of Civil Engineers, October 1991, pp 84.

30. HAYAKAWA, K, TAKESHITU, S & MUTSUI, T (1991). Reduction Effect of EPS Blocks on Ground Vibration Caused by Road Traffic and Its Evaluation. Journal of the Japanese Society of Soil Mechanics and Foundation Engineering Domestic Edition, 31(2).

31. YEH, S T & GILMORE, J B (1992). Application of EPS for Slide Correction. Stability and Performance of Slopes and Embankments - II. American Society of Civil Engineers 1992, pp 1444-1456.

32. HORVATH, J S (1992). Dark, No Sugar: A well known material enters the geosynthetic mainstream. Geotechnical Fabrics Report, October 1992, pp 18-23.

23. Relationship testing for acceptability assessment of cohesive soils

R. H. JONES, University of Nottingham, UK, and J. R. GREENWOOD, Travers Morgan Consulting Group, UK

SYNOPSIS. The procedure for relationship testing, now formally adopted by the Department of Transport is outlined. Essentially it consists of determining the relationship amongst MCV, dry density, undrained shear strength s_u, CBR and water content for soils covering the range of plasticity and gradings encountered on site. The results from three case studies are considered in the context of previously published general relationships. Good correlations were obtained from individual samples. When generalised, the results conformed to the expected trends but produced considerable scatter. Nevertheless, there was a broad correspondence between upper limits defined by MCV = 8, s_u = 50kPa, CBR = 2 and liquidity index = 0.15.

BACKGROUND AND PROCEDURE

1. Fill has to be placed within acceptable moisture content limits so that it is neither too wet to be stable and trafficable, nor too dry to be compacted satisfactorily.

2. Increasingly for cohesive fill, Specifications quote limiting values of moisture condition value (MCV) but other parameters such as undrained shear strength (s_u), California Bearing Ratio (CBR), relative density, air voids or moisture content (possibly related to plastic limit, w_p) have been adopted. For a particular contract it is essential that only one criterion is chosen for each class of fill.

3. Relationship testing, promoted in the Eastern Region of DOT[1] and formally introduced into DOT procedures in HA44/91[2], is a programme of tests in which the variation of compacted dry density, MCV, s_u and CBR is determined over a range of moisture contents.

4. Advantages claimed for relationship testing[2] are that it:

(a) provides information on soil characteristics at a range of moisture contents,

(b) indicates susceptibility to a reduction in strength due to wetting and increased strength due to drying,

(c) helps determine wet and dry acceptability limits for the optimum use of available materials,

(d) enables acceptability assessments to be carried out by different methods and approaches, and correlations to be made between them,

(e) aids site control testing where sometimes only limited data, ie moisture content or MCV, are immediately available,

Engineered fills. Thomas Telford, London, 1993

(f) helps the Contractor assess plant requirements and assess haul distances.

Also, it enables the designer to relate an MCV limit to values of s_u and CBR which are required for stability and pavement thickness calculations.

5. Typically two or three relationship tests are performed on each material type. These should represent the two extremes and perhaps the average for the material. Some 50kg (say 3 bags) of material is required. Two or three tests have been considered sufficient to give a good insight into the behaviour of the soil without incurring excessive costs.

6. The tests recommended are MCV calibration, dry density/moisture content (2.5kg rammer), CBR/moisture content, s_u/moisture content, sample description, Atterberg limits and particle size distribution (wet sieving and sedimentation). s_u may be determined either by hand vane tests on the compacted CBR and MCV specimens or by undrained triaxial tests on remoulded specimens. The latter method is necessary for drier specimens. CBR tests should be unsoaked with a surcharge of 10kPa. The preferred procedure is to start with the sample at natural moisture content and perform tests both wetting up and allowing it to dry. Alternatively it may be air-dried to begin with. Additional data, such as heavy (4.5kg rammer) compaction tests or MCV densities can be obtained if required. Except for the hand vane all the above tests are described in BS1377(3).

7. Although both wet and dry limits are important, relationship testing has tended to concentrate on the wet limit. Testing must include material which is manifestly too wet so that the wet acceptability limit is interpolated, not extrapolated. Thus it should extend to CBR < 2 (1.5 preferable), MCV < 6, s_u < 30 kPa. It was proposed that compaction tests should include ideally one point dry of optimum, one close to optimum and three wet of optimum. Thus for plastic clays, moisture contents of 0.7, 1.0, 1.2, 1.4 and 1.6 x w_p were thought appropriate(1). It may however be preferable to test over a range of liquidity indices (say up to 0.25). It should be noted that additional points may well be required to investigate the dry limit, so that the normal 5 point compaction test may not be adequate.

8. The choice of limits is influenced particularly by:-

- a. earthworks balance
- b. availability of import
- c. trafficability,
- d. ability to compact to satisfactory density,
- e. susceptibility to moisture content change.

Other factors include frost susceptibility, improvement by drying, chemical composition, availability for landscaping bunds and the use of more plastic material in core of embankments. Whilst the designer has to consider stability, settlement and pavement thicknesses, for many schemes the wet limit is effectively decided by trafficability.

9. Typically, limits (2,4) are in the ranges:-

	Wet	Dry
s_u	30 - 50 kPa	150 - 200 kPa
MCV	7 - 9	12.5 - 15 MCV

10. Cohesive soil will normally be compacted by Methods 1 or 2 (Table 6/1)(5). The Notes for Guidance(4) imply that the dry limit should be the

moisture content (less than optimum) at which the 10% air void line intersects the standard compaction curve. Method 3 (for silty materials) should give 95% of the maximum dry density (MDD) by the 2.5kg rammer method.

GENERAL TRENDS

11. For saturated remoulded soils there is an approximate linear relationship between liquidity index, I_L, and log c_u (Fig. 1)(6). In earlier practice, the upper limit of moisture content was often specified as $1.2w_p$. The lower limit of shear strength for trafficability (30- 50kPa), corresponds to a liquidity index of about 0.15. In view of the overall approximations, the effect of the small air content in the wet compacted soil is neglected.

12. Many British soils(7) obey the relation $I_p = 0.84 (w_L - 17)$ but some glacial tills(8) are characterised by the relation $I_p = 0.80 (w_L - 10.6)$. These relations imply that the wet limit of moisture content should be x^*. w_p where x^* varies with I_p (Fig. 2). However, since w_p is measured on the fraction

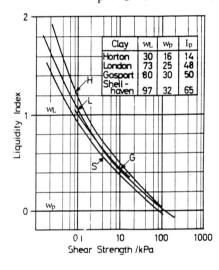

Fig. 1. Liquidity Index vs Shear Strength (after Ref. 6)

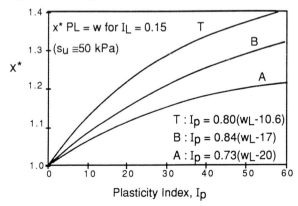

Fig. 2. Inferred Relationship between Limiting Moisture Content and Plasticity

of soil passing a 425μm sieve, there are fundamental difficulties in applying this method to soils containing significant proportions of coarser materials.

13. Moisture condition value is essentially a measure of the effort required to compact a soil to refusal. Specifically, it is $\log_{10}N$ where N is the number of standard blows required - so that MCV is essentially a log scale. Moisture content, w, (and thus liquidity index) usually has a linear relationship with MCV which is sample specific. Published trends suggest that MCV8 corresponds to an s_u of about 40kPa(9).

14. The relations amongst CBR, moisture content and plasticity index have been considered in detail(7, 10). Many empirical relations exist, including

$$CBR = \frac{s_u}{23} \tag{1}$$

15. The significance of correlations between various parameters is affected by the reproducibility of tests. The reproducibility of MCV is 1.6 for clays(11) but for other tests, there does not appear to be any post-1975 published information.

CASE STUDIES

16. Data from relationship tests was obtained from three sites in Eastern England. For a fourth site, N, only compaction data was available. The three sites were chosen because they covered the range from low to high plasticity (CL to CH) materials and the data was readily available. They included some early applications of relationship testing when the shear strength was obtained by triaxial tests on remoulded soils rather than by hand vane tests. An unusual feature was the dependence on heavy compaction tests.

17. Classification, plasticity and grading data are shown in Table 1 and Figs. 3a and b. Separate silt and clay fractions were often not determined. For the present detailed study, it would have been helpful to have had complete data. Typical relationship data are tabulated in Table 2 and plotted in Fig. 4.

Table 1 - Materials studied

Site	No of sets of data	Material	BS Class[n]	DTp Class[n]
B	3	Glacial Till	CI	2A
	1	Weathered Oxford Clay	CH	2A
E	10	Oxford Clay	CI- CH	2A
I	4	Glacial Till	CL- CI	2A
N	12	Glacial Till	CL-CI	2A/2C

18. MCV calibrations were generally good straight lines of the conventional form

$$w = a - b \; (MCV) \tag{2}$$

Whilst it was possible to distinguish zones of low, medium and high plasticity materials on a plot of 'a' vs 'b', general contouring was not possible. This accords with previous findings(12).

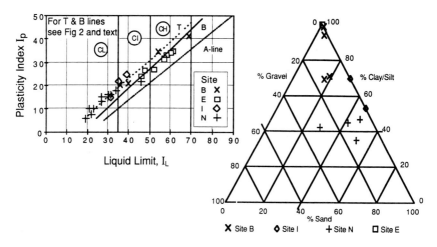

Fig. 3(a). Plasticity (b) grading data

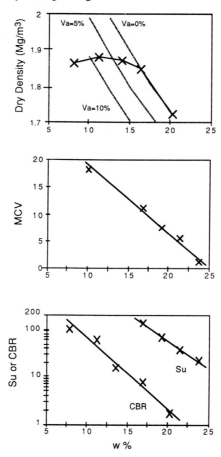

Fig. 4. Typical results from relationship testing

Table 2 – Relationship testing

LAB TESTING BY: Geotechnical and Ground Associates	PROJECT: Beetown Bypass	Form T(i)

Form T(i)

HOLE No(s) 210
SAMPLE No(s) 3
DEPTH(s) 4.00-4.80M
DATE January 1993
SOIL TYPE
Glacial Till

SOIL DESCRIPTION
Grey brown, slightly sandy CLAY with a little fine to coarse gravel

>425μm removed:-
PL 16
LL 36
Classn CI
SG 2.65 assumed

Sample Preparation and Procedure Details:
> 20mm removed:- 2%

GRADING %

Clay	Silt	Sand	Gr	Cob.
27	44	15	0	0

FIG. NO. T(i) 147
REL TESTING Sheet 1 of 1

SAMPLE No	HOLE No
3	210

COMPACTION
Heavy/Light*
CBR/BS mould*
.... Percentage air voids
* delete as appropriate

w %	Dry Density Mg/m³
8.1	1.867
11.2	1.885
14.0	1.872
16.8	1.831
20.2	1.720

CALIFORNIA BEARING RATIO
T CBR at top and
B base
A Average CBR if top and base results differ by <10%
Surcharge pressure 10kN/m²

	w %	CBR %	Dry Density Mg/m³
T	7.7	96	1.867
B	8.1	109	
T	11.0	61	1.885
B	11.4	63	
T	13.2	14	1.872
B	13.7	18	
T	16.9	10	1.831
B	16.7	6.1	
T	20.2	2.0	1.720
B	20.2	1.6	

MOISTURE CONDITION VALUE
Sample mass 1.5kg
Hammer mass 7kg
Hammer drop 250mm

w %	MCV	Dry Density Mg/m³
10.0	18.4	1.730
16.8	11.3	1.813
19.2	7.7	1.732
21.4	5.7	1.650
23.8	1.4	1.576

SHEAR STRENGTH
Hand Vane
mean of 3 on each end of CBR
mean of 3 on MCV
Vane Type   Pilcon

w %	s_u kPa (CBR)	s_u kPa (MCV)
7.7	>140	
8.1	>140	
11.0	>140	
11.4	>140	
13.2	>140	
13.7	>140	
16.9	134	
16.7	125	
20.2	50	
20.2	51	
		>140
		140
		70
		38
		23

19. For individual samples, linear relationships were obtained between $\log s_u$ and MCV and also between moisture content and both $\log s_u$ and log CBR. Combining the results of all the individual tests, (where necessary by converting moisture contents to liquidity indices), broadly gave the expected trends(9) but with wide scatter bands. There was no discernible dependence of the coefficients obtained from regression equations with plasticity index. Furthermore no relation was found between MCV results and plastic limits. Direct comparison of s_u and CBR results, possible only on site B, see Fig 5, supported equation 1.

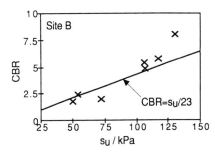 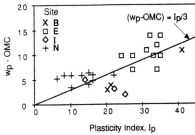

Fig. 5. CBR vs s_u Relationship for Site B.

Fig.6.(w_p -OMC) vs I_p Relationship for Heavy Compaction

20. The relationship between OMC, w_p and I_p for heavy compaction is shown in Fig. 6. There is a general trend for the difference between OMC and Ip to increase with plasticity index but the correlation is not strong.

21. In summary, the relationship testing from these three sites conforms to the general trends discussed in paragraphs 11 and 13 although there is a fair amount of scatter.

22. In order to look specifically at the effects of alternative methods of specifying the wet limit, the following moisture content parameters are introduced. (d) is included for comparison purposes. It is not proposed as an alternative for contractual purposes.

Criteria	Associated limiting moisture content
a. MCV $\lessgtr$ 8	w_{m8}
b. CBR $\lessgtr$ 2	w_{c2}
c. s_u $\lessgtr$ 50 kPa	w_{s50}
d. $I_L \gtrless$ 0.15	w_{IL15}

23. Plots of moisture contents relating to various limits are shown in Fig. 7(a) to (f) together with equality and linear regression lines. From the graphs, it is apparent that there is close correlation between w_{s50} and w_{IL15} and between w_{s50} and w_{m8} although the former classifies marginally more material as acceptable. The poorest correlations involve the CBR and also w_{m8} *vs* w_{L15}. Nevertheless in view of the general scatter of the results over the full

moisture content range and the assumptions built into the predictions of equality (see paras 11-13) the correlations are good.

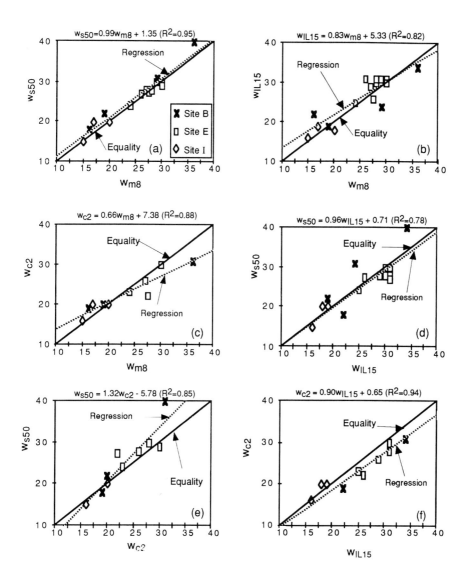

Fig. 7. Relationships between limits set by various criteria (showing regression and equality lines)

24. The selection of the extreme and average soils for relationship testing is likely to be based on plasticity index and gradings. Although this is the best available method it is not foolproof since many relationship parameters (para 19) are largely independent of I_p. Fig. 8a shows how well various criteria picked out the extreme samples with respect to the MCV results for a particular horizon. The alternative criteria considered are I_p, w_p, w_L and "offset". "Offset" is the difference between the actual liquid limit and the liquid limit obtained from equation B (Fig. 2) for the same plasticity index (see Fig. 8b). It is Δw as defined by Black (7). The actual distribution of w_{m8} ranges from 24 to 30% as shown in the lower part of Fig. 8a. The horizontal lines in the upper part of the figure show the range of w_{m8} values which would have been

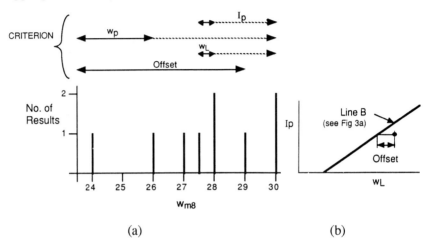

(a)

(b)

Fig. 8(a). Selection of 'Extreme' Soils by various criteria for Site 'E'. (b) Definition of Offset.

revealed by testing samples having extreme values of other criteria; the dashed portions indicate that different samples, each having the same maximum value of the alternative criterion gave a range of values of w_{m8}. Although not observed with this data set, a similar problem could occur with the minimum values of alternative criteria. The solid horizontal lines in Fig. 8a thus give the minimum range of w_{m8} values which could be revealed by selecting samples on the basis of alternative criterion.

25. For these particular samples the 'offset' method proved best. However for picking out extreme performances in the compaction tests, plasticity index appeared better. Perhaps the best scheme is to choose samples having the most extreme offsets together with those having the greatest and least plasticity indices. Additionally, for coarser soils, the grading needs to be considered, perhaps with the aid of a triangular diagram (Fig. 3b) which can be annotated with plasticity data. In view of the difficulty of identifying the extreme soils, it may be advisable to increase the number of relationship tests. Careful attention must be given to the selection of the extremes without giving undue weighting to outliers. If the range of variability is large, it is important that "average" (typical) soils are tested.

310

CONCLUDING REMARKS

26. Relationship testing gives a formal framework for considering the relations amongst MCV, s_u, CBR and plasticity. Plant performance on site is not considered specifically, but experience shows that MCV is a good indicator of this and is also probably the best method of site control of acceptability.

27. Relationship testing has a particularly useful role for the designer in that it relates MCV to design parameters such as s_u and CBR. Although it is fairly difficult to select the extreme soils, it gives a warning of anything untoward. It also increases confidence in interpreting in terms of acceptability, the many natural moisture content and plasticity index determinations undertaken for classification purposes during ground investigation.

28. Correlations amongst MCV, s_u, CBR and I_L tend to be sample specific, even for soils of the same horizon. However broad general correlations exist.

29. Even so, for the three sites studied, there was broad agreement amongst the upper limits defined by MCV = 8, s_u = 50kPa, CBR = 2 and I_L = 0.15.

ACKNOWLEDGEMENTS

G W Herrick, Eastern Region, Department of Transport
H U Baba & G P Barnes, Department of Civil Engineering, University of Nottingham for data processing.

REFERENCES

1. EASTERN REGION, DEPARTMENT OF TRANSPORT (1984) Transport Note 1/84. (Amplified by 1987 Note prepared for design/contractor advice (unpublished)).
2. DEPARTMENT OF TRANSPORT (1991) Earthworks: Design and Preparation of contract documents. Departmental Advice Note HA44/91.
3. BRITISH STANDARDS INSTITUTION 1990. Method of test for engineering soils. Part 2 Classification tests, Part 4 Compaction related tests. Part 7 Shear strength tests (total stress). BS1377:1990.
4. DEPARTMENT OF TRANSPORT 1991. Specification for Highway Works Vol. 2 Notes for Guidance.
5. DEPARTMENT OF TRANSPORT 1991 Specification for Highway Works. Vol. 1.
6. SKEMPTON, A.W. & NORTHEY, R.D. 1953 'The sensitivity of clays', Geotechnique 3, p3-53.
7. BLACK, W.P.M. 1962 'A method of estimating the California Bearing Ratio of Cohesive Soils from Plasticity Data' Geotechnique Vol. 12, No. 4, p271-282.
8. BOULTON, G.S. & PAUL, M.A., 1976 'The influence of genetic processes on some geotechnical properties of glacial tills. QJ Eng. Geol. Vol. 9, pp159-194.
9. PARSONS, A.W., 1981 'The Assessment of Soils and Soft Rocks for Embankment Construction'. Q.J. Eng. Geol. Vol. 14, No.3. p219-230.
10. BLACK, W.P.M. & LISTER, N.W. 1979 The strength of clay fill subgrades: its prediction in relation to road performance. Clay Fills ICE London p37-48.
11. PARSONS, A.W. & TOOMBS, A.F. 1987. 'The precision of the moisture condition test', TRRL Res Report 90.
12. EWAN, V.J. and WEST, G. 1983 Appraising the moisture condition test for obtaining the Casagrande classification of soils; TRRL Sup. Rep. 786.

24. The effect of stone content on the determination of acceptability for earthworking

M. G. WINTER, Transport Research Laboratory, Scotland, and
SUHARDI, formerly University of Strathclyde, UK

SYNOPSIS.   Test   procedures   for   the   determination   of
acceptability of fills for earthworking require the removal of
particles greater than a predetermined maximum size. In
particular, use of the Moisture Condition Apparatus requires
the removal of stones retained on a 20mm test sieve. Many tills
contain a significant proportion of stones. The application of
test results carried out on the residual matrix material can
thus be questioned. A technique to assess the maximum stone
content, for which tests on the matrix material will indicate
the acceptability of the excavated material, is described. The
preliminary results presented indicate that this maximum stone
content is approximately 45%.

INTRODUCTION
  1. The Moisture Condition Apparatus (MCA) is routinely used
for the determination of soil acceptability for earthworking
in Scotland [1 & 2]. The removal, prior to test, of the
particles retained on a 20mm test sieve (the stones) is an
integral part of the MCA test. The removal of such large
particles, often referred to as scalping, can have a profound
effect on measured properties.
  2. If the stone content of a given soil is such that the
stones determine behaviour, then acceptability tests on the
material passing a 20mm test sieve (the matrix) do not
represent the bulk soil properties.
  3. This paper presents the results of a preliminary study to
determine the transition between matrix and stone determination
of behaviour. This is expressed in terms of the maximum stone
content at which tests on the matrix material are valid. In
cases where the stones determine behaviour an alternate
approach to acceptability determination is required.

THE EFFECT OF SCALPING
  4. The limits of use of the MCA are governed by limits on a
ternary diagram (Fig. 1). Potential usage is evaluated by
reference to the particle size distribution (PSD) of the
excavated material, in terms of fines, sand and gravel [1 &
2]. The gravel fraction includes particles in the range 2 to
60mm. Thus, scalping of stony materials prior to MCA testing
may produce a specimen at variance with the excavated sample.

Engineered fills. Thomas Telford, London, 1993

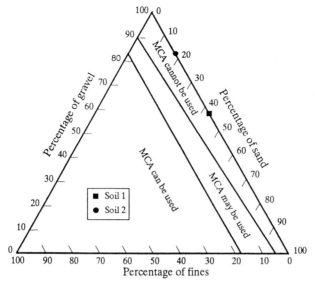

Figure 1 - Limits of use of the MCA (after [1 & 2]) showing
particle size distributions of Soils 1 and 2.

5. Barnes [3] illustrated the importance of scalping prior
to test. Fig. 2 shows PSDs of excavated samples. Sample A is
finer than sample G. After scalping sample A is coarser than
sample G.

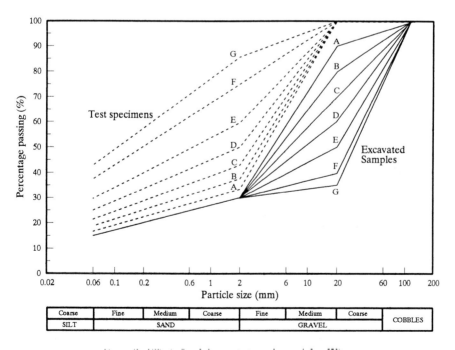

Figure 2 - Effect of scalping on test specimens (after [3]).

313

6. A number of methods are available for the calculation of densities of unscalped materials, from tests on scalped laboratory samples [4 to 7]. These methods relate to the standard Proctor-type compaction tests as specified, for instance, in the UK in BS 1377 [8]. None of the methods of which the author is currently aware are applicable, without the use of large scale in-situ density tests, to soils with more than 30% stones. Limited success can be had by replacing small amounts (5 to 10%) of large particles (>37.5mm) with smaller particles (20 to 37.5mm) of similar characteristics [8 & 9]. Simple correction of MCA results for stone content is not as yet possible.

ACCEPTABILITY OF STONY SOILS

7. In selecting an appropriate method for determining the acceptability of stony soils, it is necessary to know whether the matrix or the stones determine behaviour. If the matrix material determines behaviour then acceptability is determined by conventional means. On Scottish trunk road projects this is by either MCA testing or by conventional Proctor-type compaction testing. The choice of method depends on the position of the PSD, of the excavated material, on the ternary diagram (Fig. 1).

8. If the stones determine behaviour then there are two options. Firstly, the material may be free-draining. In general, such materials can be placed regardless of moisture condition, provided that adequate drainage is possible. PSD testing is required to check that the excavated material is indeed free-draining. Secondly, if the material is not free-draining then acceptability is determined by comparing the measured moisture content with compaction curves determined on the representative excavated material. Thus ensuring that the material has the potential to compact to low air voids.

9. For the above procedures to be followed, a reliable method is required of determining the stone content at which the transition from matrix to stone determination of behaviour occurs. This indicates the maximum stone content at which the results of tests on the matrix material represent the behaviour of the excavated material.

10. Barnes [3] demonstrated a clear transition between fines and sand determination of behaviour for a till with measured proportions of fines and sand (Fig. 3). The results for fines, sand and gravel mixes are not so clear cut.

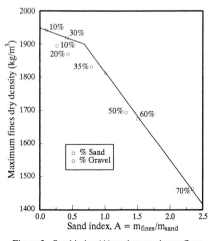

Figure 3 - Sand index (A) against maximum fines dry density (after [3]).

314

However, the results indicate that it might be possible to adopt a similar approach for matrix (up to 20mm) and stone (20 to 37.5mm) mixes.

11. Up to 40% sand (Fig. 3), the reduction in fines dry density is a result of the compaction energy being used to compact the fines around the sand particles. At sand contents greater than 40%, particle interference becomes a more dominant effect [3] and the sand determines soil behaviour.

DEFINITION OF STONE CONTENT

12. Barnes interpreted his results using $A = m_{sand}/m_{fines}$, where m is dry mass. For the purposes of the current work this is termed the "sand index" (Fig. 3). An analogue for the present work would be stone index, B, defined as m_{stones}/m_{matrix}. This compares with the stone content defined as (m_{stones}/m_{total}) x 100% (Figs 5-7). The stone index has values of 0 to 3.0 for the range of stone content 0 to 75%. The use of the stone index distorts data on the x-axis scale. The stone index also has the disadvantage of being counter-intuitive. That is, the meaning of a stone index of 3.0 is not clear. In contrast the meaning of a stone content of 75% is immediately clear. The stone content is used for all interpretations in this paper.

EXPERIMENTAL WORK

Test Materials and Techniques

13. The experimental work was carried out on two granular tills from the South of Scotland. Classification and index tests results are given in Tab. 1. Soil classifications are in accordance with BS5930 [10] and the Specification for Highways Works [11]. Test results are in accordance with BS1377 [12]. PSD curves, for the excavated material, are shown in Fig. 4 and the corresponding positions are plotted on the ternary diagram (Fig. 1).

Table 1 - Classification and index test results.

MATERIAL	Soil description to BS5930: [10] (Soil classification to Specification for Highways Works: [11])	Specific Gravity, Gs	Liquid Limit, LL (%)	Plastic Limit, PL (%)	Plasticity Index, PI (%)
Soil 1	Well graded very sandy GRAVEL with occasional cobbles (Well graded granular material)	2.70	22	12	10
Soil 2	Well graded sand GRAVEL with some cobbles (Well graded granular material)	2.73	25	14	11
Barnes [3]	BS5930 description not available (Stony cohesive material)	2.71,2.65, 2.63 (Fines, Sand, Gravel)	34	15	19

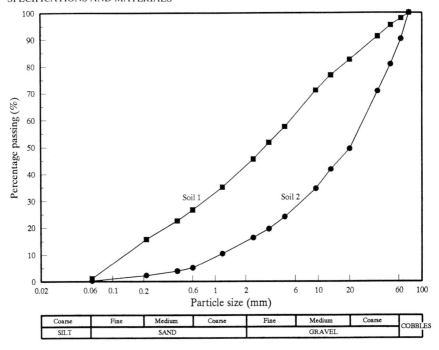

Coarse	Fine	Medium	Coarse	Fine	Medium	Coarse	
SILT		SAND			GRAVEL		COBBLES

Figure 4 - Particle size distribution curves of Soils 1 and 2.

14. Compaction tests were carried out in a large CBR-type mould on material passing a 37.5mm test sieve. Stone contents were varied from 0 to 75%. A 2.5kg rammer was used and the number of blows increased to 62 [6] to ensure that the energy levels applied during compaction were comparable with the British Standard test [12]. The current edition of BS1377 [8] sets a limit of 30% stone content for routine compaction tests in the CBR mould.

15. The effects of crushing were evaluated by means of PSD determinations both before and after compaction. The observed decreases in particle size were found to be within the limits of experimental error (i.e., around 2% variation in the percentages passing a 20mm test sieve).

The Effect of Stone Content on Total Dry Density

16. Figs 5 and 6 show dry density-moisture content curves for Soils 1 and 2, respectively. These curves are for stone contents in the range 0 to 75%.

17. In each case, for an increase in stone content up to 35%, there is an increase in maximum dry density and a decrease in the optimum moisture content. Constant air voids are maintained. As Barnes [3] observed, these increases in density do not actually represent improvements in the stability of the soil. They reflect the replacement of high moisture content, low density matrix material with low moisture content, high density stones. For stone contents in the range 50 to 75% a decrease in maximum dry density is apparent with a slight

316

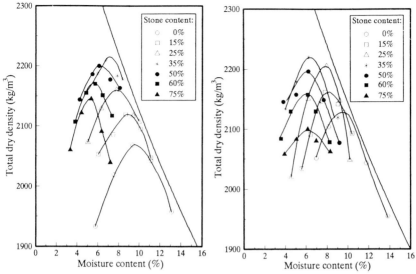

Figure 5 - Moisture content against total dry density for Soil 1.

Figure 6 - Moisture content against total dry density for Soil 2.

decrease in optimum moisture content. Large reductions in air voids are a consequence.

The Effect on Stone Content on Matrix Dry Density

18. Fig. 7 shows stone content plotted against the matrix dry density at optimum moisture content for both Soils 1 and 2. Up to approximately 45% stone content, the matrix dry density exhibits only very small reductions. Plainly, the stones act in isolation within the matrix with little or no stone-to-stone contact.

19. At a stone content of around 45% the matrix dry density begins to decrease rapidly with increasing stone content. As the stone content increases the number of stone-to-stone contacts inceases. As a consequence, compaction energy, which was previously used in densifying the matrix material, is used to overcome the effects of friction between stones. The stone dry density is thus increased.

20. These changes in the relations between stone content and matrix dry density mark the transition from matrix to stone determination of behaviour (see paras 7 to 9)

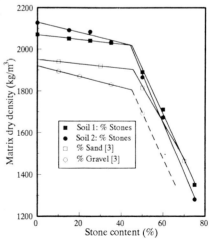

Figure 7 - Stone content against matrix dry density at optimum moisture content for Soils 1 and 2. Results from Barnes [3] are shown for comparison.

317

SPECIFICATIONS AND MATERIALS

21. Data from Barnes [3] have also been included in Fig. 7 (basic soil properties are given in Table 1). These data are used to plot sand and gravel content against fines dry density at optimum moisture content. The transition between fines and sand, and fines and gravel, determination of behaviour occurs at approximately 45% sand and gravel content. (c.f. 40% for the sand index: see para. 12 & Fig. 3).

22. Describing sand, gravel and stones as inclusions, and fines as matrix, all the data presented in Fig. 7 is for matrix/inclusion mixes. Thus, a consistent pattern begins to emerge. The transition between matrix and inclusion determination of behaviour consistently occurs at approximately 45% inclusion content. Ongoing experimental work and other published data [13 & 14] give support to this conclusion.

SUMMARY AND FURTHER WORK

23. If stones determine the behaviour of a soil, then acceptability tests on the matrix material do not represent the behaviour of the excavated material.

24. A technique has been described for the determination of the transition between matrix and stone determination of behaviour. This allows an appropriate technique to be selected for the determination of acceptability for earthworking.

25. The technique is based upon earlier work by Barnes [3]. However, the interpretation of results is in terms of stone content rather than stone index.

26. From the preliminary experimental results, and published data [3], the transition from matrix to inclusion (sand, gravel or stones) determination of soil behaviour consistently occurs at around 45% inclusion content. Ongoing work gives support to this conclusion.

27. This corresponds to the maximum stone content at which test results, such as MCV, on the matrix material represent the behaviour of the excavated material.

28. Further testing work is ongoing to confirm the proposed technique, on a wide variety of tills, and to evaluate possible variations in the transition from matrix to stone determination of behaviour for different soil types.

ACKNOWLEDGEMENTS

29. Crown Copyright 1993. The views expressed in this paper are not necessarily those of the Department of Transport or the Scottish Office Roads Directorate. Extracts from the text may be reproduced, except for commercial purposes, provided the source is acknowledged.

30. The work described in this paper forms part of a University of Strathclyde funded research programme supervised by the Transport Research Laboratory Scotland (Director, Dr GD Matheson), and the paper is published by permission of Professor KZ Andrawes and the Chief Executive of TRL.

REFERENCES

1. MATHESON, G.D. & OLIPHANT, J. 1991. Suitability and acceptability for earthworking with reference to glacial tills in Scotland, In: Forster, A., Culshaw, M.G., Little, J.A., Cripps, J.C. and Noon, C.F. (Eds), Quaternary Engineering Geology, 239-249. Engineering Geology Special Publication 7, Geological Society, London.

2. ANON 1989. The use and application of the Moisture Condition Apparatus, Applications Guide No. 1, Scottish Development Department, Edinburgh. (1989 Revision of SDD Technical memorandum SH7/83.)

3. BARNES, G.E. 1987. The moisture condition value and compaction of stony clays, Compaction Technology, 79-90, Thomas Telford, London.

4. DAY, R.W. 1989. Relative compaction of fill having oversize particles, Journal of Geotechnical Engineering, ASCE, 115(10), 1487-1491. (Also Discussion 1991, 117(10), 1635-1639.)

5. ANON 1982. Specification for transportation of materials and method of sampling and testing: Part 2. Methods of sampling and testing, AASHTO Designation: T 224 (Correction for coarse particles in the soil compaction test), American Association of State Highway and Transportation Authorities, Washington.

6. HEAD, K.H. 1980. Manual of Soil Laboratory Testing: Volume 1. Soil classification and compaction tests, Pentech Press, London.

7. HSU, T.-S. & SAXENA, S.K. 1991. A general formula for determining density of compacted soils with oversize particles, Soils and Foundations, 31(3), 91-96. Japanese Society for Soil Mechanics and Foundation Engineering.

8. ANON 1990. British Standard methods of test for soils for civil engineering purposes: Part 4. Compaction-related tests, BS 1377, British Standards Institution, London.

9. PARSONS, A.W. 1992. Compaction of soils and granular materials: A review of research performed at the Transport Research Laboratory, HMSO, London.

10. ANON 1981. Code of practice for site investigations (formerly CP2001), BS 5930, British Standards Institution, London.

11. ANON 1986. Specification for highways works: Part 2; Series 600 - Earthworks, Department of Transport, Scottish Development Department, Welsh Office, Department of the Environment for Northern Ireland, HMSO, London.

12. ANON 1975. Method of test for soils for civil engineering purposes, BS1377, British Standards Institution, London.

13. ANON 1952. Soil mechanics for road engineers, HMSO, London.

14. MADDISON, L. 1944. Laboratory tests on the effects of stone content on the compaction of soil mortar, Roads and Road Construction, 22(254), 37-40.

25. Behaviour of compacted earth fill embankments, constructed during winter and summer conditions

O. MAGNUSSON and S. KNUTSSON, Luleå University of Technology, Sweden

SYNOPSIS

A field test has been carried out in northern Sweden to find out the behaviour of an earth-fill embankment constructed during winter and summer conditions respectively. Two materials were used, a coarse-grained till and a fine-grained till. Settlements, compaction results, temperatures and frost depth were measured. The movements in the embankment were closely related to thawing and the magnitudes were highly dependant upon amount of fines, water content, temperature and degree of compaction at the time of construction. Movements going on for longer periods of time, i.e. during the following summer were also detected.

INTRODUCTION

1.   It is known from experience and laboratory tests, that compaction of earth-fill embankments during winter months gives less reliable and more variable results than if the work is done during non-freezing periods. Experience shows, that compaction during winter periods can be performed with acceptable results for some coarse-grained and dry soils. If fine-grained soils are used under these conditions, the result can be poor compaction and sometimes the work needs to be repeated.

2.   Recently, more emphasis has been paid to the question of deformations in the soil during loading, and this includes both man made earth-fills and natural soil deposits. We know for example, that even when the compaction has been done during summer conditions, some deformations will normally take place in the fill after some time of use. A number of studies have been performed in order to find methods of determining these future settlements during the construction of the fill. With winter construction this question is even more important, as frequently we have situations where the deformations might be very large.

3.   The aim of this study was to investigate the behaviour of compacted earth-fill embankments constructed during winter- and summer conditions. The aim was to study the development of deformations during and after the completion of the construction.

Engineered fills. Thomas Telford, London, 1993

The study consisted of a full-scale field test, with laboratory compaction tests.

MEASURING METHODS AND EQUIPMENT

4.   The road embankment was separated into four test sections, two of which were constructed during winter conditions and the other two during the following summer. In both test embankments, two different materials were used. A coarse-grained, free draining soil (A) and a fine-grained till with a large silt content (D).

5.   In all the test sections the following parameters were measured: density, water content, CMV-value (Compactometer Measurement Value), deformations, frost depth and temperatures in the air and in the ground.

6.   The embankments were constructed in five separate layers, in which the deformations were measured separately, by using the EMM-gauge (Earth Movement Measurement), three levels of flexible hoses and by surface surveying.

7.   The temperatures were measured by thermocouples in each one of the five layers, thus giving a vertical temperature profile for each test section as a result. In each of the four test sections a frost front indicator was also installed.

8.   In laboratory, the compaction properties of all the soils involved, were determined at above zero temperatures as well as at freezing temperatures.

Compactometer

9.   During recent years, electronic, roller-mounted gauges aimed for surface documentation of the compaction results, are being used more intensively.

10.   In the compactometer, the accelerometer is the principal gauge, mounted directly on the vibrating drum. The dynamic response from the compacted soil, is an indicator of the soil density and strength. From empirical relationships, the so-called CMV-value, measured during compaction, can be translated to a dry density value of the soil, see Fig. 1.
The main advantage with the equipment is that the driver of the roller can follow the development of the compaction results directly on a computer display. The measured CMV-values are shown continuously on the display as a function of the position of the roller.

11.   In contrast to other methods for compaction control, it is with this method possible to get a continuos graph of the compaction result during the construction work. The method is well proven for compaction control of coarse granular material, i.e. sand and gravels.

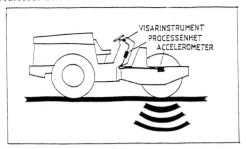

Fig. 1. The compactometer schematically depicted.

Gauges for vertical deformations

12. The vertical deformations were measured in each layer after the completion of the embankment. The equipment used, called EMM (Earth Movement Meter) consists of a number of pipes telescopically linked together. Each pipe can move individually, thus giving an opportunity to follow the movements in each layer. By using a measuring rod with a scale on the top, driven down to a firm bottom layer, it is possible to detect the movements of each layer. The equipment can be used for detection of horizontal, as well as vertical movements within a maximum length of 15 m, see Fig. 2. The equipment is capable of detecting both permanent and elastic deformations, including deformations due to frost heave. The accuracy of the detected deformations are in the order of ± 0.5 mm.

13. The vertical deformations in the embankment were measured in 8 different sections and the horizontal deformations were measured in one section.

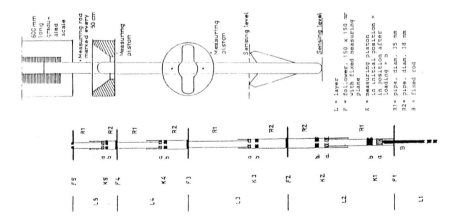

Fig 2. The EMM (Earth Movement Meter) schematically shown.

322

Deformations detected by the use of flexible hoses

14. The settlements in profiles perpendicular to the direction
of the road were followed by the use of flexible hoses placed on
the terrace level and within the fill on two different levels
respectively.

15. The principle of the equipment is shown in Fig. 3 (ref. 1).
The flexible hoses were placed on the ground, and within the
embankment during construction. Due to the flexibility of the
hoses, they follow the movements of the fill. By determining the
hose's position, in relation to a fixed reference point, it is
possible to measure the settlements of the embankment. From
readings taken at different times, it was possible to follow the
settlement as a function of time.

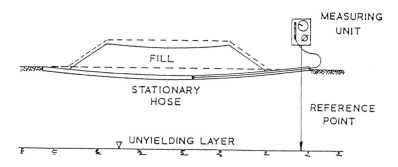

Fig. 3. Settlement gauge with flexible hoses according to
Bergdahl and Broms (ref. 1).

Temperatures and frost front position

16. The air temperatures were continually recorded during the
construction period starting from February 15 1989, and it was
found, that the winter was milder than normal. The temperatures
were never lower than -15 $^{\circ}$C and were as high as 0 $^{\circ}$C during
February and March. In April the temperatures were close to 0 $^{\circ}$C
and from mid May the thawing period started.

17. Thermocouples were installed in each layer of the fill and
the measured temperatures were in the "winter embankment" close to
0 $^{\circ}$C throughout the winter. They were never lower than -2 $^{\circ}$C in
any position and were consequently evenly distributed throughout
the embankment height.

18. A frost front indicator was placed in each section in order
to follow the changes in the frost front position. In the coarse-
grained material the maximum recorded frost depth was 1.1 m and in
the fine-grained material 0.8 m (April 7). The coarse-grained
material was completely thawed in early May, in comparison with
the fine grained material which was not completely thawed until
the end of June.

FIELD TESTING

19. The field test was carried out on road 1052, Risbäck-Borga close to the village Borgafjäll in the mountain area in the interior of the district Västerbotten, Sweden. The road was chosen for the test in co-operation with Vägverket (department of Transportation) as the road was going to be re-built during both summer and winter conditions due to employment reasons.

20. Two types of soils were used for the test:

- a coarse-grained, free draining soil; type A according to the Vägverket (dept. of Transportation) classification system

- a fine-grained till with a large silt content; type D according to the Vägverket (dept. of Transportation) classification system

21. However, due to problems in finding enough material of each type, the test was carried out with two types of material , denoted A and D, with similar grading curves, see Fig. 4. This situation was most pronounced for the summer embankment, where the two types of material looked very similar and visually gave the impression of being D-material. Despite of this circumstance, clear differences could be noticed between the test results obtained from each type of material.

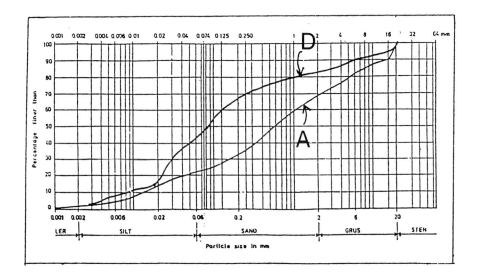

Fig. 4.Grain size distribution curves for the two materials used in the field test.

RESULTS AND DISCUSSION

Laboratory results

22.   Modified Proctor tests were conducted in the laboratory at different temperatures on the two materials used. The results are given in Fig. 5 for the two temperatures $+20\ ^{\circ}C$ and $-20\ ^{\circ}C$. It can be seen that the obtained dry densities are very sensitive for sub-zero temperatures and water content. The coarse-grained material shows higher dry densities at $+20\ ^{\circ}C$ than the the fine-grained material, but the result at $-20\ ^{\circ}C$ is approximately the same.

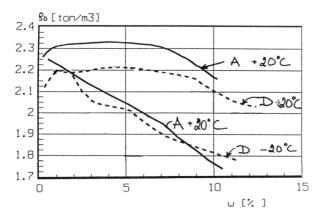

Fig. 5. Compaction curves from Modified Proctor tests at $+20\ ^{\circ}C$ and $-20\ ^{\circ}C$ for material A and D respectively.

Obtained compaction in summer and winter embankments evaluated from CMV-testing

23.   In the fine-grained material (D-material), no increase in the CMV-values could be found from the first pass to the sixth, neither for the summer embankment nor for the winter embankment. It was not until consolidation during a number of months had taken place, as well as re-compaction of the final layer, that the embankment could serve as an acceptable support for the road subbase. At this stage the winter embankment had been constructed for 17 months and the summer embankment for 12.

24.   In the coarse-grained material (A-material), no increase in the CMV-value could be found in the summer embankment between the first and the sixth pass, in contrast to what normally is found for this type of material. Normally the CMV-value increases with the number of passes. In contrast to what was found in the summer embankment, the CMV-value obtained in the winter embankment increased from pass number one to pass number six.

25. After the completion of the winter embankment, in the end of January, the road subbase was put in place the following summer. However, during compaction of this layer it was found, that the same high CMV-value obtained during the winter compaction of the embankment could not be reached by compaction of the base course under summer conditions.

26. Therefore it is concluded, that the CMV-values obtained cannot be regarded as reflecting the degree of compaction of the soil or the dry density value. It just reflects the changes of these properties taking place during the compaction process. Therefore, if we are going to determine the dry density and water content of a soil, it is the opinion of the authors, that we must rely on conventional methods, e.g. the Troxler isotope instrument.

27. When comparing the compaction result for the winter embankment, measured by the CMV and the results obtained from the Troxler, Figs. 6 and 7, it is found that the CMV-values are low even after six passes and that the Troxler shows very high, and to some extent unrealistic, values for the dry density of the soil.

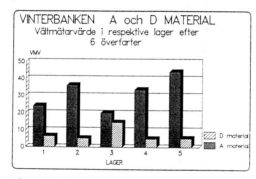

Fig. 6. The CMV-values after six passes, in each one of the five layers in the winter embankment for the two materials used.

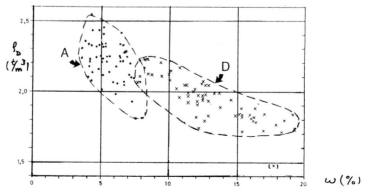

Fig. 7. The obtained relation between the dry density and water content from field measurements with the Troxler method in the winter embankment.

326

28. It can therefore be concluded, that neither of these methods give a correct picture of the degree of compaction in an embankment made up by a fine-grained till and this is independent of the climatic conditions. This is most obvious if the fill is constructed during winter-time, when the water content is high and the pores are filled with ice and super-cooled water.

Deformations

29. The aim of the study was to study the deformations in earth-fill embankments constructed during winter and summer conditions respectively, and as a function of the soil materials used.

30. The winter embankment, with the two types of material (A and D), was constructed during the end of January 1989 and was compacted according to common practise. In August 1989, it was re-compacted from the top of the terrace surface. During the intermediate period of time it was used for as a haul road.

31. After the re-compaction was completed, the subbase layer was placed and compacted. The readings of the deformations started in early February 1989 and went on until May 1990, when the last complete set of readings were taken. The total length of the measuring period was 440 days.

32. The deformations in the winter embankment constructed of the fine-grained material (D-material) is shown as a function of time in Fig. 8.

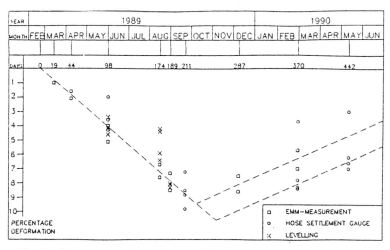

Fig.8. The compression of the winter embankment as function of time. The section was constructed by the use of the fine-grained material (D-material). The compression is related to the total height of the embankment after the compaction of the fifth layer (terrace surface).

327

33.   As long as the soil material is frozen, the measured deformations are generally small, less than 1% of the total height of the embankment. When the embankments thaw, consolidation settlements took place. The compression increased during the first summer to 4-7% and reached after re-compaction 7-9%. After re-compaction, the compression increased further, to about 10%. The first winter after the completion of the embankment (winter 1989/90), some frost heave took place. This was in the order of 5% of the final embankment height.

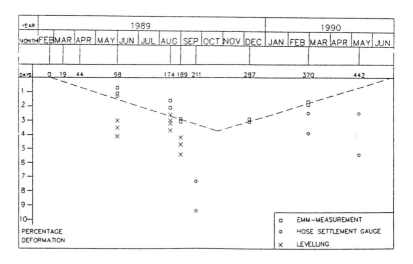

Fig. 9. The compressions in the winter embankment as function of time. The test section is constructed by using the coarse-grained material (A-material). The deformation is related to the total height of the embankment after the compaction of the fifth layer (terrace surface).

34.   In the coarse-grained material the total compression was in the order of 4% after the re-compaction. During the following first winter after completion of the fill, the recorded frost heave was in same order of magnitude as the settlements, i.e. 3%.

35.   The measurements of the deformations in each one of the five individual layers, of which the summer embankment was constructed, could not be carried out as planned, due to major failures in the EMM-gauges. The failures were caused by the heavy haul traffic on the embankment surface, causing large soil movements and breakage of the gauges. Therefore, we only have information of the total embankment compression, measured as settlements on the final embankment layer.

36.   From all the readings taken, it can be concluded that most of the compression can be correlated to the thawing of the initially frozen embankment. But not all of the compression.

328

Deformations due to consolidation and creep, seem to take place in the embankment for some time after completion of the re-compaction.

37. Similar observations have been found in other projects where the degree of compaction has been found to increase as a function of time. The mechanism behind this phenomena, as well as other similar observations, have been discussed in several papers (refs 2-5). However, it is the authors opinion that the process is not yet entirely understood and therefore not fully explained.

CONCLUSIONS

1. Soil embankments compacted during winter time, experience deformations and compression during the thawing process. The magnitude of these movements are to a major degree dependant upon the type of soil, water content, temperature and degree of compaction at the time of construction. The amount of fines in the soil is important, and if this is high, the deformations will be high and vice versa.

2. If the result of compaction work, is measured by the use of CMV-values in a fill constructed during winter-time, this will only give information of the CMV-value of the frozen material, and **not** of the compaction results for the thawed embankment.

3. The thawing of a frozen, fine-grained embankment is a very slow process.

4. In earth-fill embankments constructed during winter time, movements directly related to the thawing of the fill will take place as the thawing proceeds. In addition to this, there will also be movements going on for long periods of time i.e. during the following summer.

REFERENCES

1. Bergdahl, U. and Broms, B. Ny metod för mätning av sättningar hos bankfyllnader på kompressibla jordlager. *Väg- och Vattenbyggaren*, nr 3, 1966, Stockholm (in Swedish)

2. Clough, W. et. al. Influence of Cementation on Liquefaction of sands. *Journal of Geotechnical Engineering*, vol 115, no 8, 1989, New York

3. Mesri, G. et. al. Postdensification Penetration Resistance of clean Sands. *Journal of Geotechnical Engineering*, vol 116, no 7, 1990, New York

4.   Mitchell, J. and Solymar, Z. Time Dependent Strength Gain in freshly deposited or densified Sand. *Journal of Geotechnical Engineering*, vol 110, no 11, 1984, New York

5.   Mitchell, J. Practical Problems from surprising Soil Behaviour. *Journal of Geotechnical Engineering*, vol 112, no 3, 1986, New York

26. Importance of air void for compacted loam soils

N. MOROTO, Hachinohe Institute of Technology, Japan

SYNOPSIS.   It has been well known that applying different
sample preparation methods or compactive energy yields
different compaction curves for volcanic cohesive soils
called loam soils in Japan. This investigation clarified
that the air void ratio becomes an important parameter when
determining optimum moisture condition and mechanical
properties of compacted loam soils.

INTRODUCTION
   1.   About 16% of the total land area of Japan is covered
with volcanic soils most of which are found on the flat land.
These volcanic soils are generally categorized into the
following: volcanic cohesive soils called 'loam soils',
volcanic sandy soils canned 'Shirasu' and volcanic black
soils called 'Kuroboku'. The word loam originates from a
grading of soil which contains a relatively equal mixture of
sand, silt and clay. Dr D Brown, who lectured at the
University of Tokyo in Japan from 1881, first named the red
soils distributed widely in terraced areas of the Kanto
District as 'loam soils' because these red soils contained a
relatively equal mixture of sand, silt and clay like the
'Yellow soils' in China. The name 'loam' did not originally
refer to volcanic soil. It just happens that the red soils
are volcanic soils with a grading of loam. Thus, the word
'loam' is used to describe volcanic cohesive soils in Japan.
In this paper, the author calls volcanic cohesive soil, loam
soil.
   2.   The ejecta from volcanoes thickly covers the hills and
terraced areas of the Aomori Prefecture. This ejecta
includes loam soils and hence earthworks in this prefecture
are likely to be constructed with loam. It is recognised
that different sample preparation methods yields different
compaction curves for loam soils. It has been also
recognized that applying different compactive energy results
in different compaction curves. The mechanical properties of
compacted loam soils are dependent upon the sample
preparation methods or compactive energy even at the same
moisture content. In this paper, the author will attempt to

clarify the parameter that could be used to uniquely correlate with the optimum moisture condition and mechanical properties of compacted loam soils.

OPTIMUM MOISTURE CONDITION AND AIR VOID

3. Soil is composed of solid, water and air. The total volume of soil $V(=1)$ is given by

$$1 = Vs + Vw + Va \tag{1}$$

where

Vs:volume of solid
Vw:volume of water
Va:volume of air

A constant air void line can be expressed as

$$Vs + Vw = 1 - Va = constant \tag{2}$$

The degree of saturation Sr is defined by
$$Sr = Vw/(Vw+Va)$$
A line of constant degree of saturation, Sr can be obtained from

$$Vs + Vw/Sr = 1 \tag{3}$$

The air void ratio, e_a, is defined by

$$e_a = Va/Vs \tag{4}$$

A line of constant air void ratio can be obtained from

$$Vs + (1/(1+e_a))Vw = (1/(1+e_a)) \tag{5}$$

The three equations, Eqs. (2), (3) and (5), produce three straight lines on a plot of Vs and Vw. The gradient of a degree of saturation line, $1/Sr$, is greater than one. The gradient of an air void line is unity. The gradient of an air void ratio line, $1/(1+e_a)$, is less than one.

4. The author has accumulated much data for loam soils taken throughout Aomori Prefecture. Physical properties of the loam soils are shown in Figs. 1, 2, 3 and 4. The soils cover a wide range of plasticity as shown in Fig. 1. The relationship between the optimum moisture content, Wopt, and the corresponding degree of saturation, Sr, is shown in Fig. 2. The relationship between the volume of air, expressed as a percentage, and the maximum dry density, ρ_{dmax}, is shown in Fig. 3. The relationship between maximum dry density and the optimum moisture content is shown in Fig. 4.

5. The volume characteristics of these loam soils at the optimum moisture content are plotted on the graph of Vs and Vw as shown in Fig. 5. Linear regression analysis yields an average trend expressed as

332

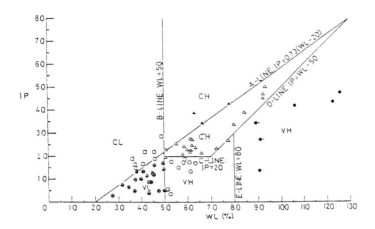

Fig. 1. Plasticity chart for loam soils

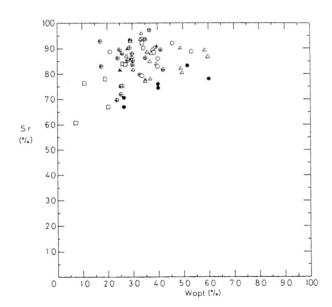

Fig. 2. Relationship between degree of saturation and
optimum moisture content

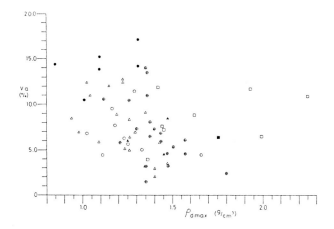

Fig. 3. Relationship between air porosity Va and
maximum dry density at optimum moisture condition

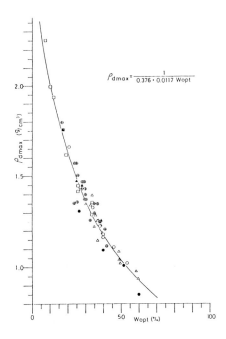

Fig. 4. Relationship between maximum dry density and
optimum moisture content

$$Vs + 0.822Vw = 0.893 \qquad\qquad (6)$$

Data from areas other than the Aomori Prefecture together
with the data in Fig. 5 are shown in Fig. 6. A revised
linear relation between the volume of solids and the volume
of water is given by

$$Vs + 0.830Vw = 0.851 \qquad\qquad (7)$$

From Figs 5 and 6 it can be seen that there is an approximate
unique relationship between the volume of solids and volume
of water at the optimum moisture content for a variety of
loam soils.

MECHANICAL PROPERTIES OF COMPACTED LOAM SOIL AND AIR VOID
    6. It is well known that different sample preparation
methods yield different compaction curves even though the
same compactive effort is used. In this investigation,
Takadate loam from this city of Hachinohe was tested.
Unconfined compression tests were used to obtain mechanical
properties of the compacted soil.
    7. The physical properties of the soil used for the
compaction test are as follows:

In-situ moisture content	: 40-50(%)
Liquid limit	: 31-39(%)
Plastic limit	: 15-21(%)
Specific gravity	: 2.69-2.74
Grading Gravel	: 2-3(%)
Sand	: 44-50(%)
Silt	: 40-48(%)
Clay	: 3-9(%)

    8. The soil was compacted using a 2.5 kg rammer in three
layers with twenty five blows per layer. The mould had
internal dimensions of: 10cm in diameter, 12.7cm in height
and 0.001 cubic meters in volume.
The sample preparation methods were as follows:

Method-A   The soil is air dried before each test and the
            moisture content is increased. The same soil is
            used throughout the series of tests.
Method-B   The soil is air dried before each test. The soil
            is not reused.
Method-C   The soil is reused for each test but it is not
            air dried. The moisture content is reduced
            starting at the in-situ condition
Method-D   New material is used for each test.

    9. The samples used for the unconfined compression tests
were formed by trimming the compacted soils after the
compaction tests were complete. The compacted samples were

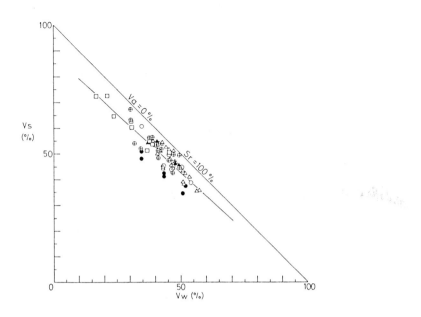

Fig. 5. Optimum moisture condition on Vs vs Vw diagram
for loam soils in Aomori Prefecture, Japan

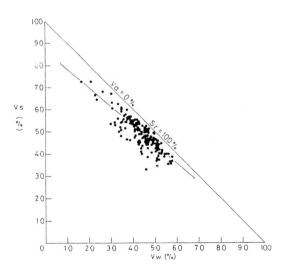

Fig. 6. Additional data to Fig. 5

not always uniform because the compacting operation tends to produce an inhomogeneous material. Therefore, the author has had to be selective in choosing only the data of relatively uniform samples. Samples that could not be trimmed were rejected.

10. The results observed from the unconfined compression tests were expressed in terms of the parameters q_u, ϵ_f and E_{50}

where

q_u : unconfined compressive strength
ϵ_f : axial strain at failure (%)
E_{50} : modulus of deformation defined as the secant modulus at $q_u/2$

The results of the compression tests on samples prepared using Methods-A and -D are shown in Figs. 7, 8 and 9 in which the mechanical properties are plotted against the moisture content. These figures indicate that soils compacted by Method-D are stronger, stiffer and more brittle than soils compacted by Method-A for the same moisture content. Method D is more representative of the conditions in situ, that is during placing of a fill. It is the recommended procedure for evaluating the mechanical properties of loam from the Hachinohe district.

11. Fig. 10, 11 and 12 show the same data from the compression tests plotted against the air void ratio, e_a, which is defined by Eq.(4) and can be written as

$e_a = e(1-Sr/100)$

e: void ratio   (8)

Use of this air void ratio, e_a, in place of the moisture content becomes very effective, as can be seen in Figs. 10, 11 and 12 since there is a unique line irrespective of the method of sample preparation. Test data presented in this section has been reported in the author's paper (ref. 1).

CONCLUDING REMARKS

12. The data for loam soils shows that:

(a) The volume of air within a sample is a simple function of the optimum moisture content.
(b) The air void ratio is uniquely related to the mechanical properties obtained from unconfined compression tests on loam soil even if the samples are prepared by different methods.

13. In conclusion, the air void ratio rather than the water content should be used when comparing data for loam soils.

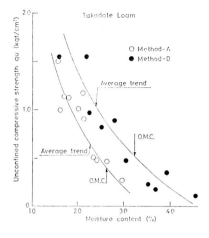

Fig. 7. Unconfined compressive strength and moisture content obtained with Methods-A and D for Takadate loam

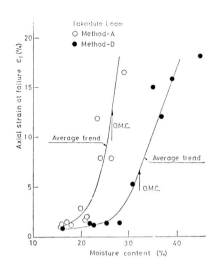

Fig. 8. Axial strain at failure and moisture content obtained with Methods-A and D for Takadate loam

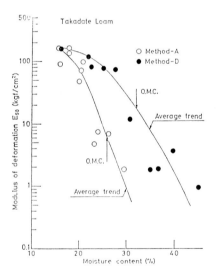

Fig. 9. Modulus of deformation and moisture content obtained with Methods-A and D for Takadate loam

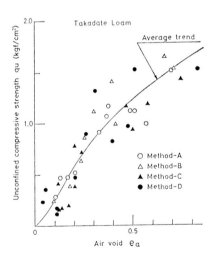

Fig. 10. Unconfined compressive strength and air void obtained with various sample preparation methods for Takadate loam

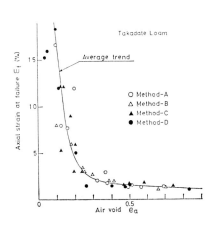

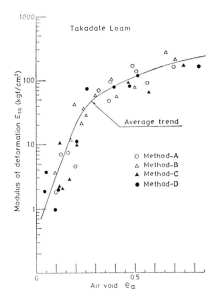

Fig. 11. Axial strain at
failure and air void
obtained with various
sample preparation
methods for Takadate loam

Fig. 12. Modulus of
deformation and air void
obtained with various
sample preparation methods
for Takadate loam

REFERENCES
1. MOROTO N. Mechanical behaviour of two typical compacted
volcanic soils in Hachinohe, Japan under different sample
preparation methods. Soils and Foundations, Vol. 31, No. 2,
1991, pp. 108-116.

27. Volcanic ash treated with foam and cement for lightweight filling materials

J. NISHIKAWA, Y. MATSUDA and N. MIHARA, Hokkaido Development Bureau, Japan, and M. KUWABARA, M. MURATA and F. KUSAKABE, Fudo Construction Co. Ltd, Japan

SYNOPSIS. Foam mixed solidified soil, which is a mixture of volcanic ash, water, cement and a foaming agent, has been developed as a foam of light weight fill. This soil attains a typical 7-day strength of 1.5 kgf/cm² with a cement quantity of 85-110 kg/m³. The density of the foam mixed solidified soil may be adjusted by varying the quantity of foam used in the unit. In experiments conducted in a concrete pit, it has been shown that target densities can be achieved reliably. Lightness and a relief in horizontal soil pressure are two major features of the foam mixed fill material. Horizontal soil pressure of this soil became almost zero after consolidation. The material can be used as an embankment material or as fill materials.

INTRODUCTION
1. The development of the road network in Japan requires major road widening work using retaining walls, fill and bridge construction work in the numerous valleys which is a feature of the Japanese topography. Technical and economic benefits can be achieved if the horizontal soil pressures applied to retaining walls and abutments, and the vertical pressures applied to embankment culverts together with the confining pressures applied to underground installations such as pipe utility conduits can be reduced. One way of reducing soil pressure is to lighten and solidify the fill materials used and examples of this material are expanded polystyrene (EPS) and foam cement.
2. Disposal of spoil materials generated by construction has become a problem in terms of environmental preservation, a potential method of solving this problem is to transform them into lightweight and solidified materials.
3. Laboratory experiments on lightweight fill materials have been conducted using sand, volcanic ash and fly ash (Noto et al., 1992) (ref. 1). Further experiments on a lightweight fill material, formed from a mixture of volcanic ash found throughout Hokkaido, water, foam and cement are reported in this paper. Model tests, using this soil, are discussed.

Engineered fills. Thomas Telford, London, 1993

LABORATORY EXPERIMENTS

4.   Foam mixed solidified soil is a mixture of volcanic ash, water, cement and foaming agent.   It is light, strong and workable.   Material lightness is thought to depend primarily on foam quantity, the strength and quantity of cement and the fluidity of the water.   In order to establish these relationships a number of experiments have been conducted.

5.   The volcanic ash used was Shikotsu pyroclastic fall deposit which is found around Sapporo, Hokkaido.   The grade composition, specific gravity and maximum dry density of this ash are given in Table 1.   Agents used to generate the foam were surface active agents.   The volcanic ash was put through

Table 1.   Physical properties of volcanic ash used in the experiment

specific gravity (particle)		2.402
max. compaction density		1.603 g/cm³
grain size distribution	gravel	17.5%
	sand	50.2%
	silt	26.0%
	clay	6.3%

a 5mm sieve prior to use.   Preliminary experiments showed that bleeding does not occur when the adjusted moisture ratio of the mixed materials is slightly below 100%.   The basic characteristics of the foam mixed solidified soil specimens used in the experiments are shown in Tables 2 - 4.

Table 2.   Mix proportion (1) relationship between adjusted moisture ratio and strength

adjusted moisture ratio (%)	water-cement ratio	cement quantity (kg/m³)	remarks
50	3.7	83	target density
	3.8	80 *	= 1.0 g/cm³
55	4.1	80 *	initial moisture
60	3.7	92	ratio = 31.0%
	4.3	80 *	* 27.7%
70	3.7	100	
	3.7	100 *	● not prepared
85	3.7	110	sample
100	3.7	119	because of
120	3.7	120 ●	bleeding

Table 3. Mix proportion (2) relationship between density and strength

target density (g/cm³)	cement (kg/m³)	foam (%)	remarks
1.60	160	0	water-cement ratio = 3.7
1.20	120	28.1	adjusted moisture ratio = 70%
1.00	100	40.1	
0.80	80	52.1	initial moisture ratio of volcanic ash = 31.0%
0.60	60	64.1	

Table 4. Mix proportion (3) relationship between cement quantity and strength

cement (kg/m³)	adjusted moisture ratio(%)	water-cement ratio	remarks
60	50	5.2	target density =1.0g/cm³
	60	5.9	initial moisture
		5.9 *	ratio of volcanic
	70	6.5	ash = 27.7%
	50	3.8	(* 47.4%)
	55	4.1	
	60	4.3	
		4.3 *	
	70	4.7	
100	55	3.4	
	60	3.4	
		3.4 *	
	70	3.7	
120	55	2.6	
	60	2.8 *	
150	60	2.1 *	

Figure 1 shows the process of mixing foam solidified soil and the experiments used to classify the soil.

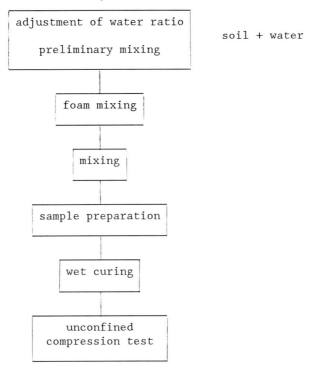

soil + water

Fig. 1.   Experimental flowchart

6.   Figure 2 shows the relationship between the unconfined comprehensive strength of the foam mixed solidified soil, and the adjusted moisture ratio at an age of 7 days.   The unconfined compressive strength reached a maximum value when the adjusted moisture ratio was between 55 and 60%.   Figure 3 shows the relationship between density and unconfined compressive strength of the foam mixed solidified soil when the quantity of foam varied and the water cement ratio and the adjusted moisture ratio were kept constant at values of 3.7 and 70% respectively.   The density of the soil is that of a slurry when materials were mixed.   As the foam quantity is decreased or the density increased, so the unconfined compressive strength increased.   As the water cement ratio is constant, the basic structure of solidified mixed soils (without foam) can be considered to be the same regardless of variations of density.   Variations of the unconfined compressive strength with different densities is thought to be a function of the quantity of foam.

7.   Reference to established standards shows that this foam mixed solidified soil used as a filling material requires a minimum unconfined compressive strength at 7 days of 1.5 kgf/cm², (Geotechnical Section of Civil Engineering Research Institute, 1985) (ref. 2).

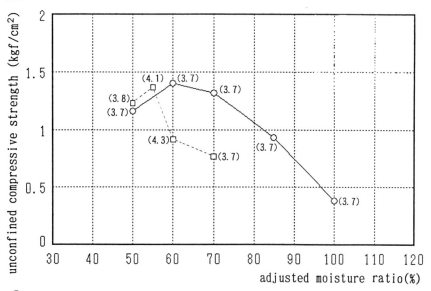

○:31.0% , □:27.7% initial moisuture ratio of volcanic ash
number in brackets = water-cement ratio
target density = 1.0g/cm³

Fig. 2. Relationship between adjusted moisture ratio & strength

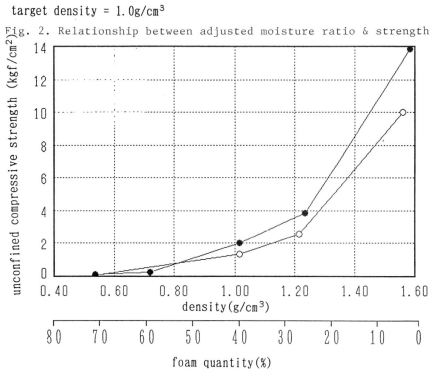

○:strength at 7days later, ●:strength at 28days
adjusted moisture ratio = 70%, water-cement ratio= 3.7

Fig. 3.   Relationship between density and strength

344

8. The relationship between the quantity of cement used and the unconfined compressive strength of the material is shown in Figure 4. In the case of volcanic ash with an initial moisture ratio of 27.7%, the unconfined compressive strength increased with an increase of cement quantity. The cement quantity which satisfies the required strength is 85-110 kg/m³ when the adjusted moisture ratio is 55-60%. When the initial moisture ratio of the material soil is 47.4%, the unconfined compressive strength tended to increase with an increase of cement quantity. However, the strength did not increase when the quantity of cement used exceeded 100 kg/m³. In addition to the cement quantity, the initial moisture ratio may affect the strength of the mix.

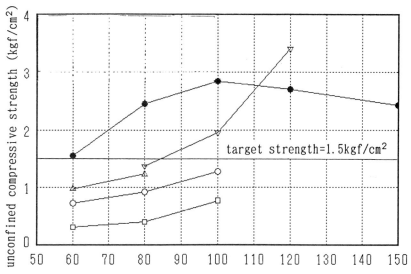

Wa:adjusted moisture ratio

cement quantity(kg/m³)

Wm:intial moisture ratio of volcanic ash

△:Wa=50%, ▽:Wa=55%, ○:Wa=60%, □:Wa=70% & Wm=27.7%

●:Wa=60% & Wm=47.4%   target density = 1.0g/cm³

Fig. 4. Relationship between cement quantity and strength (7 days)

9. The relationship between the adjusted moisture ratio and the flow values of materials with a density of 1.0 g/cm³ is shown in Figure 5. It can be seen that as the adjusted moisture ratio increases so the flow values increase. The rate of increase is larger in cases of smaller adjusted moisture ratios. Flow values differ with different moisture ratios and flow values are smaller in cases of larger initial moisture ratios. The study suggested that moisture which forms part of the initial moisture ratio, may not influence fluidity.

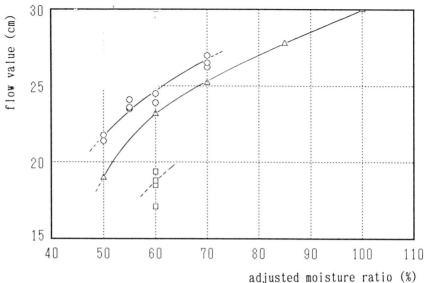

adjusted moisture ratio (%)

Wm:intial moisture ratio of volcanic ash
○:Wm=27.7%, △:Wm=31.0%, □:Wm=47.4%
target density = 1.0g/cm³

Fig. 5. Relationship between adjusted moisture ratio
and flow values

MODEL EXPERIMENTS
    10. Following the laboratory work, experiments were
carried out to investigate the relief of soil pressures and
the workability provided by the foam mixed solidified soil.
Figure 6 shows a flowchart of the experiments and Figure 7
shows the layout of the experimental pit and the
instrumentation. Three experiments were conducted; the first
case covered a sand embankment and case two and three covered
the placement of foam mixed solidified soil on model ground.
For the latter two cases, the expected density of the foam
mixed soil was 1.0 g/cm³ and 0.8 g/cm³ respectively. Both
adjusted moisture ratios were 60% and the cement quantities
were 100 kg/m³ and 130 kg/m³ respectively.
    11. The sand embankment and model ground on which the foam
mixed solidified soil was placed were fully rolled and their
compaction confirmed by density measurements and the Swedish
type penetration test.
    12. Before and during placement of the slurry, the density
and flow values were measured every 30 minutes and the slurry
was adjusted to maintain the target density. Placement of
the slurry was undertaken in two parts over two days. After
the slurry had been placed up to the planned height, its
surface was levelled. Soil pressures and temperatures were
measured from the start of placement for a period of 28 days.
At the end of 28-days further curing was conducted without

346

watering or covering with sheets, and a range of experiments
were undertaken.

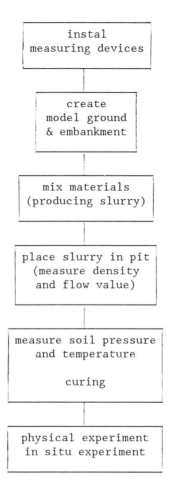

```
┌──────────────────────┐
│        instal        │
│   measuring devices  │
└──────────────────────┘
            │
┌──────────────────────┐
│        create        │
│     model ground     │
│      & embankment     │
└──────────────────────┘
            │
┌──────────────────────┐
│     mix materials     │
│   (producing slurry)  │
└──────────────────────┘
            │
┌──────────────────────┐
│   place slurry in pit │
│    (measure density   │
│     and flow value)   │
└──────────────────────┘
            │
┌──────────────────────┐
│  measure soil pressure│
│    and temperature    │
│                       │
│        curing         │
└──────────────────────┘
            │
┌──────────────────────┐
│  physical experiment  │
│   in situ experiment  │
└──────────────────────┘
```

Fig. 6.   Construction tests

PRODUCTION PLANT FOR FOAM MIXED SOLIDIFIED SOIL
    13.   The foam mixed solidified soil slurry was produced
using a 2-shaft paddle-type mixer, Photo 1.   The production
process was as follows;

(a) Volcanic ash was mixed with water to produce the
    planned adjusted moisture ratio, and cement and foam
    was added to the mixer.
(b) The slurry was removed from the mixer and its density
    and flow values measured.
(c) The slurry was pumped into the experiment pit

347

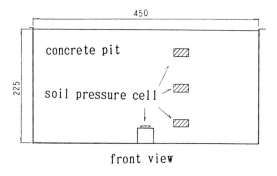

front view

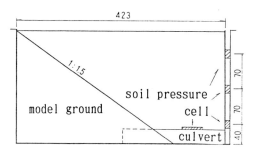

longitudinal section profile

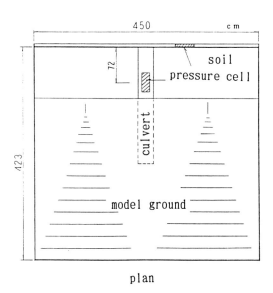

plan

Fig. 7. Layout of the experimental pit and measuring devices (Note dimensions in cm)

RESULTS OF EXPERIMENTS AND DISCUSSION

14. The results of the experiments are reported in this paper, these covered:
Case I (sand embankment) and Case II (foam mixing solidified soil with density of 1.0 g/cm³).

Photo 1. A 2-shaft paddle-type mixer used in this construction tests

15. Figure 8 shows density and moisture ratio of the embankment produced in Case I and model ground in Case II. Their densities ranged from 1.60 to 1.65 g/cm³ with an average of 1.63 g/cm³. The Swedish type penetration test shows that NSW is 10-40 at depths of over 20 cm. The embankment and model ground produced for these experiments were considered to be almost homogeneous and fully compressed.

16. Figure 9 shows the density and flow values of the slurry measured during its placement. The densities ranged between 0.86-1.11 g/cm³ with an average value being 0.997 g/cm³. The average density of the placed slurry was close to the initially planned value.

17. The range of flow value dispersion is larger than that of the variations of the densities. However, there were no problems during construction.

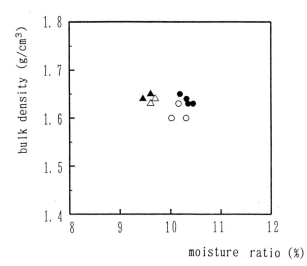

○:lower part of case I, ●:upper part of case I,
△:lower part of case II, ▲:upper part of case II

Fig. 8.　Bulk density and moisture ratio

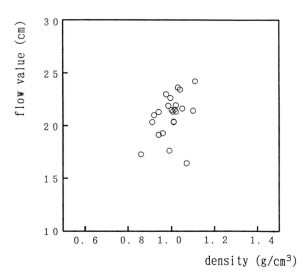

Fig. 9.　Density and flow values of slurry

(a) <u>Soil pressure by filling</u>.　Figure 10 shows the vertical
and horizontal soil pressures recorded during filling.　As
the embankment height increased the vertical and horizontal
soil pressures increase.　The vertical soil pressure on a
2.25m high embankment was close to the theoretical value of
0.293 kgf/cm². The soil pressure cells are thought to have
indicated almost exact values.　The horizontal soil pressure

350

was 30% higher than the theoretical value. It is considered
that the effects of rolling the soil around the soil pressure
cells affected the results.

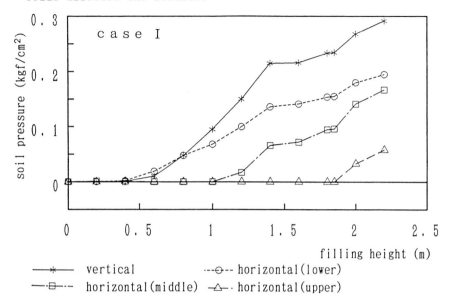

Fig. 10. Soil pressure by filling Case I

(b) <u>Soil pressure by foam mixed solidified soil</u>. Figures 11
and 12 show the vertical and horizontal soil pressures and
curing temperatures for the slurry used in Case II.

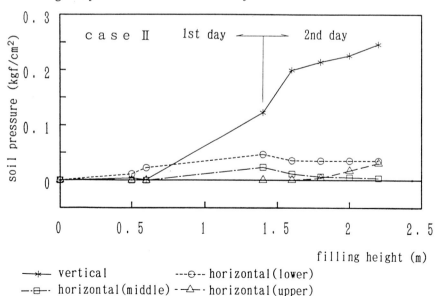

Fig. 11. Soil pressure by placing slurry in Case II

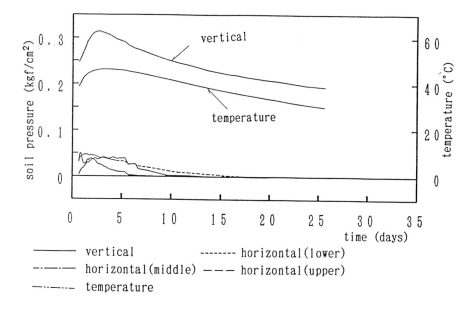

Fig. 12. Soil pressure and curing temperatures by
curing in Case II

Initially, the vertical and horizontal pressures increased as
the slurry level increased. However, the horizontal pressure
on the second day did not increase, even though the slurry
level increased. This phenomenon is considered to have been
caused by the slurry being initially in a state close to the
liquid state. After initial placement the foam mixed soil
started to solidify. The vertical soil pressure recorded at
28 days was 0.194 kgf/cm^2, which is close to the soil
pressure calculated from the density of the slurry, 0.18
kgf/cm^2. The tendency of the vertical soil pressure to vary
after slurry had been placed corresponds to the tendency of
the curing temperatures to vary. This is thought to occur in
the following manner;

   i. an increase in the curing temperature results in an
      expansion of the foam in the solidified soil, and the
      soil pressure temporarily increases with the
      development of solidification.

   ii. later, a decrease of the curing temperature results in
      the contraction of foam, and soil pressure decreases.
      Horizontal soil pressures became almost zero after 20
      days. This is considered to be a result of the
      solidified soil becoming self-supporting. After the
      construction was complete, the finished surface
      subsided about 2 mm.

(c) <u>Physical experiments</u>. Figure 13 and Table 5 show the strength and density of the specimens of the material at 7 days and 28 days. The specimens were wrapped in film and cured. Their strength exceeded the 7-day unconfined compressive strength of 1.5 kg/cm², and reached 3.3-5.0 kgf/cm² at 28 days. The increase inbetween the 7 day and the

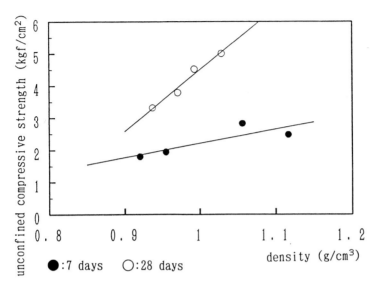

●:7 days   ○:28 days

Fig. 13.   Strength and density at 7 and 28 days

Table 5.   Average strength and density at 7 and 28 days

	7 days	28 days	28 days/7 days
density (g/cm³)	1.01	0.98	0.97
strength (kgf/cm²)	2.26	4.16	1.84

28 day results was approximately 80%. The average density was almost the same at 7 days and 28 days and remained close to the density of slurry. The results of other physical experiments are shown in Table 6. In Case II, the water absorption did not affect the strength and the foam mixed solidified soil had an acceptable permeability. The CBR and frost heave experiments produced good results.

(d) <u>In situ experiments</u>. The results of in situ experiments are shown in Table 7. The density of specimens extracted from the pit was equal to the planned values and the density of slurry did not vary. It can be conoluded that there was no decrease in foam quantity following placement. The strengths of specimens extracted from the pit were lower than that of the specimens produced in the experiments, and this is thought to be the result of different curing conditions. The CBR values measured in situ were larger than the values in the experiments. The thermal conductivity measured was close to that of sand.

Table 6. Results of physical experiments

test		results			
unconfined compressive (water absorption)	strength (kgf/cm²)	3.58(3.32) 4.40(4.52)	density (g/cm³)	1.06(0.94) 1.07(0.99)	
permeability	coefficient of permeability (cm/s)	1.74×10^{-4} 2.44×10^{-4}	density (g/cm³)	1.29(1.07) 1.19(0.98)	
C.B.R.	C.B.R. (%)	10.7			
frost heave	frost heave rate (%)	2.3			

*figure in brackets shows data before absorption or permeability test

Table 7. Results of in situ experiments

		lower part	upper part	average
unconfined compressive strength of in situ samples	kgf/cm² (28 days)	2.52 3.74 3.58	3.62 2.77 3.21	3.240
	density (g/cm³)	0.90 0.97 1.02	1.02 1.00 1.09	0.998
thermal conductivity	(W/m·K)	0.37 0.36		
in situ C.B.R.	C.B.R.(%)	24.0 30.0		

CONCLUSIONS

18. The study of the use of volcanic ash to produce a lightweight foamed fill material produced the following conclusions:

(a) Volcanic ash was used in these experiments. The foam mixed solidified soil, formed from a mixture of volcanic ash, water, foam and cement attained a 7-day strength, of 1.5 kgf/cm^2 with a cement quantity of 85-110 kg/m^3. The unconfined comprehensive strength reached a maximum when the adjusted moisture ratio was 55-60%.

(b) In the construction experiments, the target density of the slurry was obtained and was seen to be stable.

(c) The density of the foam mixed solidified soil 28 days after slurry placement was the same as that of the initial slurry.

(d) The vertical soil pressures measured on the base of the experimental pit agreed with the soil pressures calculated from the material density. The horizontal soil pressures decreased with solidification and eventually became almost zero.

(e) Foam mixed solidified soil can be used as an embankment material and as fill. Foam mixed solidified soil will relieve soil pressures because of its lightness, and steep embankments can be constructed with the material because it is highly self-supporting.

REFERENCES
1. NOTO S., KUSAKABE H., YAMAGUCHI H. and MURATA M. Strength of foam soil cement using local soil. Proc., 47th Ann. Conf. JSCE., 3, 1992, pp. 970-971. (in Japanese)
2. Geotechnical Section of Civil Engineering Research Institute. Countermeasure manual for problem soils in Hokkaido. 142p. 1985 (in Japanese)

28. Engineered fills on dolomitic terrain in South Africa

F. VON M. WAGENER and P. W. DAY, Jones and Wagener, South Africa

SYNOPSIS.   Structural distress and foundation problems have occurred extensively on dolomitic rock in South Africa. Apart from the danger of sinkholes, problems arise due to non-uniformity of the in situ materials. Engineered fills, or mattresses, constructed of selected granular materials have been used extensively on the dolomite below structures and other development. The paper reviews the types of mattresses used, material selection, design, methods of construction and advantages over other founding methods on this geological environment.

INTRODUCTION

1.   Damage to structures and loss of life has been far more severe on dolomite than on any other geological formation in South Africa. Apart from the danger of sinkholes and subsidence, problems arise due to the non-uniformity of the in situ material particularly where pinnacles and boulders occur at shallow depth.

2.   Dolomitic rocks cover 14 percent of the industrialised Witwatersrand area around Johannesburg in the Transvaal province of South Africa.   The demand for land is leading to ever increasing residential and industrialised development on these subsidence-prone areas.

3.   A number of foundation solutions (ref. 1) can be applied for a site on dolomite of which engineered fills or "mattresses" constructed of selected granular material have been used extensively. This paper reviews the typical profile on dolomite, the types of mattresses used, material selection, design and methods of construction. In addition the advantages of the engineered mattress over other founding methods on this geological environment are discussed.

TYPICAL PROFILE ON DOLOMITE

4.   Dolomite consists of the mineral Calcium/Magnesium carbonate ($CaCO_3$ $MgCO_3$). The rock contains impurities in the form of chert (SiO_2), manganese oxide (MnO_2), iron oxides (Fe_2O_3) and others. Solid dolomite is impervious and relatively insoluble in pure water. However, where the rock is jointed and accessible to water charged with carbon dioxide, solution of the dolomite takes place.

5.   The movement of water through a dolomite mass is illustrated in Figure 1. Erosion and solution by percolating water create voids or cavities within the rockmass. These voids form receptacles which may become filled or partially filled with residuum derived from the insoluble impurities within the

rock. The extent of the voids (or cavities) and the type of residuum has a marked effect on the engineering behaviour of the weathered rockmass.

6.   The typical profile on dolomitic rocks as it occurs in large parts of the Transvaal is shown on Figure 2 (ref. 1). The site is often covered by a mantle of transported material which is usually underlain by a thin pebble marker before residual material is encountered. In a chert rich dolomite the residuum usually grades from a coarse angular chert gravel to the finer insoluble components of dolomite namely wad (MnO_2) and clay. This is followed by the highly irregular dolomite rockhead consisting of pinnacles and boulders. Cavities may be found in the residuum between the pinnacles with solution caverns in the upper layers of "solid" dolomite. The thickness of cover over pinnacles and boulders can vary from zero to more than 100 m. Intrusives such as syenite dykes and sills are also encountered.

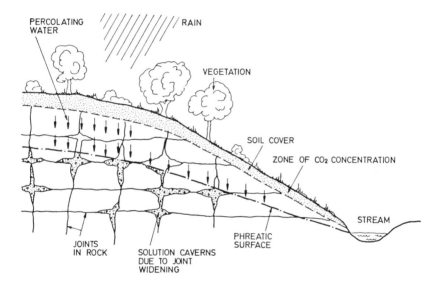

Fig. 1 Movement of water through soluble rocks.

7.   On a dolomite site, the consistency or in situ strength of the material usually reduces with depth from relatively competent chert gravel to compressible, lightweight wad before it increases sharply as the "solid" rock is encountered. This is contrary to the situation on most geological formations encountered in the South African context where the consistency improves with depth. In addition the overburden varies markedly in thickness and properties over short distances. These aspects make dolomitic terrains difficult to investigate and develop by conventional means (ref. 2).

CLASSIFICATION OF SITE AND SELECTION OF FOUNDING METHOD
8.   The geotechnical investigation of a dolomite site concentrates, in the first place, on defining the risk of the development of sinkholes and subsidence (ref. 3). Should the risk be found to be acceptable, and the site is declared fit for development, it is classified and a founding method can be selected

9.   The site can be classified according to the average thickness, c, of overburden above the top of pinnacles and boulders as follows:

357

SPECIFICATIONS AND MATERIALS

Class A : Pinnacle and boulder dolomite at or near surface
c < 3m

Class B : Pinnacle and boulder dolomite overlain by moderately thick
overburden
3m < c < 15m

Class C : Pinnacle and boulder dolomite overlain by thick
overburden
c > 15m

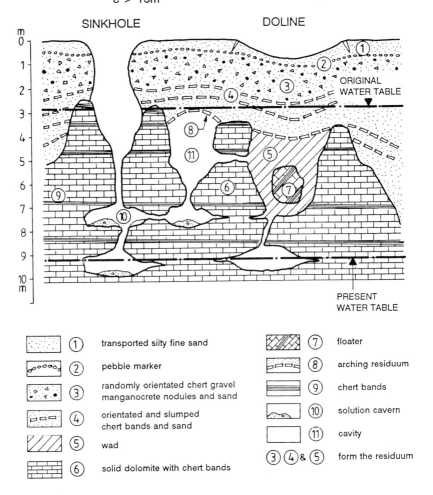

Fig. 2 Typical profile on dolomite

10.   The founding method depends on the type of structure and thickness
and type of residuum over "solid" dolomite. Piles cannot be used on a Class
A site where dolomite pinnacles and boulders are close to the surface
(Figure 2). Furthermore, the use of conventional strip footings could lead to
severe differential settlement and structural distress where loose material is
found between outcropping pinnacles.

11.   As mentioned above, the consistency of the soil cover on a dolomite
site, in addition to being extremely variable within short distances and difficult

to assess by conventional means, often deteriorates with depth. It has been found that a mattress of engineered fill can often be used for a structural foundation with success on such a site.

ENGINEERED MATTRESS

12.   If a certain thickness of soil cover is removed and replaced under controlled conditions, a mattress of known thickness and known strength parameters is formed.

13.   The functions of a mattress on a dolomite site are:
- To control total and differential settlement.
- To reduce the bearing pressures to an acceptable limits at the level of the underside of the mattress.
- To reduce the risk of the formation of small sinkholes and dolines.

14.   The thickness of the mattress will depend on a number of factors, the most important of these being the thickness and properties of soil cover and the sensitivity of the proposed structure to settlement.

15.   The material available on a site for mattress construction depends largely on whether the dolomite is chert-rich or chert-poor. Chert gravel, mixed with the transported soil cover, is often an excellent construction material as illustrated in Figure 3. Typical grading and strength curves for chert gravel are given in Figures 5 and 6. Residuum in the form of wad, which usually has a clayey silt grading, is unsuitable as a fill.

Fig. 3   Chert gravel for use in mattresses.

Fig. 4. Dumprock in mattress construction.

16.   Should the dolomite be chert-poor, selected granular material has to be imported. Fortunately, some of the dolomite areas are located in a mining environment where dumprock is usually available and is often successfully used for the construction of mattresses (Figure 4). This material is a waste product from mining and consists of angular fragments of very hard rock quartzite ranging in size from sand to 300mm.

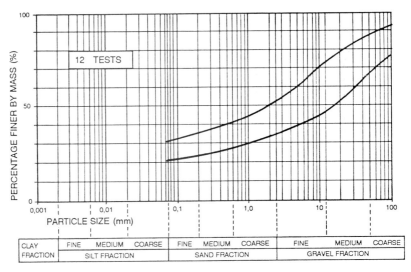

Fig. 5   Grading envelope for Swartkops chert gravel.

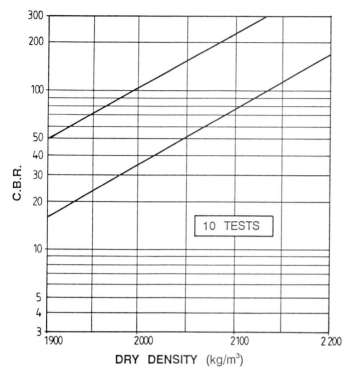

Fig. 6   Strength/Density envelope for Vaal Reefs chert gravel.

METHOD OF CONSTRUCTION

17.   On a Class A site with shallow pinnacles and boulders where chert gravel and dumprock is available as backfill, the method of construction would be as follows (Refer Figure 7):

- Remove material to a depth of (say) 1m below tops of pinnacles and large boulders. It may be necessary to blast off the tops of a few pinnacles which project above the general pinnacle level.
- Inspect the bottom of the excavation and order additional excavation where pinnacles are spaced further apart or where loose material is present. Level the bottom of the excavation, moisten and compact with a small vibratory roller or by hand.
- Place selected waste dumprock into the excavation to about 200mm higher than the tops of pinnacles. Compact by means of a vibratory roller with a static mass of at least 8 tons. Use ample water during the compaction process but avoid uncontrolled water ingress into the underlying dolomite. Ten passes of the roller are usually sufficient.
- The remainder of the terrace is built up to the required elevation by using selected chert gravel or other suitable material placed under controlled conditions. Chert gravel is usually compacted to 95% of modified AASHTO maximum dry density at ± 2% of optimum moisture content. Layer thickness is generally 200 - 250mm.
- The finished terrace should be above natural ground level to promote good drainage.

361

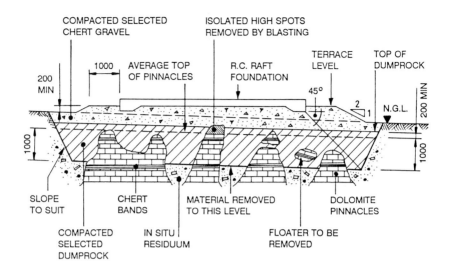

Fig. 7   Mattress on a Class A site.

18.   Where heavy settlement sensitive structures such as bridges, headgears, mine winders or compressors have to be founded on a Class A site, mass concrete could replace the dumprock between pinnacles. In this case it may be necessary to distinguish between pinnacles and loose boulders (known as floaters) and have the floaters removed.

19.   For light structures such as houses, the thickness of mattress can be reduced considerably. In this case the mattress is usually placed entirely above the pinnacles and boulders and the house built on a light raft with thickened edge beams.

20.   A cross section through a typical light raft on a Class A site is shown in Figure 7.

21.   The treatment of a site with a thick cover over pinnacles and boulders ie. Class B and C, is shown in Figure 8. Backfill could consist of selected chert gravel placed under controlled conditions. Dumprock could also be used provided it is capped by a substantial layer (1,0m or more) of the less permeable chert gravel to limit water ingress into the underlying dolomite.

22.   The foundations of structures founded on the mattress should be as shallow as possible. There is no point in providing a mattress and then excavating through it to found the structure. For this reason the raft type foundation shown in Figures 7 and 8 is often used. Not only is this foundation founded high in the terrace but it also distributes the applied loads as widely as possible.

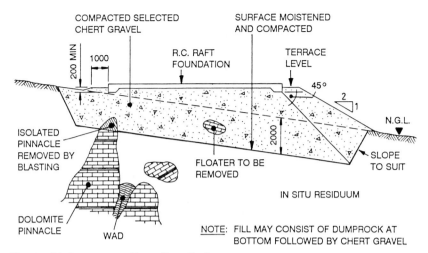

Fig. 8   Mattress on a Class B or C site.

DESIGN OF MATTRESS

23.   A mattress will be the most effective where its thickness is greater than or comparable to the size of the footings to be built on it, or where the depth of cover over the tops of pinnacles is small (ref. 4).

24.   The effectiveness of the mattress in reducing both total and differential settlements is shown in Figure 9.   The plot of settlement versus mattress thickness shows settlement to reduce sharply as the mattress thickness increases to 2m. Thereafter further increase in thickness has little effect. The principal compressive stress vectors below the foundation on a 2m thick mattress are plotted on Figure 10. The plot shows how the mattress tends to shed the load into the pinnacles by arching.

25.   Figure 11 shows the settlement analysis of a similar footing on a Class B site, with 10m cover over rock. In this case there is a more gradual reduction in settlement with increased mattress thickness.

26.   If one assumes an elastic (Boussinesq) stress distribution within the mattress, only 10% percent of the bearing pressure reaches the base of a mattress for a square footing with sides half the mattress thickness. A mattress thus spreads the load very effectively over weaker underlying layers.

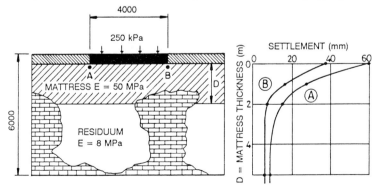

Fig. 9   Finite element settlement predictions for spread footings on pinnacle dolomite.

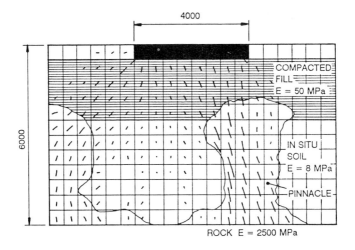

Fig. 10   Plot of principal compressive stress vectors for a 2,0 m thick mattress.

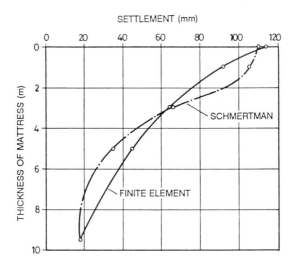

Fig. 11   Settlement of a 4,0 m wide strip footing on a thick cover over dolomite.

ADVANTAGES AND DISADVANTAGES OF A MATTRESS

27.   The main advantages of a mattress over other methods of founding, such as piles, caissons, strip footings etc., on a dolomite site are:
*   If suitable fill material is available, it is highly cost effective.
*   It exposes the whole construction area enabling an in situ inspection of the founding material. Problem spots such as paleo-sinkholes and slumped chert bands, which could indicate a potential subsidence problem, can be identified and addressed before backfilling operations commence.

- A mattress spreads the load fairly uniformly over the underlying material.
- For certain combinations of footing size and mattress thickness the bearing pressure on the underlying material is reduced considerably.
- The mattress will form a relatively strong roof over any cavities which may form below the structure. By reinforcing the mattress with, say, metal strips additional flexural strength may be obtained.

28. Disadvantages of a mattress construction are:

- The reduction of bearing pressure and hence settlement for large footings on mattresses on Class B & C sites is not very effective.
- A mattress becomes more expensive if suitable material has to be imported.
- The mattress solution is most effective for isolated footings. Closely spaced footings or large distributed loads, result in reduced efficiency of the mattress.

GENERAL CONSIDERATIONS WHEN USING MATTRESSES

29. The following aspects should be borne in mind when using engineered mattresses on dolomitic terrain.

- The foundations of the structure should be as shallow as possible and raft foundations are often preferred (see Figures 7 and 8).
- Where possible, dumprock should be used between pinnacles because of the ease with which it can be compacted in thick layers. Other materials such as stabilised sand or chert gravel can also be used provided the compaction process is modified accordingly.
- The chert gravel is usually compacted to a density of 95% Mod. AASHTO maximum dry density at optimum moisture content ± 2%.
- The bearing pressures below foundations on compacted chert gravel are generally limited to 200 - 250kPa. Bearing pressures of 300 - 400kPa may be used on compacted dumprock.
- If the chert gravel is too coarse to permit compaction control by sand replacement tests, or where dumprock is used, compaction is controlled by means of a method specification.
- The greatest benefits of a mattress are achieved where the depth to pinnacles is not great, i.e. on a Class A.
- On Class B and C sites a mattress can offer considerable benefits when the footings are small and widely spaced compared to the thickness of the mattress.
- In order to limit water ingress and its detrimental effects on the stability of the dolomite, a dumprock mattress should not extend to the surface but should be topped with chert gravel. The chert gravel not only provides a relatively impermeable cover over the dumprock but also makes it easy to cut foundation or service trenches into the mattress.
- The excavation for services which are placed into a completed mattress should be kept as shallow as possible and the material should be reinstated to the same density as the mattress after installation of the services.
- The top of the mattress should be slightly higher than the surrounding ground to prevent the ponding of water.

- The mattress should be constructed by an experienced civil engineering earthworks contractor using suitable equipment and control.

REFERENCES

1. WAGENER F. von M. Engineering Construction on Dolomite. PhD thesis, University of Natal. Published by Geo. Div. SAICE Johannesburg, 1983.
2. WAGENER F. von M. Dolomites - Problem Soils in South Africa - State of the Art. The Civil Eng. in S.A., Vol. 27, July 1985.
3. DAY P.W. and WAGENER F. von M. Investigation techniques on dolomites in South Africa. Sinkholes: Their geology, engineering and environmental impact. Edited by Barry F. Beck, University of Central Florida, Orlando, 1984.   .
4. DAY P.W. Dumprock and chert gravel mattresses. Seminar on engineering geology of dolomite areas. Dept. of Geology, University of Pretoria. 1981.

29. A classification of pulverised fuel ash as an engineered fill

Y. YANG, B. G. CLARKE and C. J. F. P. JONES, University of
Newcastle upon Tyne, UK

SYNOPSIS. Pulverised fuel ash is recognised as either selected cohesive fill or
general fill. This is based on the age of the ash since selected cohesive fill can only be
fresh ash. Triaxial tests show that all ashes exhibit cohesive properties but only
certain ashes exhibit a permanent increase in strength. These include fresh and
stockpiled ash. Evidence from microscopic work confirms that chemical and physical
permanent changes do take place. A modified classification is suggested based on the
worst credible effective strength parameters determined at one and twenty eight days.

INTRODUCTION
    1. Pulverised fuel ash (pfa) is formed of silt sized particles which are carried from
the furnace of a coal fired power station with the flue gases. A significant quantity of
pfa is used in the cement industry but the majority is either stored or used for land
reclamation projects, or general and structural fills. In this context structural fills
include embankment fills and fills behind retaining structures.
    2. The early use of pfa as an engineered fill material was pioneered by Local
Government agencies who developed their own specifications and expertise during the
rapid expansion of highway construction in the 60's. In 1966, the then Transport and
Road Research Laboratory began a research programme to investigate the properties
of pfa and methods that could be used to compact the material. The UK Department
of Transport (DoT) first accepted pfa for earthworks in 1966. Design was based on
total stress parameters. The current DoT Specification for Road and Bridgeworks,
published in 1986, permits the use of effective strength parameters for design.
    3. Specifications prior to 1986 were based on the assumption that, if the material
was workable, then the desired density could be achieved which implied that the design
strength was mobilised. This was acceptable since it was known that the strength of
the ash would increase with time leading to an increase in the factor of safety provided
the ash was protected from inundation.
    4. The latest specification (ref 1)is more rigorous because of the concern of a loss
in strength with inundation. It is based on the premise that ash is a consistent material
and therefore the target densities can easily be achieved. End product control is used
to ensure that these requirements are met. It is often difficult to comply with this
specification, yet experience has shown that design strengths are easily achieved even
for poorly compacted material. As a result, a research programme has been instigated
to investigate the effects of compaction and time on the properties of ash. One

conclusion reached from this programme is that the existing classification may not be strictly correct.

SOURCE OF PFA AND THE CURRENT CLASSIFICATION

5. Two ashes are produced during the burning of coal; pulverised fuel ash (pfa) and furnace bottom ash (fba). Pulverised fuel ash, which forms about 80% of the UK ash produced, is defined as the material extracted from the flue gases of a furnace fired with bituminous coal. It is the most common ash in use. After extraction it is known as fresh pfa. Fresh pfa can be conditioned with water at about the optimum to produce conditioned ash (referred to as conditioned or fresh ash). This is either used directly as fill or stored. Conditioning pfa allows it to be transported more easily and reduces the dust problem. Ash stored as conditioned ash is known as stockpiled ash. Some ashes are mixed with water to form a slurry and are pumped into lagoons to be stored as lagoon ash. This ash can later be retrieved as stockpiled ash.

6. There are two classes in the DoT Specification which apply to pfa; selected cohesive fill (7B) for structural purposes and normal cohesive fill (2E) for general fill. Conditioned fresh ash can be used as structural and general fill, stockpiled ash can only be used as general fill. This suggests that conditioned ash is better than stockpiled ash.

7. The properties of pfa in its fresh state are dependent on the type of coal, the source of the coal, the degree of pulverisation, the efficiency of combustion, the variation in firing conditions and the extraction process. Thus, it may be expected that fresh pfa will vary from source to source and from one source (ref 2). Further changes take place with time depending in the water content, the fineness of the particles and the chemical composition of the ash.

8. Pfa is a non plastic, predominantly silt sized material. The specific gravity of pfa particles varies from 1.90 to 2.72 depending on its source. Many ashes exhibit cohesive properties hence the DoT classification. There is evidence that some ashes show an increase in strength with time (ref 3). There is a concern that the cohesive properties of the ashes are due to suction forces created during compaction and that these may dissipate with time and inundation leading to loss of strength. A DoT Memorandum BE 3/78 (revised 1987) covering the use of pfa as a structural fill in reinforced soil, limited the cohesion intercept to 5 kN/m^2.

THE STRENGTH OF PFA

9. The UK specified procedure to determine the strength of ash is based on the 300 mm direct shear box test. This is used to test representative samples since pfa, especially stockpiled ash, contains lumps of pfa formed by self-hardening. It is assumed that the larger specimen gives a lower, more representative design value of strength. These large shear boxes are not commonly available. Clarke et al (1993) showed that tests using a 95 mm direct shear box produced more conservative results than those from a 300 mm shear box. The results from direct shear tests on pfa are very dependent on sample preparation (ref 4). Further, it is difficult to guarantee full saturation which represents the worst conditions following inundation.

10. Clarke et al (1993) demonstrated that, for practical purposes, triaxial results give similar results to those from the 95 mm shear box test. The results are less than those obtained from the large shear box, which implies that they are more conservative in respect of design values. Triaxial tests have the advantages that drainage and saturation can be controlled.

11. Pfa could be classed as a free draining material since its permeability ranges between 10^{-7} and 10^{-3} cm/sec. Pfa is placed at about the optimum water content. The detailed design of a fill includes drainage systems to help prevent possible saturation of pfa fill. It is expected that a pfa fill would remain partially saturated. However, there may be instances when in situ pfa becomes inundated. Thus, drained triaxial tests on saturated pfa, with the saturation taking place after compaction, may be more representative of the worst case. These tests would give the worst credible effective strength parameters suitable for design. Tests on partially saturated pfa would give the most likely parameters for design provided the design included adequate provision to prevent inundation.

THE TESTING PROGRAMME AND PROCEDURE

12. A series of consolidated drained triaxial tests were carried out on fresh and stockpiled ashes from Tilbury, Blyth and Fiddlers Ferry power stations. Identical specimens were tested at varying ages from one day to one year. The specimens were statically compacted to the maximum dry density and at 95% of the maximum both on the wet and dry side. These are referred to as 95%D, 100% and 95%W respectively. Static compaction was used since it was possible to guarantee the required density. The dry density was chosen from a compaction curve produced using the 2.5 kg test. A single delivery of ash was used in each case because of the variations noted above. Specimens of 38 mm diameter were stored at $20^{\circ}C$ and 40% RH in plastic liners and sealed with paraffin wax. It is critical during storage to ensure that there is no loss of water due to evaporation.

13. Partially and fully saturated specimens of each ash were tested. The specimens were saturated immediately prior to testing. Saturation was carried out in two stages maintaining the effective stress at about 10 kN/m^2. During the first stage, a back pressure of 50 kN/m^2 was applied at the base of the specimen and maintained until water flowed out of the top of the specimen. This pushed some of the air out of the specimen. During the second stage, any remaining air was dissolved by increasing the back pressure in increments of 50 kN/m^2 up to 250 kN/m^2. The cell pressure was then increased to the consolidation pressure and the change in pore pressure measured. Effective isotropic pressures of 0, 25, 50, 100 and 150 kN/m^2 were used representing a maximum height of embankment of ten metres.

14. Consolidation took place in under ten minutes no matter the age or source of the specimen although it was usual to isotropically consolidate the specimen for twenty four hours. Typically, the B value after saturation was less then 85%. For some ashes, the B values reduced with the age of the specimen suggesting that some change was taking place resulting in a stiffer material. B values for Tilbury and Blyth conditioned ashes were about 30% after six months.

15. The majority of tests showed that all the ashes exhibited peak and post rupture strength with the peak strength occurring at under 5% axial strain. In the case of the Tilbury and Blyth conditioned ashes the axial strain at peak stress reduced with the age of the specimen.

16. Fig 1 shows that there is an increase in peak strength with age for partially saturated ash compacted at the optimum water content. For example, for a consolidation pressure of 150 kN/m^2, the compression strength increased from 300 kN/m^2 (point A) at one day to 950 kN/m^2 (point B) at six months.

17. There is a trend toward an increasing φ' and c' with age (Fig 1). Fig 2 simply shows the change in φ' with age for an ash prepared at three water contents. There is scatter in the data making it difficult to clearly define the change in φ'.

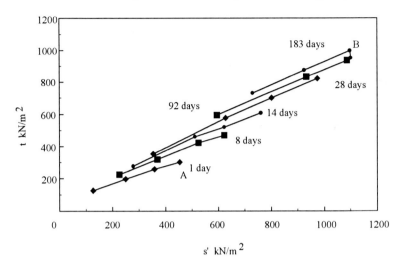

Fig 1   Increase in compression strength with time for a partially saturated ash (Tilbury fresh pfa)

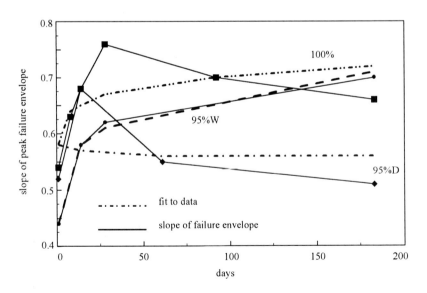

Fig 2   The variation in peak angle of friction with time (Tilbury fresh pfa)

18. Evidence from electron microscopic studies (ref 5) have shown that cementitious actions take place which could account for the increase in cohesion. Cohesion is made of two components; cementitious action between particles and

370

suction. The fluctuation in the peak strength parameters may be a result of age, density, different amount of pozzolanic action within a set of specimens and, for partially saturated specimens, a variation in the suction within a set of specimens. It is anticipated that suction pressures will change with time to reach an equilibrium condition. In general, it was found that the variation in c' with age was much greater than the variation in ϕ' but since they are interrelated a reduction in c' could be accompanied by an increase in ϕ'. It would require a significant number of tests on specimens of the same age prepared at the same water content and subject to the same consolidation pressure to determine the change in effective strength parameters with time.

19. The electron microscopic studies also show that the particles of fresh pfa change shape which would lead to an increase in the angle of friction. It is likely that this change is due to the presence of water only and therefore the change with age is more consistent.

20. The variation in ϕ' with time is given by:-

$$\phi' \quad = \quad C1 + C2 * \log_{10}(\text{time}) \tag{1}$$

where   C1 represents ϕ' of the ash at one day
          C2 represents the change in ϕ' with time

21. The fit to the data is shown in Fig 2. Values of C1 and C2 for the peak and post rupture failure envelopes are given in Table 1. Negative values of C2 are associated with significant increases in cohesion.

Table 1   The trend of the change in the peak and post rupture angles of friction with age for ashes compacted to the maximum dry density

Station	Ash	Saturation	Peak Strengths		Post Rupture Strengths	
			slope (C1)	intercept (C2)	slope (C1)	intercept (C2)
Blyth	fresh	S	0.06	0.58	0.11	0.32
		U	0.11	0.52	0.07	0.41
	stockpile	S	0.00	0.65	-0.04	0.56
		U	0.06	0.62	-0.01	0.53
Tilbury	fresh	S	0.14	0.54	0.09	0.49
		U	0.18	0.42	0.07	0.48
	stockpile	S	0.01	0.61	-0.07	0.59
		U	0.03	0.67	0.00	0.59
Fiddlers Ferry	fresh	S	0.05	0.44	0.06	0.27
		U	0.09	0.40	-0.07	0.59
	stockpile	S	0.01	0.53	0.02	0.40
		U	0.03	0.55	-0.02	0.50

22. The saturated specimens should contain no air and, therefore, suction forces will not exist. Note that saturation was carefully carried out to ensure this condition; soaking will not remove all the air which implies that specimens in shear boxes may retain some suction and therefore apparent cohesion giving non conservative results.

371

23. The constants in Table 1 are a result of a smoothing operation and do not necessarily represent the actual increase in angle of friction with age. They do give realistic values of friction. The fluctuation in the data may be a result of differing amounts of suction or cementitious action within a set of specimens and varying densities and water contents.

24. The smoothing operation used for the peak strengths does not produce such clear distinctions between the different ashes for post rupture strengths. In some cases the angle of friction is reducing with age. The data shown in Table 1 are based on the best fit lines to the failure points. The specimens all failed by shearing. Shearing will break any form of bonding and may cause loss of suction. Thus, there should be no cohesion following rupture.

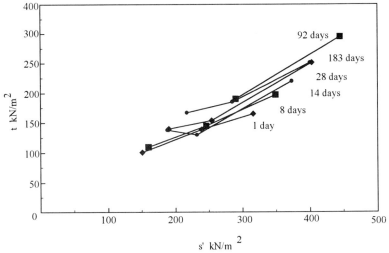

Fig 3   Increase in compression strength with time for a saturated ash
              (Tilbury fresh pfa)

25. Fig 3 shows that, for practical purposes, given the variability in the material, that this is the case. Table 2 gives the constants C1 and C2 for post rupture strengths based on a total loss of cohesion

26. Twenty eight days was chosen because, in general, most of the self hardening occurs within that period and, if the benefits of self hardening are to be realised, then tests at twenty eight days are practical.

27. The results given in Table 2 show several trends.

a)   All tests on the ashes show that saturated ash is weaker than unsaturated ash supporting the hypothesis that inundation leads to a loss of strength.

b)   Tilbury and Blyth fresh ashes show that strength increases with time even if the ash is saturated before testing.

c)   Fiddlers Ferry fresh ash does not show a significant increase in strength with time.

d)   Tilbury stockpile ash increases in strength with time.

e)   Blyth and Fiddlers Ferry stockpile ashes do not show a significant increase in strength with time.

28. This suggests that some ashes are able to develop a permanent self-hardening component of strength while others do not. Note that stockpiled ash is aged fresh ash therefore it may be expected that tests on stockpiled ash show no increase in strength with time. This is not always the case. Blyth stockpiled ash shows a continuing increase in strength with time.

Table 2 The change in the post rupture angle of friction with age for ashes compacted at different water contents (it is assumed that there is no cohesion)

Station	Ash	Saturation	Nominal Density	Post Rupture Strengths		Angle of Friction	
				slope (C1)	intercept (C2)	1 day	28 days
Blyth	fresh	S	95%D	0.088	0.42	25	33
			100%	0.072	0.43	25	32
			95%W	0.068	0.41	24	30
		U	95%D	0.031	0.53	32	35
			100%	0.045	0.57	35	39
			95%W	0.033	0.54	33	36
	stock	S	95%D	0.060	0.49	29	35
			100%	0.004	0.56	34	35
			95%W	0.055	0.49	29	35
		U	95%D	0.009	0.63	39	40
			100%	0.021	0.63	39	41
			95%W	0.026	0.61	37	41
Tilbury	fresh	S	95%D	0.072	0.50	30	37
			100%	0.111	0.51	31	42
			95%W	0.047	0.55	34	38
		U	95%D	0.025	0.64	40	42
			100%	0.053	0.61	38	44
			95%W	0.068	0.59	36	44
	stock	S	95%D	0.008	0.58	35	36
			100%	-0.015	0.60	37	35
			95%W	0.025	0.54	34	35
		U	95%D	0.101	0.57	35	46
			100%	0.008	0.66	41	42
			95%W	0.014	0.64	40	41
Fiddlers Ferry	fresh	S	95%D	0.047	0.33	19	23
			100%	0.014	0.42	25	26
		U	95%D	0.002	0.53	32	32
			100%	0.013	0.55	35	36
			95%W	0.019	0.52	31	33
	stock	S	95%D	0.010	0.42	25	26
			100%	0.013	0.44	26	27
		U	95%D	0.004	0.60	30	31
			100%	0.012	0.30	31	32
			95%W	-0.004	0.49	31	30

29. The permanent increase in peak strength for Blyth and Tilbury ashes is more dramatic. However, the displacement required to reach peak stress reduces with time.

Thus, it may not be prudent or practical to rely on this cohesion since deformations of about 1% could lead to the mobilisation of post peak strength. This high cohesion is the reason that vertical cuts through pfa fills can remain standing unsupported for a considerable time.

THE DEVELOPMENT OF THE STRENGTH OF PFA WITH TIME

30. Simons and Jeffery (ref 6) suggested that some form of chemical binding occurs with pfa to the growth of ettringite. Raymond (ref 7) and Sutherland et al (ref 8) proposed that the presence of free lime contributed to the development of cohesion. Briam (ref 9) considered the increase in strength was probably due to the interlock of gypsum crystals.

(a) 0 day x10000

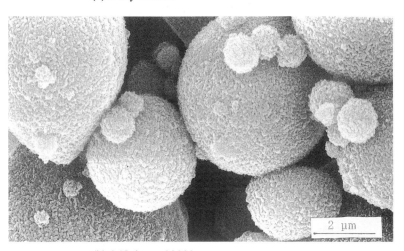

(b) 365 days x10000

Fig 4   The change in particle shape with time for Fiddlers Ferry fresh ash

31. Figs 4, 5 and 6 show examples of the results of SEM studies on the ashes from the three stations. Fig 4 shows that Fiddlers Ferry fresh ash particles retain their basic spherical shape but the surface of the particle changes. This could lead to an increase in suction, hence the apparent increase in strength of partially saturated specimens, and an increase in the angle of friction due to the change in surface roughness.

(a) 365 day x2000

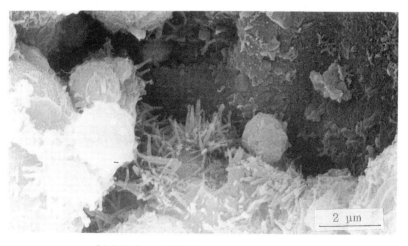

(b) 365 days x10000

Fig 5   Randomly distributed needle like crystals in Tilbury fresh ash

32. Fig 5 shows that Tilbury fresh ash develops randomly distributed needle like crystals. This leads to an increase in the angle of friction and an increase in peak cohesion.

33. Fig 6 shows that Blyth ash develops a form of gel which contributes to a permanent increase in ϕ' and c'.

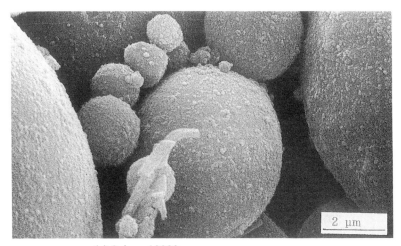

(a) 0 day x10000

(b) 180 days x10000

Fig 6   Gel like structures in Blyth fresh ash

A SUGGESTED CLASSIFICATION

34. Fig 7 shows a suggested classification based on the results of the triaxial tests and the qualitative data from the SEM studies.

35. In order to classify an ash it is necessary to determine its effective strength parameters at one an twenty eight days. These are assessed from triaxial tests on ash saturated at the time of testing. It is important to store the specimens correctly before testing.

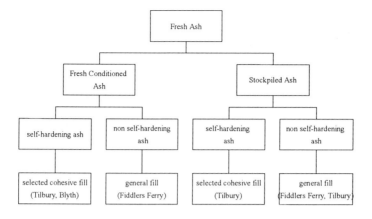

Fig 7   Classification of pfa based on self-hardening properties

CONCLUSIONS

36. The current DoT Specification for pulverised fuel ash classifies ash according to its age, in that fresh ash is assumed to be selective cohesive fill and stockpiled ash is general fill. This leads to the assumption that fresh ash is better than stockpiled ash.

37. Consolidated drained triaxial tests on partially saturated and saturated ash at varying ages show that all ashes increase in strength with time but the percentage increase varies. It is a function of the source of the ash, though this can be further complicated because of the influence of the source of the coal and the method of burning.

38. Ashes can be classified according to their self-hardening abilities. Not all fresh ashes are self-hardening and some stockpiled ashes are.

39. To establish the classification it is necessary to conduct tests at one and twenty eight days. The tests should be consolidated drained triaxial tests on saturated specimens and the results should be taken from the post peak behaviour assuming no cohesion. This gives the worst credible conditions for design. If the angle of friction has only changed by 1 or 2 degrees then the ash is not self-hardening.

40. Peak effective strength parameters are well in excess of design values but to be confident in choosing values for design, several tests have to be carried out to obtain mean values because of the fluctuation in data due to variations in density and pozzolanic action within a set of specimens.

ACKNOWLEDGEMENTS
The Authors would like to thank National Power and Powergen for their permission to publish data in this paper and the staff at the University of Newcastle upon Tyne who assisted in the laboratory work.

REFFRENCES
1. Department of Transport (1986) Specification for Road and Bridgeworks

2. CLARKE, B G, YANG, Y and COOMBS, R (1993) The effect of time and compaction on the strength and stiffness of coal ash. EPRI, to be published

3. RAYMOND, S (1961) Pulverised fuel ash as embankment material, Proc Inst Civ Engnrs, Vol 19 pp 515 - 536

4. TRI UTOMO, H and CLARKE, B G (1992) The effect of time on the properties of pulverised fuel ash. Geotropika'92, Malaysia

5. YANG, Y (1992) Study of the mechanical properties of pulverised fuel ash for use in geotechnical applications, PhD Thesis, University of Newcastle upon Tyne

6. SIMONS, H S and JEFFERY, J W (1960) An X-ray study of pulverised fuel ash, J App Chem, No 10, pp 328-333

7. RAYMOND, S (1965) Pulverised fuel ash in highway engineering structural fill

8. SUTHERLAND, H P, FINLAY, T W and CRAM, I A (1968) Engineering and related properties of pulverised fuel ash, J Inst Highway Engrs, pp 1-9

8. BRIAM, M (1981) Pulverised fuel ash structural fill; properties and behaviour, PhD Thesis, University of Leeds

30. Recent experience with soft rockfill in dam engineering

R. A. READER, C. G. HOSKINS and H. T. LOVENBURY, Rofe,
Kennard & Lapworth, and M. FLETCHER, RKL Geotechnical, UK

SYNOPSIS. Embankment dams worldwide are now often
constructed of materials classed neither as 'earthfill' nor
'rockfill', but of 'soft rockfill'. Quarries frequently
yield this potentially useful material in large quantities
during a search for sound hard rockfill; in the past it has
been discarded as unsuitable overburden. Its use as an
attractive and economic alternative to other engineering
fills can greatly reduce the environmental impact of
embankment dams and borrow areas. The paper sets out some
experience gained from successful construction over the last
20 years.

INTRODUCTION
1. High quality rockfill is usually quarried at depth with
much weathered overburden being rejected. The widespread
excavations and spoil tips have a large impact on the
surrounding area and cannot be confined within the reservoir
basin. With careful selection, winning and placement, this
'waste' material performs well as an engineered fill and
makes the search for sound rockfill unnecessary.
2. Soft rockfill is a transition material between sound
(or hard) rockfill and earthfill. It is free-draining,
granular material with a grading such that the fine fraction
always 'overfills' the voids between the coarser particles
giving near maximum density when thoroughly compacted. It
can be used to form embankments with relatively steep slopes
and few zone boundaries, more akin to true rockfill than
earthfill because no positive pore pressures will be
developed by changes in total stress. Settlement is less
than for sound rockfill.
3. Penman and Charles (ref. 1) published some information
on the early use of soft rockfill in Great Britain. More
recent experience both in this country and abroad has been
given by Wilson and Evans (ref. 2) and Williams and Reader
(ref. 3). These have detailed the use of soft rockfill in
various dams but have generally not discussed the more
general issues involved in the identification and
appreciation of the material. Brown (ref. 4) provides a

detailed case history and refers briefly to the wider aspects of the assessment of soft rockfill. Experience at Balderhead Dam (completed 1965) described by Kennard, Knill and Vaughan (ref. 5) and at Roadford Dam (completed 1988) by Wilson and Evans (ref. 2), shows that soft rockfill is a very variable material, depending on its origin. There are two broad categories; that derived from non-argillaceous rocks which breaks down into relatively fine but granular material, and that derived from mudstones or shales that can break down into silt or clay. Mudstones and shales need careful handling if they are to retain their properties as soft rockfill without degrading to a cohesive low permeability 'earth' fill.

SOFT ROCKFILL - WHAT IS IT?
4. Hard rockfill specifications call for dense sound rock without weathered or weakened material or fines. Handling causes little breakdown. When compacted it has high intergranular stresses at point-to-point contacts, a large void ratio, a high permeability and relatively low density. Hard rockfill is sluiced during placing to reduce subsequent deformation.
5. Soft rockfill, by comparison, has a wide grading with a large percentage of finer material. Handling is deliberately rough to break up the rock into a well-graded fill with sufficient finer material to 'overfill' the voids between larger fragments. Compaction produces much lower intergranular stresses, a low void ratio, a lower permeability but a higher density. Water is added only to achieve better compaction.

HOW DO YOU RECOGNISE IT?
6. Potential soft rockfill quarry sites must be adequately investigated and assessed at an early stage. Competent logging of natural exposures and sufficient exploratory boreholes must be complemented by trial excavations and visual observation of the material during handling. The strength, hardness, toughness and durability of the rock and the likely required strength and grading of the rockfill must be considered during extensive laboratory tests. The nature of the fine fraction is fundamental; whilst low plasticity, mainly silt-sized rock flour would be acceptable, a plastic chemically-altered rock-derived silt or clay would not.
7. Overburden usually represents the end stage of weathering and is likely to have similar properties to the fully-degraded rockfill. Soft rock will break down appreciably during excavation, transportation and placement. Pockets of fresher material should be deliberately broken up to improve its grading and reduce subsequent weathering or breakdown after compaction into the embankment. Some soft rockfill may degrade to plastic material if it has been chemically altered. Any small quantity that escapes notice in the quarry should not affect embankment stability so long

as it is either locally contained or dispersed within the
fill. Once buried in the structure further breakdown is
unlikely to proceed at a rate comparable to the life of the
dam. Ground investigation and laboratory testing can
investigate all these properties, but a comprehensive visual
assessment of the material is essential.

8. Rockfill has a non-linear strength envelope [Marsal
(ref. 6), Charles and Watts (ref. 7)]. At high effective
confining stresses the strengths of hard and soft rockfill
are similar. At lower confining stresses the strength of
soft rockfill is lower than hard rockfill. In addition:

(a)   The strength of rockfill is usually measured under
      plane strain conditions in a large shear-box. This
      cannot be compared directly with the strength of soft
      rockfill measured under triaxial stress conditions.
(b)   Fragment strength controls the strength of rockfill and
      the amount of breakdown in shear and dilatancy.
(c)   Well-graded rockfill exhibits less fragment breakdown
      and greater dilatancy giving higher strengths than
      uniformly-graded rockfill.
(d)   Scaled-down tests of large-sized rockfill indicate that
      strengths increase when the larger fragments are
      removed so that laboratory tests tend to overestimate
      field strengths.

9. Fig. 1 compares the required design strengths (with
appropriate factors of safety applied) with the available
strengths of soft rockfill from several sites over a range of
normal stress. Results from elsewhere [(Marsal (ref. 6) and
Charles and Watts (ref. 7)] are also included. Specification
of a conservative strength envelope will obviate the need for
selection within the quarry or rejection of dubious weathered
upper layers.

10. The performance of soft rockfill is governed by the
properties of its fine fraction, particularly permeability
and strength, when there is an excess of granular finer
material to 'overfill' the voids. If the fine fraction is
insufficient to 'overfill' the voids, as shown on Fig. 2,
then the fill will have the properties of sound rockfill and
require sufficient strength to support point-to-point
contact. Soft rockfill with more than a few percent of
cohesive plastic fines from chemical weathering or breakdown
to clay or from the inclusion of a clay band in the quarry
will have a greatly reduced permeability and drained strength
and will begin to approach the properties of an earthfill
needing flatter slopes and causing placing problems during
wet weather. Vaughan and Soares (ref. 8) suggested a method
of cohesive testing which, combined with gradings, double
hydrometer tests, small shear-box and permeability tests
related to clay content and plasticity tests, will determine
whether the material will act as soft rockfill or degenerate
into earthfill.

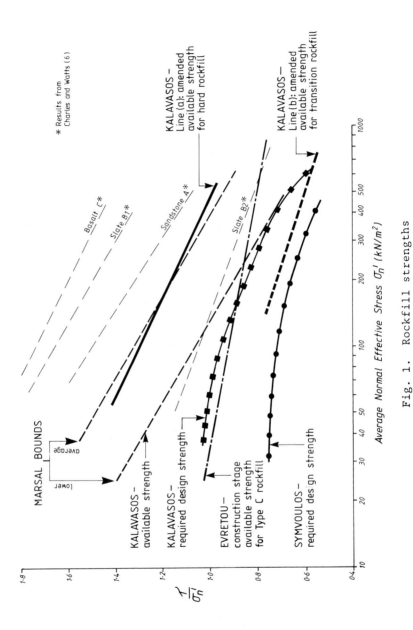

Fig. 1. Rockfill strengths

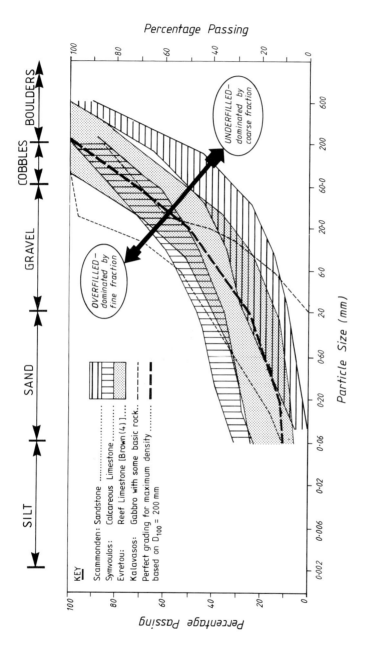

Fig. 2. Comparison of soft rockfill gradings

HOW DO YOU USE IT?

11. Soft rockfill can be successfully specified for use in embankment dams but its properties need continually re-assessing against the design assumptions. The zone boundaries within the dam or even the slopes themselves may need revision in some cases and space should be allowed for such changes. Use of virtually all material arising from the quarry (preferably within the reservoir basin) should be the aim; material moved other than into the dam is a costly luxury.

12. A clear specification should detail the method of working the quarry to produce material within the required grading envelope and define a maximum plasticity. The placing specification (layer thickness, compactive effort and density control) should be confirmed by site trials. Normally the soft rockfill should be tipped some metres back from the edge of the advancing layer and dozed forward whilst being wetted with 5-7% by volume, more in hot climates, to form a homogeneous fill. Some British soft rockfills may not need extra water but compaction in too dry a state may lead to collapse settlement later as noted by Wilson and Evans (ref. 2). Over-watering is not a problem in such permeable fill although compaction by smooth-wheeled vibrating rollers can skin over the surface of each layer causing ponding. This can be cured by scarifying, but trial pits through completed layers have shown little evidence that the skin survives compaction of a subsequent layer, when larger fragments of rock appear to penetrate the skin and maintain vertical permeability.

13. Development of the quarry may alter the quality of the soft rockfill and require a change to the placing specification. Regular compaction trials with increasing roller passes to refusal are useful to determine optimum layer thickness for minimum compactive effort. In-situ density and permeability tests should be carried out in the trial area together with shear-box checks on strength.

14. A layer of coarser rock is frequently placed on the downstream slope to prevent surface erosion.

RECENT EXPERIENCE

Scammonden Dam

15. A moderately well-graded material with an adequate fines content derived from adjacent motorway cuttings in Carboniferous Sandstone forms Scammonden Dam with upstream slopes of 1V:3.1H steepening to 1V to 1.8H and a similar 1V to 1.8H slope downstream. The embankment has an upstream inclined clay core. The specified grading is shown on Fig. 2.

16. Excavation and compaction trials [Williams and Stothard (ref. 9)] investigated the best method of winning the fill material and placing it to minimise segregation. The rock was won by blasting. By tipping the rockfill some 3-4 metres back from the advancing face of each layer and

dozing forward, the larger fragments tended to roll out and be fully covered by the smaller fragments and fines which filled all the voids to create a dense layer with a smooth top surface.

17. Five passes of an 11.5t vibrating roller compacted the 0.9m max. sized rock in 0.9m layers to a dry density of 2.0 t/m^3 at about 7% moisture content. At higher moisture contents the fines congealed and were not distributed within the fill, leading to lower densities. Rockfill placing was suspended if the material passing the 20mm sieve had a moisture content above 12%.

Kalavasos Dam - Cyprus

18. Gabbro rockfill was available locally to form shoulders with slopes of 1V:2H upstream and 1V:1.75H downstream either side of a central clay core. An alternative diabase quarry 3km distant was too costly. Toe berms incorporated spoil from the tunnels, tailbay and dam foundations.

19. Shear-box tests on the gabbro gave a wide scatter of strengths dependent on the degree of weathering. The Marsal (ref. 6) lower bound was adopted for a conservative design that allowed for the inclusion of some highly weathered material. The material was considered free-draining and without drawdown pore-pressure problems.

20. The quarry proved to contain great depths of highly weathered material with little rock of specified quality available without the removal of huge quantities of overburden. Initial quarry development had dozed this overburden into tips that appeared to have a remarkably consistent grading of stable, silty sandy gravel. Shear-box tests indicated no further breakdown during shear and gave strengths only marginally below the design values. Visual inspection confirmed that further loading, placing and compaction would not cause further breakdown. The rock was won by ripping and dozing. Shear-box tests on the originally specified rockfill gave strengths well in excess of the design values.

21. A redesign incorporated the lower strength soft rockfill [Fig. 1, Line (b)] in inner zones of the dam shoulders, the outer zones being constructed of the specified rockfill [Fig. 1, Line (a)] found at greater depth in the quarry. The soft rockfill grading is shown on Fig. 2 requiring maximum breakdown to be achieved in the quarry. Chemically-weathered material was excluded by a 10% plasticity index limit. A 13.6t vibrating roller compacted the fill in 500mm layers, determined by site trials, adding about 8% water by volume.

22. The redesign allowed the dam to be completed early despite the soft rockfill being more weather-susceptible.

Symvoulos Dam - Cyprus

23. A quarry in thin-bedded, fine-grained variable strength calcareous limestones of the Pakhna Formation yielded homogeneous soft rockfill for an embankment with 1V:2.7H upstream and 1V:2.3H downstream slopes, sealed with an upstream HDPE membrane.

24. The rockfill was prone to breakdown by mechanical abrasion to a well-graded, granular material which strength tests proved to have high shear strength at low confining stresses. Los Angeles abrasion tests showed a loose trend between the degree of weathering of the rock and the strength of the rockfill. The high percentage wear of many of the samples was considered to reflect breakdown of the cementing material rather than the strength of the individual rock fragments. A design strength envelope just below the Marsal lower band indicated that spoil from the foundations and the spillway could be incorporated in the toe berms and the slope beneath the membrane would be stable.

25. The quarry was opened in small sections to reduce degradation of the rock during handling or weathering. Fill placing trials established an optimum layer of 500mm with a 10% by volume addition of water. Four passes of a 10t vibrating roller compacted the soft rockfill to a dry density of 1.78 t/m^3 at 13.8% moisture content, the low density value reflecting the low bulk specific gravity of this rock.

26. The embankment was completed on programme from a quarry largely below reservoir top-water level.

Reva Dam

27. The stability of Reva dam, near Bradford, was improved by construction of a downstream rockfill berm which needed to be free draining as seepages emerged from the downstream slope. Rockfill had the added advantage of allowing placement in winter.

28. Water replacement density tests at different depths in the trial embankment constructed within the completed berm, together with grading tests, investigated the effects of compaction on successive layers of rockfill to establish a suitable specification.

29. Spreading the rock with a D3 dozer and compacting it in 200mm layers with 8-12 passes of a 5.8t vibrating roller broke down the rock sufficiently for it to pass the 200mm maximum size specified without engendering too many fines. The fill achieved a dry density of 2.1 t/m^3 without adding water although wet weather prevailed during placement.

30. The added berm was completed to programme during the winter of 1990-91 without problems.

EXPERIENCE ELSEWHERE

31. Evretou Dam in Cyprus was constructed just after Kalavasos, being completed in 1987. The rockfill was derived from reef limestone, a crystalline limestone with zones of powdery rock debris. The design and construction are

discussed by Brown (ref. 4). Problems with rock quality similar to those at Kalavasos were evident on opening up the quarry. In the revised design better quality rockfill was restricted to the upper levels of the upstream shoulder whilst well graded material with up to 40% passing the 2mm sieve was used elsewhere. The results of the boreholes and surface trial blasts were interpreted too optimistically so trial excavations at an earlier stage would have been beneficial. The gradings of the material, included on Fig. 2, are again typified by a high proportion of sand and silt to produce an 'overfilled' grading.

32. Following trials the soft rockfill was placed in 500mm thick layers with the addition of 10% of water and compacted by six passes of either a 12t or 15t vibrating roller. An average dry density of $1.97t/m^3$ was achieved with a moisture content of 8.5% in the material up to 50mm in size and 4.7% in the larger sizes. The upstream slope was flattened from 1V:1.9H to 1V:2.4H with the addition of a toe berm of the very finest rockfill, whilst the downstream slope was unchanged at 1V:2H. Full upstream filters were retained between the soft rockfill and the core, although not apparently necessary. The flattening of the upstream slope led to problems with encroachment onto the diversion works due to their proximity to the dam.

CONCLUSIONS

33. The identification of soft rockfill and an appreciation of its behaviour, by visual inspection, laboratory testing and site trials, are essential at an early stage. The effects of excavation, transportation and placement on the engineering behaviour of the material need careful study.

34. The benefits of using soft rockfill are:-

(a)   Advantages of hard rockfill but at lower cost
(b)   Overburden and waste removal from quarry and selective working significantly reduced
(c)   Depth of quarry reduced
(d)   Quarry working more economic
(e)   Smaller fill quantities than for earthfill
(f)   Potential omission of filter/transition zone between upstream fill and core
(g)   Placement not greatly weather dependent
(h)   Faster construction
(i)   Reduced environmental impact

REFERENCES
1. PENMAN A.D.M. and CHARLES J.A. The quality and suitability of rockfill used in dam construction. Transactions of the 12th International Congress on Large Dams, Mexico, 1976, vol. 1, 533-556.
2. WILSON A.C. and EVANS J.D. The use of low grade rockfill at Roadford Dam. Proceedings of the 6th Conference of the

British Dam Society on The Embankment Dam, Nottingham, 1990, 21-27.

3. WILLIAMS O.P. and READER R.A. Design and construction of Kalavasos Dam, Cyprus. Proceedings of the Institution of Civil Engineers, Part 1, 1991, vol. 90, October, 975-992.

4. BROWN A.J. Use of soft rockfill at Evretou Dam, Cyprus. Geotechnique, 1988, vol. 38, no. 3, September, 333-354.

5. KENNARD M.F., KNILL J.L. and VAUGHAN P.R. The geotechnical properties and behaviour of Carboniferous shale at the Balderhead Dam. Quarterly Journal of Engineering Geology, 1967, vol. 1, no. 1, September, 3-24.

6. MARSAL R.J. Mechanical properties of rockfill. Embankment Dam Engineering, Casagrande Volume, Wiley, New York, 1973, 109-200.

7. CHARLES J.A. and WATTS K.S. The influence of confining pressure on the shear strength of compacted rockfill. Geotechnique 1980, vol. 30, no. 4, December, 353-367.

8. VAUGHAN P.R. and SOARES H.F. Design of filters for clay cores of dams. Journal of the Geotechnical Engineering Division, Proceedings of the American Society of Civil Engineers, 1982, vol. 108, no. GT1, January, 17-31.

9. WILLIAMS H. and STOTHARD J.N. Rock excavation and specification trials for the Lancashire - Yorkshire motorway, Yorkshire (West Riding) section. Proceedings of the Institution of Civil Engineers, 1967, vol. 36, March, 607-631.

31. Some aspects of the use of skip load tests to determine the settlement behaviour of fills

C. S. ECCLES, Soil Mechanics Limited, UK

SYNOPSIS. The use of skip load tests to determine the settlement characteristics of fills has increased in the UK since publication of a standard specification of skip load tests in BS 1377 : Part 9 : 1990. Some suggested modifications to this procedure are discussed. The results indicate that in addition to measuring immediate and long term creep settlement, the differential settlement and the potential of the ground for collapse settlement can also be assessed. It is recommended that the ground around all skip load tests is inundated to investigate the potential of the fill for collapse settlement.

INTRODUCTION

1. The use of skip load tests to determine the settlement characteristics of made ground and fills has increased in recent years in the UK. This is probably due to a number of factors. The skip load test has been included in the recently revised publication of BS 1377[1] and hence there is now a standard specification for carrying out skip load tests. The test has been named "The Shallow Pad Maintained Load Test" and the skip load test is the simplest method of carrying out the test. In recent years there has been growing recognition that when testing heterogeneous fill, simple large scale in situ test methods (such as skip load tests) will be more appropriate than small scale in situ and laboratory tests which can only be carried out on recompacted materials causing some uncertainty in their relevance. There is also an increasing desire to re-use land, particularly in inner cities, for housing and light industrial purposes. Many of these sites have a surface layer of variable fill which is often of a substantial thickness.

2. Skip load tests have the advantage of being a simple in expensive form of long term in-situ test. Long term settlement is often of more concern than immediate settlement or bearing capacity therefore standard plate tests may not be useful due to the short duration of such tests and the size of plates that are used. Skip load tests can be used to approximately reproduce the stress level and distribution with depth of strip footings for lightweight structures which typically stress the ground only to depths of 1.5 to 2.5 m. Hence direct estimates of the performance of shallow footings can be made from skip load tests. Skip load tests have also been used to indicate the effectiveness

of ground improvement techniques by carrying out tests both before and after treatment [2].

SKIP LOAD TEST PROCEDURE

3.   The first paper on the skip load test was by Cox [3] however, this gave only a very brief description.   Charles and Driscoll [4] gave a more detailed account of the skip load test in 1981.   The British Standard test specification; BS1377 : Part 9 : 1990 : Clause 4.2 is based on the work by Charles and Driscoll.   Table 1 presents a brief comparison of the procedure for carrying out a skip test in accordance with the British Standard and a suggested slightly modified procedure.

4.   Charles and Driscoll described application of the load by applying an empty skip onto a prepared base, taking a set of datum levels, then filling the skip with sand and capping it with concrete within 45 minutes.   This method has a number of deficiencies.   As the load is applied over a period of time it could be difficult to differentiate between immediate, consolidation and creep settlements of the skip.   In many fills such as granular soils and cohesive soils with a significant air voids, the initial and consolidation settlement of a skip may be effectively instantaneous and hence the above loading procedure would obscure differentiation between the three types of settlement.   A second difficulty is that the load applied by the skip is generally assessed by calculating the volume of the sand in the skip and determining the density of the sand in the skip by an in situ density test   However, as the skip is filled rapidly, it may be difficult to place the sand in the skip with a uniform density and hence some uncertainty in calculating the applied load occurs.   A third but minor deficiency is under utilisation of plant.   Due to the volume of the sand, skips are usually filled using a hydraulic excavator.   As a significant amount of time is taken for levelling, few skips can be filled in a day.   Unless the excavator is required for other purposes on a site, it cannot be fully utilised and may be required for several days if a number of tests are to be carried out.

5. Alternatively if the skips are filled on site at a position remote from the testing area and then moved onto the loading pad using a skip delivery truck an instantaneous load can be applied. This method also means that many skips can be filled in a more controlled manner in a single day giving better utilisation of plant.   To get a datum reading, the skip has to be lowered while empty, briefly onto the loading pad, surveyed, then moved away from the test location to await filling.   One disadvantage of this method of applying load is that on some sites where the ground surface is relatively soft, a skip truck with a fully loaded skip may not be able to place the skip on the loading pad due to the trafficability of the ground.   Where a site is in an urban area another possible improvement, for little extra cost, is to weigh the full skips on a public weighbridge if there is a weighbridge close to the site.

6.   Fully loaded, nominal 8 cubic yard (6 m^3) skip will apply a bearing pressure of 27 to 31 kNm^{-2}.   However it should be noted

Table 1 : Comparison of British Standard and Suggested
Test Procedures

BS 1377 Procedure	Suggested Procedure
1. Install levelling datums and prepare base for skip.	1. Install levelling datums and prepare base for skip.
2. Place empty skip on base.	2. Place empty skip on base.
3. Take initial level of skip.	3. Take initial level of skip.
4. Fill skip with sand.	4. Remove skip from base and fill both first and second skip with sand in an area remote from the base.
5. Monitor the settlement of the first skip as the it is filled.	5. Move the first skip and carefully lower it on to the base, monitor settlement.
6. Apply second skip by carrying out stages 2 to 5.	6. Apply second skip and monitor settlement.
7. Remove the skips monitoring movement.	7. Flood test location and monitor movement.
	8. Remove the skips monitoring movement.
	9. Excavate trial pit at the location of the test to confirm the ground conditions.

that many skip delivery vehicles cannot lower a second skip onto the top of a fully loaded skip due to height restrictions, hence the first skip has to be only partially filled so that the bottom skip may only apply a pressure of about 17 to 21 kNm^{-2}.

7.   When carrying out skip load tests the four lifting lugs on the skips are usually used as levelling datums. However, when the second skip is applied only two of these datums can be seen from any one location hence requiring the tripod and level to be moved to take a full round of readings. It is recommended that two levelling datum stations are located close to each test to enable a full round of levels to be made quickly and accurately. These levelling datums will probably be in the area of fill therefore they should be surveyed on a regular basis to check if all the fill on the site is undergoing long term settlement.

8.   A final suggested addition to the skip load test is to carry out a trial pit below the skip location on completion of the test to confirm the ground conditions. This is particularly important if the ground conditions at the site have been proved to be variable.

CASE HISTORIES
Site A
9.   This site is a former clay pit which has been backfilled with domestic waste and capped with approximately 3 m of clay. Dynamic compaction had been carried out on the site some six years after it was capped and about two years prior to carrying out three skip load tests using the suggested alternative test procedure discussed above without Stage 7. The results of skip tests at Site A and Site B are summarised in Table 2 and the full settlement data for one of the tests from Site A is presented in Fig 1. Total settlements at two months at Site A range from 36 to 163 mm. Initial and consolidation settlements both occured almost instantaneously. The variability of compressibility at this site can be seen in Fig 2a.

10. By analysing the settlement of the four monitoring points on the skip separately the differential settlement across a skip can be assessed. Fig 1 indicates that the compressibility of the ground after the dynamic compaction was still highly variable even over relatively short distances. The range in differential settlements at 2 months are listed in Table 2 in terms of relative deflection as defined by Burland et al [5]. These relative deflections are very high at Site A. It can be concluded that dynamic compaction has not been effective in reducing settlement and variability to an acceptable level at this site.

Site B
11.   Site B is an area which has been reclaimed and was previously inter-tidal mud flats. Ground conditions comprise 4 m of slightly to very clayey, sandy to very sandy GRAVEL, this fill overlying approximately 10 m of soft normally consolidated clays. An envelope of particle size analyses is presented in Fig 3. The site is currently a storage area, mainly for cars

Table 2 : Summary of Skip Load Tests

Test	Net Applied Pressure (kNm⁻²)	Average Immediate Settlement (mm)	Creep Settlement Per log Cycle (mm)	Total Settlement at 2 months (mm)	Proportion of Settlement at 2 Months			Relative Deflection at 2 months
					Immediate (%)	Creep (%)	Collapse (%)	
Site A								
Test 1	18.11 46.6	58 99	1.8 4.1	163	96	4	–	1/11
Test 2	17.5 45.8	5 20	1.1 4.1	36	71	29	–	1/215
Test 3	16.9 44.0	9 30	1.3 8.6	51	76	24	–	1/80
Site B								
Test 1	31.6	2.2	2.4,3.7	8.1	27	54	19	1/360
Test 2	21.5 53.2	0.3 4.6	0.7 0.8	6.7	73	27	–	1/255
Test 3	21.8 49.3	0.5 4.2	0.4 0.9	8.1	58	42	–	1/400

Note: 1. Immediate settlement for second skip is the additional immediate settlement when second skip applied not total immediate settlement.

2. Two creep rates for Site B Test 1 : first is prior to collapse, second is after collapse.

393

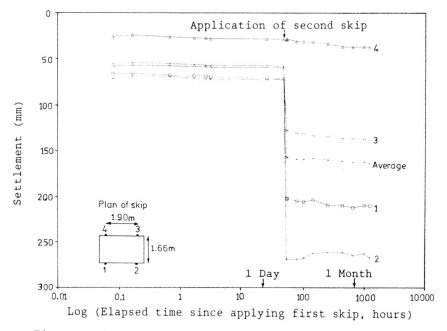

Fig. 1.   Site A Test 1, Settlement against logarithm of elapsed time

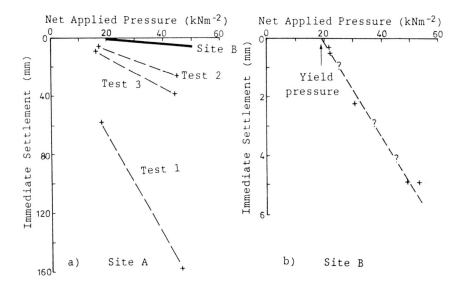

Fig. 2.   Average settlement of the skips

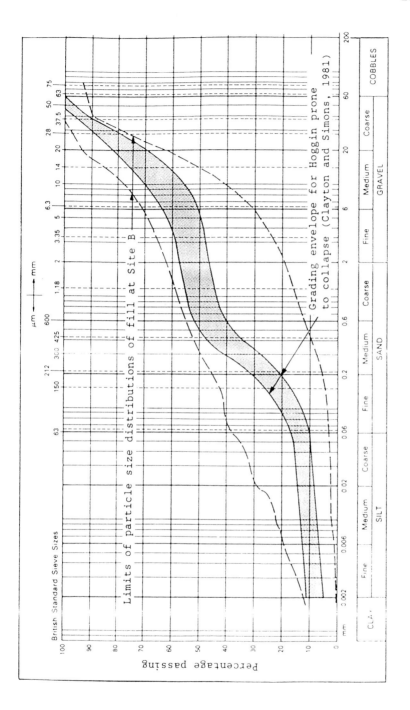

Fig. 3. Site B, Particle size distribution of fill

(Fig 4). Three skip load tests were again carried out using the suggested alternative procedure. In Test 1, a single skip was used to load the ground. Total settlements of the skips at 2 months varied from 6.7 to 8.1 mm. A comparison of the immediate settlements of the skips (Fig 2b) indicates that the ground conditions are more uniform at this site than at Site A. An approximately linear load/settlement curve can be inferred in Fig 2b together with an approximate value of the yield pressure.

12. The general impression of the fill at Site 2 is that it has desirable properties; it appears to be fast draining, relatively incompressible and to have a low creep rate. However, during the period of monitoring there was a period of heavy rain and the area around Test 1 was inadvertently flooded as it was located in an area of slightly lower ground on what was essentially a flat site. This inundation caused collapse settlement and the rate of creep settlement increased after inundation (Fig 5). The particle size distribution of the fill at Site B encompasses the grading envelopes of Hoggin which is known to be a material which can be prone to collapse, Clayton and Simons [6]. A number of in situ density measurements were made and air voids of 5 to 25% were calculated indicating the potential for collapse settlement.

CONCLUSIONS

13. Skip load tests are simple in situ tests which are being used increasingly to assess the settlement characteristics of fills. Modifications to the British Standard test procedure are suggested which should help to differentiate between the different stages of settlement whilst improving the quality and cost effectiveness of the test. Skip load tests can also be used to investigate the effectiveness of ground improvement techniques. Collapse settlement is potentially an important mode of deformation of fills of many different compositions, including rockfill, sandy gravels with a significant fines content, stiff clays and chalk. Therefore it is recommended that an additional stage of inundating the ground around the loaded skips is included in all skip load tests to assess the potential of the fill for collapse settlement.

REFERENCES

1. BS 1377 : 1990 : Methods of tests for soils for civil engineering purposes. Part 9 : In situ tests. British Standards Institution.
2. WATTS, K S and CHARLES, J A : 1991 : The Use, Testing and Performance of Vibrated Stone Columns in the United Kingdom. Deep Foundation Improvements : Design, Construction and Testing, ASTM SPT 1089, M I Esrig and R C Bachus Eds, American Society for Testing and Materials, Philadelphia. pp 212 - 223.
3. COX, D W : 1978 : Contribution to discussion : Conf on Clay Fills. Instu Civ Engrs, London, pp 271.
4. CHARLES, J A and DRISCOLL, R M C : 1981 : A Simple In Situ Loading Test for Shallow Fill. Ground Engineering, Vol 14, No 1, pp 31 - 36.

Fig. 4.   Precise levelling of skip load tests at Site B

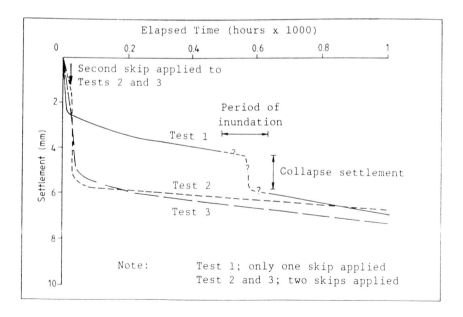

Fig. 5.   Site 2, Settlement of skips against time

5.   BURLAND, J B, BROMS, B B and de MELLO, V F B : 1977 : Behaviour of foundations and structures.   State of the Art Report, 9th Int Conf Soil Mech and Foundn Eng, Tokyo, Vol 2, pp 495 - 546.

6.   CLAYTON, C R I and SIMONS, N E : 1981 : The collapse of slabs on a compacted hoggin fill.   Proc 2nd Conf on Ground Movements and Structures, Cardiff, pp 655 - 670.

398

32. Initial assessment of new rapid ground compactor

K. S. WATTS and J.A. CHARLES, Building Research Establishment, UK

SYNOPSIS. The effectiveness of a new ground improvement technique using a rapid impact compactor in improving the properties of fills has been assessed and compared with other techniques currently available. Detailed investigations have been carried out at four sites; the depth and degree of ground improvement have been measured and other features of the technique have been monitored.

INTRODUCTION

1. A rapid impact ground compactor or dynamic tamper has been jointly developed by BSP International Foundations Ltd. and the Defence Research Agency (formerly RARDE) for repairing explosion damage to military airfield runways. The compactor comprises a modified hydraulic piling hammer acting on an interchangeable articulating compacting foot. A 7 tonne weight falls 1.2 m onto a 1.5 m diameter foot at a frequency of about 40 blows per minute. The plant was developed to re-compact crater debris 2 - 3 m deep as well as reinstate explosion damage to underlying soils.

2. With good building land becoming increasingly scarce in the urban environment many sites with former industrial usage are being redeveloped. Such sites often contain miscellaneous fills unsuitable for supporting structural loads without some improvement of their engineering properties. Effective densification through the full depth of fill will improve load carrying capacity, but more importantly will minimise or remove adverse long term behaviour such as excessive creep settlement and susceptibility to collapse compression when inundated. Currently two techniques are principally used, dynamic compaction and vibrated stone columns (vibro).

3. The dynamic compaction technique (Menard, 1972)[1] has been used mainly to densify loose partially saturated fills. It was introduced into the United Kingdom (UK) from France in 1973. Deep compaction of a fill is effected by repeated impacts of a heavy weight onto the ground surface. The mechanism of compaction is simply to reduce the air voids, thus increasing density and improving the overall engineering properties of the

fill. This should lead to an increase in bearing capacity and a reduction in long term settlement potential. BRE has investigated the effectiveness of dynamic compaction on a number of sites in the UK (Charles et al, 1981)[2].

4. The ground improvement techniques most commonly adopted in the UK are the various deep vibratory processes collectively described as vibro (St John et al, 1989)[3], in which ground treatment is effected by a powerful torpedo shaped vibrating poker. The cylindrical hole formed by the poker is backfilled and compacted in stages with gravel sized stone. Improvement in ground properties may be due to a densification of the in situ material and/or the stiffening effect of the stone columns within the composite soil/column structure.

5. The rapid impact compactor's potential for ground improvement in the building and civil engineering industries was soon recognised. Although the principle of this treatment is similar to dynamic compaction, the rapid impact compactor is likely to be cheaper to mobilise and can therefore operate economically on much smaller sites, where vibro might usually be chosen as the technique for improving a shallow fill. A number of potential advantages of this technique over dynamic compaction and vibro were perceived, both from a technical and an economic standpoint.

6. The effectiveness of ground improvement techniques carried out from the surface will largely depend on the depth to which the ground can be treated. Techniques such as drop weight dynamic compaction and surcharging may be limited by this

Fig. 1. The rapid impact compactor treating old ash fill

consideration. With penetrative methods such as vibro and piling the depth of treatment can be chosen to suit the ground conditions and structural considerations. The main aims of the BRE research in assessing the rapid impact compactor were to establish the depth to which treatment would be effective and the degree of improvement achievable. A programme of four site trials was carried out. The successful outcome of the first two field assessments prompted BSP to re-design the compactor for civilian work (Figure 1).

LOOSE BUILDING WASTE LANDFILL (WATERBEACH)

7. When BRE investigations began the compactor was in the military format for the rapid repair of airfield runways, with the compacting hammer mounted on a Terex wheeled loader. A licensed landfill site, consisting mainly of loose inert building waste was chosen for the first trial. Fill has been deposited in terraced layers without any engineering compaction in a disused sand pit. Typical constituents of this fill are wood, brick, concrete, glass and rag with some soil. At the location chosen for the compactor trial the fill was 6.5 m deep, overlying a natural clay deposit. The ground water level was about 4.5 m below ground surface.

8. Two techniques were employed to assess the properties of the fill before and after treatment. Dynamic probing was carried out at each corner and the middle of the area to be treated using a rig complying with BS 1377:1990[4] and DIN Standard 4094 (DIN, 1974)[5]. The large variation in blow counts confirmed the extreme variability of this fill. A geophysical surface survey technique was then used (Abbiss, 1981)[6]. A profile of Rayleigh wave velocity with depth was measured at two points on the centre line of the area prior to tamping and values of dynamic shear modulus were calculated for the fill. A magnet extensometer was installed in a borehole through the fill and was terminated in the underlying natural clay deposit. Magnet markers were placed at 0.5 m intervals from ground level to a depth of 4 m and at 1 m intervals below this to the base of the fill. In addition to the measurement of settlement within the fill surface levelling was carried out on a grid pattern over the area to be treated.

9. To establish a suitable energy level for the treatment of the fill a single treatment point was monitored and the number of blows noted where little or no further penetration was being achieved. This was found to be about 50 blows with a 1 m drop of the weight and this was adopted as the standard number of blows per treatment point for the whole trial. An area 7 m x 7 m was treated with abutting compaction points on a square pattern using the 1.5 m diameter foot, and a similar area with a primary treatment pass abutting on a triangular pattern followed by a secondary pass between and overlapping the primary points. The average energy input for the two treated fill surface areas was 150 tonne.m/m^2 and 290 tonne.m/m^2 respectively.

10. In Figure 2 the settlement with depth of the fill treated

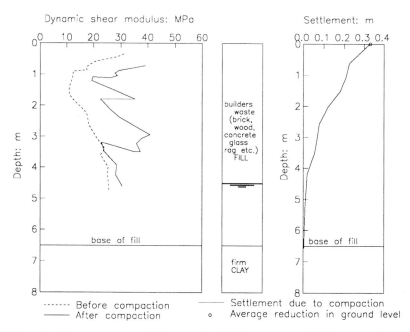

Fig. 2. Improvement of loose building waste fill (Waterbeach)

with abutting compaction points is shown against a profile of the landfill. The surface settlement indicated is the average reduction in ground level after re-grading for a total average energy input of 150 tonne.m/m^2. The higher energy input of 290 tonne.m/m^2 resulted in only a very small additional reduction in ground level. It is apparent that there was a significant effect to about 4.0 m below original ground level, the compression being fairly evenly distributed with depth. Geophysical measurements were used to calculate values of dynamic shear modulus with depth for the treated and untreated fill. The profiles obtained are also presented and confirm a substantial improvement to a depth of about 3.0 m (Butcher and McElmeel)[7]. Dynamic probing (DPH) using the standard 50 kg driving hammer and 15 cm^2 cone was carried out before and after treatment. The dynamic probe was unable to penetrate the treated fill at the majority of attempts.

NATURAL SAND DEPOSITS (WEST FREUGH)

11. The performance of the compactor has been assessed at a coastal site where an area of sand fill approximately 3 m thick lay over natural deposits of sandy peat and wind blown sand. The surface of the sand fill was level and grassed. Previous site investigation records indicated that the sand fill was a natural poorly graded wind blown sand, placed by heavy plant above the original ground surface. A borehole drilled just outside the trial area confirmed the depth of the sand fill and the existence of a shallow peat layer at what had been the original ground surface. Dynamic probing and geophysics were

(a)   (b)   (c)

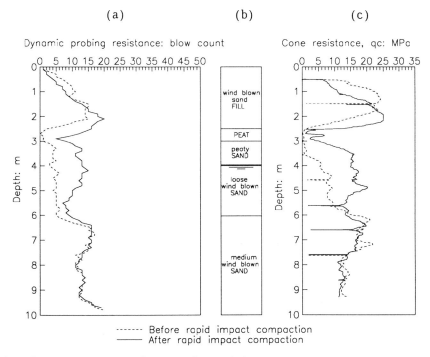

------- Before rapid impact compaction
——— After rapid impact compaction

Fig. 3. Improvement of natural sand deposits (West Freugh)

     (a) Dynamic probing
     (b) Soil profile
     (c) Cone penetration resistance

employed to measure pre-treatment ground conditions.

12. A suitable energy input level was established by
observing the point after which continued blows produced
negligible further penetration. Half the area was compacted
with 60 blows with the full 1.2 m fall at each abutting
treatment point, giving an average energy input of 225
tonne.m/m^2. The other half was treated with abutting compaction
points on a triangular grid, with a second pass compacting the
intermediate points and giving an average energy input of 420
tonne.m/m^2 over the treated area.

13. Figure 3 shows the soil profile of natural sand deposits
and measurements of penetration resistance with depth before
and after treatment at 420 tonne.m/m^2. Both dynamic probing and
static cone penetration testing showed the sand fill near the
surface was loosened by the process, although less so where a
second pass was carried out. This sand is poorly graded and is
likely to have been at or near to its maximum relative density
after placement. Severe disturbance caused by the penetration
of the compaction foot would result in an overall loosening of
this material. The peat layer varied in thickness but dynamic
probing data from one location would suggest it had been
compressed substantially. Below the original natural ground

level the increase in penetration resistance indicated significant densification of the looser natural sand deposit to a depth of 6.0 m. Dynamic probing measurements made at distances up to 3 m from the trial compaction area also indicate some improvement in the natural looser sand deposits between 3 m and 6 m below ground level. Here there was no sign of the loosening of the upper sand fill due to severe disturbance, only improvement resulting from the tendency of loose sands to densify in response to ground borne vibrations.

SELECT GRANULAR FILL (SHEFFIELD)

14. Trials were arranged between BSP and Scott Wilson Kirkpatrick & Partners on the 13 ha site of a former steel mill in Sheffield using the redesigned compactor. The site was being redeveloped for light industrial use by Sheffield Development Corporation. Redevelopment included the removal of heavy foundations, basement structures, flues and service culverts. Concrete and masonry demolition material was retained, crushed and re-used as select fill in the large excavations left in this area. This material was placed in approximately 300 mm thick layers, but received no systematic compaction other than trafficking by heavy plant. The average depth of this fill is 3.5 m, with isolated refilled deep excavations up to 8 m deep. An area about 40 m x 30 m was set aside for compaction by the rapid impact compactor.

15. The nature of the select granular fill in this trial area prevented the use of penetrative site investigation techniques. A borehole was drilled through the fill and underlying alluvial material, close to the centre of the trial area. A borehole magnet extensometer was installed down to bedrock at about 7 m below ground level. Pre-treatment ground surface levels were measured on a 5 m grid spacing against a stable off-site datum. The trial area was treated with almost abutting compaction points in seven adjacent strips. The strip widths ranged from 3 compaction points wide to 5 compaction points wide to establish the most efficient method of operation. The carrier vehicle tracked down the centre of each strip, with compaction points located on the arc described by the slew of the crane. Each compaction point received 50 blows using the full 1.2 m fall, giving an average total energy input of 170 tonne.m/m^2. The total penetration and rate of penetration for each point was recorded. Figure 4 shows the settlement with depth resulting from the compactor treatment. The settlement of the top magnet marker and the average overall reduction in ground level were measured by levelling. The treatment induced some settlement through the full depth of the fill, but compression was principally in the upper 3 m to 4 m.

OLD ASH FILL (SHEFFIELD)

16. A large part of the eastern end of the old steel works site was formerly occupied by railway sidings. Fill consisting mainly of ash, clinker, slag and some finer grained particles had been deposited over the natural alluvial valley deposits to

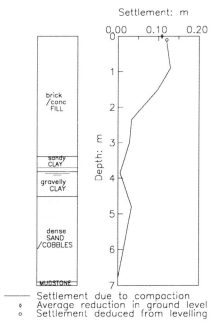

Fig. 4. Compression of select granular fill (Sheffield)

a depth of about 3.5 m. This fill was fairly loose and dynamic compaction was selected to treat most of the area, with the exception of a peripheral zone where the proximity of boundary structures and existing railway lines prohibited the use of the technique. Vibro was chosen to treat the fill within 20 m of these sensitive structures. A rapid impact compactor trial area about 40 m x 35 m was set aside within the dynamic compaction treatment area.

17. Dynamic probing and geophysics were used to assess the pre- and post-treatment ground conditions in the ash fill. Several dynamic probing profiles (DPM), using a 30 kg hammer weight with a 15 cm^2 cone, were measured prior to treatment by the rapid impact compactor and also in areas to be treated with vibro and dynamic compaction. A borehole magnet extensometer was installed in the compactor trial area. A concrete pad was cast on the surface of the ash fill and a loading test carried out on the untreated fill. Load was applied with two sand filled skips placed on the pad, with the extensometer extended up through the centre of the skips. All of the ash fill area of the site was covered with a 0.5 m working blanket of demolition waste (mostly brick and concrete fragments) prior to treatment. The complete trial area was levelled on a 5 m grid against an off-site datum prior to the treatment.

18. The loose nature of the fill in this trial area resulted in much greater penetration of the compacting foot at each treatment point. A treatment pattern of almost abutting compaction points was carried out, each point receiving 50

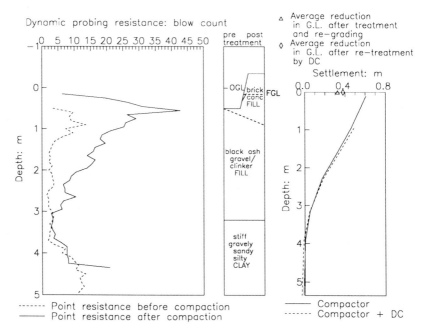

Fig. 5. Improvement of old ash fill (Sheffield)

hammer blows at the full 1.2 m weight drop, to give a total average energy input of 150 tonne.m/m^2. To avoid damaging the magnet extensometer a compaction point was located immediately over the gauge and surrounding points were treated in sequence to minimise the risk of shearing the gauge below ground level. These compaction points received some bedding blows prior to the full 50 blow treatment, resulting in the total energy input at this location being higher than the average for the rest of the trial area. The rest of the area was then treated in strips four compaction points wide.

19. This trial has provided the most comprehensive picture of the capabilities of the rapid impact compactor and the use of appropriate site control methods. The fill was fairly loose and underwent substantial compression and densification during treatment. The fill immediately surrounding the extensometer received more compaction than the average for the trial area. Figure 5 shows settlement with depth measured by the borehole extensometer. The rapid impact compactor induced significant compression within the whole depth of the fill. A heavy bulldozer was used to flatten the heavily disturbed surface of the fill and the trial area was re-surveyed to measure the average overall reduction in ground level due to rapid impact compaction. The upper fill was not systematically re-compacted, other than by the trafficking of the bulldozer, resulting in the difference between the average reduction in ground level and the settlement of the top magnet of the extensometer. The lack of compaction of fill close to the ground surface in

the trial area is an important observation which is discussed
in paragraphs 25 and 26. Dynamic probing measurements indicated
a considerable increase in resistance through the full depth of
the fill, reflecting the settlement profile.

DISCUSSION
20. The trials have shown that the compaction plant and
process can provide a considerable amount of information about
in situ ground conditions and the response to treatment. In the
trials each compaction point was identified and the rate and
total penetration of the foot was recorded for a given number
of blows. Figure 6 shows this information presented in plan for
the old ash fill trial area at Sheffield. Total penetration for
50 blows is shown on a scale of shading, a simple technique
using computer software which has highlighted a diagonal
feature across the trial area. Measurements also indicated
little or no reduction in penetration rate with increasing blow
count, confirming that the compactor had identified an area of
particularly poor fill. Site investigation borehole logs
revealed a layer of soft cohesive fill between 1.2 m and 2.4 m
below original ground level, within the mainly granular ash
fill on the line of the inferred soft zone (BH 17).
21. Dynamic probing has been used to establish depth and
degree of improvement. Comparisons between untreated and
treated ground are largely qualitative but useful information
can be obtained from a typical shallow fill site. Geophysics
can be a useful method of establishing the depth and degree of
improvement in a fill such as the extremely variable building
waste at Waterbeach. The ability of these and other techniques
to assess ground treatment is discussed further by Butcher and
McElmeel[7].
22. The initial objective of the BRE assessment of this
ground improvement technique was to establish its effectiveness
in improving the engineering properties of poor fills, with
particular emphasis on the depth to which significant
improvement could be achieved. Encouraging initial results led
to trials where the rapid impact compactor was compared
directly with other ground improvement methods currently
available. The trials suggested that the rapid impact compactor
could be used to effectively compact and improve the properties
of miscellaneous fills of a generally granular nature. Although
the depth affected is limited to about 4 m, this would be
adequate for a large number of reclamation projects where
ground improvement is currently carried out by vibro and
dynamic compaction.
23. At Sheffield it was possible to compare different
techniques used to treat an old ash fill. Dynamic penetration
tests were carried out to measure the change in penetration
resistance due to the installation of vibro stone columns,
dynamic compaction and treatment by the rapid impact compactor.
Figure 7 shows moderate improvement for vibro measured between
stone columns set out at 2.2 m centres on a triangular grid
pattern (Butcher and McElmeel)[7]. This method of assessment does

SHEFFIELD SITE LOG APPROX 35M X 40M AT 1.75M CENTRES   PHASE 2

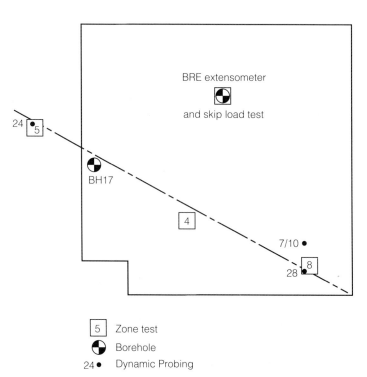

Fig. 6. Trial area on old ash fill (Sheffield)

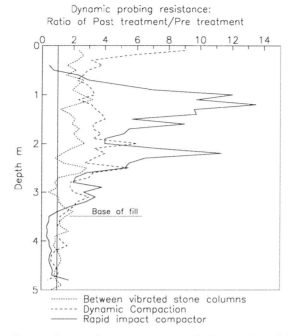

Fig. 7. Comparison of treatment techniques in old ash fill

not take into account the stiffening effect of the columns.
Dynamic compaction produced significant improvement close to
the ground surface which rapidly reduced with depth. The full
depth of fill was improved. The degree of improvement achieved
by the rapid impact compactor in the trial area was greater
than dynamic compaction, probably due to a total energy input
per m^2 of about 4 x that of the dynamic compaction. The maximum
effect of the single pass compactor treatment was at about 1 m
below the final ground level, with the fill close to the
surface showing little or no improvement. Similar observations
at all the trials indicate the necessity for a minimum of two
treatment phases, with primary compaction aimed at deep
densification and a secondary treatment to ensure adequate
compaction of the material close to the ground surface.

24. This example highlights the need for satisfactory site
control of the compaction procedure and the appropriate
measurement of the improvement of the treated ground. Ground
treatment is commonly used to provide adequate bearing capacity
for foundations in poor soils and fills, and plate or zone
loading test are often adopted as a form of proof testing to
demonstrate the performance of treated ground. These tests may
be carried out in a matter of hours and provide little, if any,
information about the long term behaviour of the treated
ground, particularly at depth, which may have a much greater
influence on the long term performance of a structure. Loose
partially saturated fills will undergo long term creep
settlement due to self weight and are likely to be susceptible

409

to collapse compression if inundated. Both of these phenomena can be reduced or eliminated by adequate densification. Treatment methods should be capable of improving poor ground at depth as well as near the surface and the testing techniques should be able to confirm the improvement. The ground conditions immediately below a test plate or pad are likely to have a controlling effect on load test results and predictions of future ground performance based on these tests. This can be demonstrated using data from the Sheffield old ash fill trial area.

25. Zone tests comprising 1.75 m square concrete pads loaded to 245 kN/m^2 over 24 hours were performed by the specialist contractor carrying out general ground improvement at Sheffield. Zone test 4 was performed after rapid impact compaction in a zone of poor fill at the location shown in Figure 6. Dynamic probing test 10 on the edge of the soft zone was also performed after the rapid impact compactor treatment, indicating a substantial increase in penetration resistance in the upper ash fill, with little improvement in or below the silty clay fill (Figure 8). Zone test 4 settled nearly 60 mm

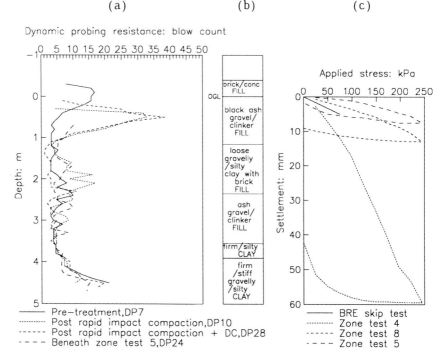

Fig. 8. Penetration resistance and zone test results in old ash fill

    (a) Dynamic probing
    (b) Soil profile
    (c) Load tests

under full load and the trial area was recompacted using
dynamic compaction to the general site specification, following
which zone test 8 was performed at the location shown in
Figure 6. Dynamic probing test 28, carried out after this
re-treatment, shows both the granular and cohesive fills were
virtually unaltered by the additional treatment by dynamic
compaction, whilst Zone test 8 indicates a remarkable
improvement. Zone test 5 was carried out in an area treated
only by dynamic compaction and settled very little although
dynamic probing test 24, in the centre of this zone test,
showed only marginal improvement when compared with
pre-treatment results in the dynamic compaction and rapid
impact compactor trial areas.

26. It is apparent from Figure 8 that the zone test results
from Sheffield did not reflect the differences in ground
conditions indicated by penetration resistance, which in
granular soils is related to relative density. The rapid impact
compactor, whilst imparting more energy and inducing greater
densification at depth, caused severe disturbance in the upper
fill. This disturbed zone probably extended about 1 m below
original ground level. Fill in the trial area was graded flat
by bulldozer but not systematically recompacted, whilst the
surface of the fill treated by dynamic compaction remained
relatively undisturbed by the primary compaction and was packed
down by a low energy secondary pass with contiguous treatment
points. It is likely that zone test 4, despite being founded
0.5 m below existing ground level, was founded on disturbed
fill. Re-treatment with dynamic compaction densified this near
surface fill, ensuring a satisfactory result for zone test 8.
The result from the BRE skip loading test carried out prior to
treatment of the ash fill in the trial area is also shown in
Figure 8. The result indicated a reduction in compressibility
of the fill near the surface due to treatment.

CONCLUSIONS
27. The BSP rapid impact compactor has demonstrated the
ability to improve the engineering properties of a range of
fills and natural sandy soils. The field trials show this to be
a very promising technique for the improvement of miscellaneous
fills of a generally granular nature up to depths of about 4 m.

28. A significant proportion of small sites being
re-developed for low rise construction comprise shallow fills
3 m - 4 m deep which are often treated with vibro stone
columns. It is likely that many of these fills could be
effectively treated by the rapid impact compactor.

29. The compaction process should include a minimum of two
treatment phases, with high energy primary compaction using
abutting compaction points to improve ground at depth and a
low energy secondary treatment, utilising a larger square foot
or heavy roller, to ensure adequate compaction of the material
up to the ground surface.

30. Site control measures such as the dynamic probing or
geophysical measurements should be selected to suit the ground

conditions. The results of surface load tests are likely to be controlled by near-surface conditions and will give little indication of ground properties at depth or the long term behaviour of the material. Data from the compaction process can give valuable information about ground conditions before treatment and changes resulting from the treatment.

ACKNOWLEDGEMENTS

31. The work described in this paper forms part of the research programme of the Building Research Establishment. A number of organisations have assisted BRE in carrying out the work described in this paper. The Defence Research Agency loaned the MoD compactor and Royal Engineers 12 Brigade the Terex carrier vehicle and support staff for the first two trials. BSP International Foundations Ltd. supplied full site support for all the trials and made their data available to BRE. Other collaborating organisations were M. Dickerson Ltd. (Waterbeach site), Test and Evaluation Establishment (formerly RAE, West Freugh) and Sheffield Development Corporation, Scott Wilson Kirkpatrick and Birse Ltd., owners, consulting engineers and main contractor respectively at the Sheffield sites. BRE staff carried out all of the site investigations and measurements. Dynamic probing (DP) measurements and the skip load test were carried out by Mr.K.McElmeel and Mr.T.Reynolds and geophysical measurement were carried out and interpreted by Mr.A.Butcher.

REFERENCES

1. MENARD L. La consolidation dynamique des remblais recent et sols compressibles. Travaux, 1972, November, pp 56-60.
2. CHARLES J. A., BURFORD D. and WATTS K. S. Field studies of the effectiveness of "Dynamic Consolidation". Proceedings of the 10th International Conference on Soil Mechanics and Foundation Engineering, Stolkholm, 1981, pp 617-622.
3. ST JOHN H. D., HUNT R. J. and CHARLES J. A. The use of 'vibro' ground improvement techniques in the United Kingdom. BRE Information Paper IP 5/89, 1989, March.
4. BS 1377: Part 9. 1990. Soils for civil engineering purposes.
5. DIN 4094 German Standard, Subsoil Dynamic and Static Penetrometers, Part 1 - Dimensions of Apparatus and Method of Operation. DIN Deutches Institut fur Normung e.V. Berlin.
6. ABBISS C. P. Shear wave measurements of the elasticity of the ground. Geotechnique 31, 1981, No 1, pp 91-104
7. BUTCHER A. P. and McELMEEL K. The ability of in situ testing to assess ground treatment. This conference.

33. Performance of an engineered fill at Lounge opencast coal site

A. K. GOODWIN and J. M. W. HOLDEN, Scott Wilson Kirkpatrick, UK

SYNOPSIS. This case history describes the controlled backfilling operations that were undertaken within Lounge Opencast Coal Site in order to allow the construction of the A42 in Leicestershire, England. After summarising the background to the site, the specification for the reclamation works and its application on site is described. The details and results of a comprehensive compaction and settlement monitoring programme are presented and analysed. The effect of periods of heavy rainfall on the movements is shown. Comparisons with previously published performance data are made and conclusions drawn.

INTRODUCTION

1. Lounge site was located approximately 3 km from Ashby-de-la-Zouch, Leicestershire, and covered approximately 231 ha. Two excavation areas, A and B, covering 136 ha, were worked north and south of the B587. The clients for the opencasting and highway were British Coal Opencast Executive and the Department of Transport, respectively. Scott Wilson Kirkpatrick were retained as consultants by both clients and were consequently involved in the works from preparation of the compaction specification to highway construction. The Contractors were A F Budge (Mining) and A F Budge (Contractors) Ltd, respectively.

2. The excavation to remove up to 8 seams at depths of up to 40 m below the A42 was progressed in a series of box cuts at an average weekly rate of about 250,000 m^3, and continued uninterrupted until completion in May 1990. The progress of the compaction works beneath the highway corridor is summarised in Figure 1. The controlled compaction typically covered a width of 45 m to 60 m, and 2.7 km of the A42 was to lie over controlled fill. The total volume of controlled backfill was approximately 3.7M m^3, deposited in the period July 1987 to June 1990.

3 The A42 runs largely at or near the opencast restored levels, as shown in Figure 1, but a 7.5 m high embankment was required to cross the B587. As construction of the A42 was split into several contracts, the works were arranged such that the period of time between backfilling and highway

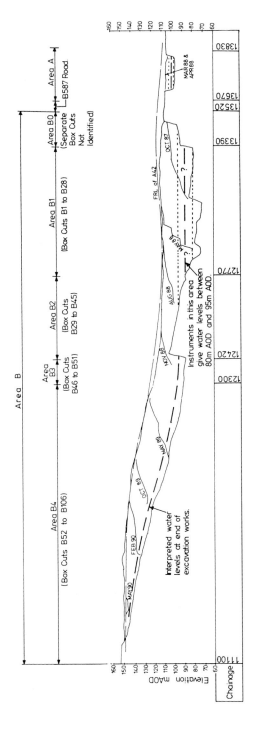

Figure 1   Cross Section of Lounge OCCS

Note: Dates on section indicate progress of controlled compaction

construction was greater over the deeper parts of the site. Over the northernmost 1.1 km of Lounge, the embankments were formed in September and October 1989, and the pavement was completed by late 1990. Over the southern part of the site, the embankments were built from July to September 1990, and pavement construction was completed in March 1991.

GEOLOGY AND GROUNDWATER

4. The solid geology of the site was Lower Coal Measures, overlain in the southern third of the site by up to 20 m of Sherwood Sandstone. Glacial Till overlay the solid geology. Typically, this Till was less than 3 m thick but in places channels had been incised through the Sherwood Sandstone into the Lower Coal Measures and infilled with up to 25 m of Glacial Till.

5. The Lower Coal Measures formed the majority of the backfill, and comprised in general highly weathered very weak mudstones and siltstones which became less weathered and typically weak to moderately weak with depth. Within the mudstone, moderately strong to strong thickly bedded sandstone horizons up to 1 m thick were encountered.

6. The groundwater conditions at the site prior to the opencasting were indicated by five deep Casagrande piezometers, and the inferred water table lay some 8.5 m to 16.5 m below the surface in Area A, and 35 m to 41 m below the ground surface in Area B which was up to 40 m higher than Area A. Above this, perched water tables in the Lower Coal Measures and Sherwood Sandstone Series were indicated by limited seepages from more permeable horizons in the excavation walls. Generally though, the site excavation was very dry.

SPECIFICATION

7. Restoration beneath the highway corridor was carried out in accordance with a specification based in part on the Department of Transport Specification for Road and Bridge Works (1). The principal features of the Specification were:

i)   It required compaction to the specification throughout a Controlled Zone beneath the highway corridor, and stipulated the minimum volume of fill beyond the Controlled Zone that was to be brought up at the same rate in order to ensure stability (see Figure 2).

ii)   It restricted allowable moisture contents to those given in Table 1.

iii) It set the compaction criteria set out in Table 2. Material compacted dry of optimum moisture content (OMC, as defined by Test 12 of BS 1377 : 1975, using a CBR mould) was required to achieve a target dry density of 2.0 Mg/m^3. The very limited amount of material compacted wet of OMC was compacted in accordance with the method specification (1).

iv)   It restricted the steepness of the final high walls and
        side walls of the excavation to reduce differential
        settlements occurring over small distances, and
        preserve the ride characteristics of the road when
        crossing such walls. Walls were generally formed at
        1v:2h or less in the weathered deposits near the
        surface, and 1v:1h through competent material (but
        1v:1.5h within 9 m of surface).

v)   It required benching of slopes within 20 m of the
        surface in undisturbed ground or in previously placed
        fill prior to placing backfill.

8. As noted above most of the controlled backfill was
compacted to the performance specification. Placement and
compaction was generally carried out by a CAT 825C (four
wheeled tamping roller with a blade), which spread 300 mm to
400 mm loose layers and compacted them with 8 to 10 plant
passes (16 to 20 compactor wheel passes). Method compaction
when used was carried out with Stothert and Pitt 72T and T182
single drum vibrating rollers (3315 kg per metre width) towed
by a CAT D8 dozer.

9. The Specification was designed to achieve a maximum air
voids for the compacted fill of either 5% (for moisture
content >OMC), or 12% (for moisture content <OMC). These
limits were considered practicable and were deemed necessary
to achieve a high standard of compaction.

Table 1   Specified Standards of Compaction

Moisture Content	Specification Type	Detail
(OMC - 4) to OMC	Performance	Target minimum dry density = $2.0Mg/m^3$
OMC and above	Method	DoT Specification for Road and Bridge Works

Table 2   Limits on Moisture Content of Backfill

Material Type	Moisture Content (%) Maximum	Minimum
Cohesive	1.0 x PL	PL - 4
Granular or dry cohesive	OMC + 2	OMC - 4

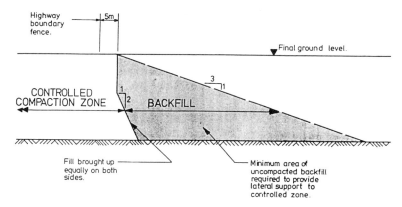

Figure 2 - Edge Detail of Compaction Zone

COMPACTION MONITORING

10. Compaction control testing was carried out several times per day using an MC3 nuclear density/moisture content gauge. The accuracy of the instrument was periodically verified by comparing MC3 results with sand replacement density tests (BS1377:1975, Test 15) and large scale water replacement tests. The calibration factor to correct the MC3 moisture content readings was checked daily using oven dried moisture content tests on samples of the fill tested with the gauge.

11. Adherence to the specification was checked also by continuous observation, site surveys, and by laboratory testing samples of the fill on a regular basis.

RESULTS OF COMPACTION CONTROL TESTS

12. The vast majority of the excavated material proved to be suitable for use in the Controlled Zone, and the backfill chiefly comprised very weak to weak mudstone with lesser quantities of weak siltstone. The mudstone had been excavated using an hydraulic face shovel (O & K RH90 or RH120), and was typically in the form of a well graded fragmented mudstone with particles up to 250 mm before compaction. OMC was typically about 10% but ranged between 9% and 11%.

13. The results of the compaction monitoring are summarised in Table 3. Figure 3 shows that the Specification was adhered to for the vast majority of the material, and Figure 4 summarises the spatial variation of air voids. As the majority of fill placed was drier than OMC and subject to the performance specification, it is evident that the air voids were consistently below the 12% upper bound intended. The small minority of materials that were shown to fall slightly below the performance criteria were accepted on site for a variety of reasons, including the uniform grading of the material and its sometimes low specific gravity.

417

Table 3 Summary of Compaction Monitoring Data

Area of Site	In situ Test Results				Lab Comp'n Test 13 (Test 12)	
	moisture content (%)	air voids (%)	dry density (Mg/m³)	relative compaction field/lab* (%)	omc (%)	mdd (Mg/m3)
Area A	8.5 [6.5-11.6]	8.4 [4-12]	2.3 [1.95-2.10]	110 (114)	11.0 (12.5)	1.85 (1.78)
Area B	8.9 [6.0-14.8]	6.3 [0-10]	2.04 [1.95-2.12]	110 (115)		

[ ] indicates range
*   Unbracketed figures are for Test 13 lab compaction; bracketed figures relate to Test 12

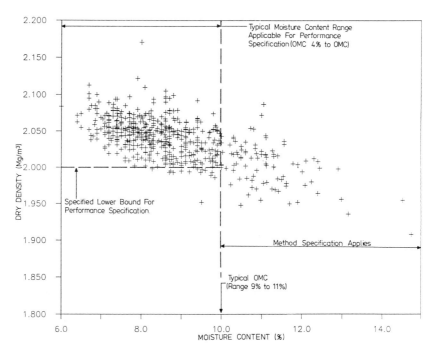

Figure 3 - Variation of Compaction Density with Moisture Content

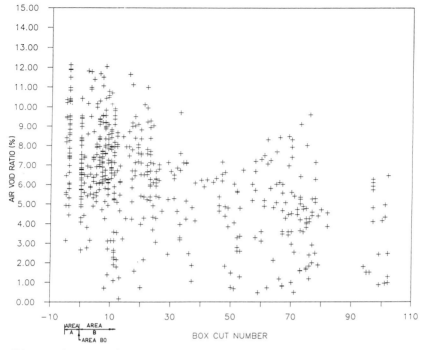

Figure 4 Spatial Variation of Air Voids Within Lounge OCCS

MONITORING INSTRUMENTATION

14.   Surface settlement markers (101 No), extensometers (9 No), piezometers (21 No) and standpipes (8 No) were installed along the length of the highway corridor within and just outwith Lounge site. A period of between 1 and 6 months elapsed between completion of backfilling and installation of instruments, with a typical interval of 2 months. All instruments were regularly read. Settlement markers were read to an estimated accuracy of ±5 mm, and extensometers to ±1 mm.

GROUNDWATER RECHARGE AFTER BACKFILLING

15.   Groundwater conditions prior to the opencasting operations were described earlier. The water levels recorded within the backfill showed that groundwater recharge does not appear to have occurred to any great extent. The instruments generally indicate that there is up to 5 m of water at the base of the excavation (see Figure 1), which dips towards worked coal seams to the north east of the site. These appear to be acting as a drain to the whole area even though pumping of the deep mines in the area ceased in 1967. Coupled with the location of the site in the vicinity of a local watershed, it is likely that the recorded depression of the water table will be a permanent feature and full groundwater recharge will not occur.

SETTLEMENT MECHANISMS

16. Settlement of opencast backfill may in general be caused by any or all of the following:

a) settlement on saturation, either due to a rising groundwater level or ingress of surface water
b) settlement due to surcharging or construction of embankments
c) long term creep settlements due to the self weight of the backfill occurring under conditions of constant effective stress and moisture content.

Creep strains and strains on saturation are proportional to the air voids in the material (2, 3, 4, 5) over the range of air voids typically encountered in opencast backfills.

SETTLEMENT RECORDS

17. The analysis of the settlement records has been divided into the six sections of the A42 identified in Table 4. Ground movements are summarised in Table 5.

Table 4 Sections of A42 Used for Interpretation of Settlements

Section	Chainage	Average Embankment Height (m)	Average Depth of Backfill (m)
Area A	13660 - 13880	5	9
Area B0	13380 - 13510	5	16
Box Cuts B1 to B28 (Area B1)	12750 - 13510	2	30
Box Cuts B29 to B45 (Area B2)	12400 - 12750	3	15
Box Cuts B46 to B51 (Area B3)	12280 - 12400	4	30
Box Cuts B52 to B106 (Area B4)	11020 - 12280	<1m fill or in shallow cut	30 m (ch 11300-12280); gradual variation from zero to 30 m (ch 11020 - 11300)

Table 5   Summary of Settlement Monitoring

| Extensometers/ Markers | Monitoring period (months after backfilling) | Strains on Inundation | | Creep compres- sion parameter (%) |
		Occurred months after backfilling/ inundation	Total strain at surface (%)	
Area A Markers 1 to 10 (8 no); Extensometer E1	1 - 49	1 - 25 after backfilling	0.56 [0.2 - 1.5]	0.10 [0.04 - 0.13]
Box cuts B11 to B18 Markers 11 to 18 (8 no); Extensometer E2	8 - 52	8 - 33 after backfilling	0.62 [0.15 - 2.8]	0.28 [0.08 - 0.50]
Box cuts B1 to B28 Markers 19 to 40 (22 no); Extensometers E3 & E4	1 - 29 and 1 - 42	1 - 24 after backfilling	0.35 [0.10 - 0.75]	0.30 [0.18 - 0.45]
Box cuts B29 to B45 Markers 41 to 50 (10 no); Extensometer E5	2 - 26 and 2 - 38	5 - 22 after backfilling	0.22 [0.0 - 0.40]	0.24 [0.05 - 0.40]
Box cuts B46 to B51 Markers 51 to 60 (9 no)	2 - 22 and 2 - 36	8 - 20	0.35 [0.25 - 0.60]	0.24 [0.12 - 0.35]
Box cuts B52 to B10 Markers 63 to 85 (8 no); Extensometers E6 to E9	1 - 22	1 - 6 after backfilling (largely complete)	0.26 [0.0 - 0.40]	0.17 [0.06 - 0.25]

[ ] indicates range

18. Examination of the results has shown that distinct settlement patterns are discernible for each zone, as exemplified by the records for box cuts B1 to B14 (settlement markers SM19 to SM28) in Figure 5. Typical forms of the "average" settlement-time records for each zone are shown in Figure 6, and from this it is apparent that there were periods of rapid settlement superimposed upon an underlying creep settlement rate. Figure 6 shows dates only up to early 1991, as monitoring showed no further periods of rapid settlement after this time (see Figure 5) and acquisition of the rainfall data ceased in early 1991 with the siteworks. The creep settlement rate may be analysed in terms of the creep compression parameter, which is defined as the amount of creep strain that occurs within one log cycle of time, usually measured from the time at which half the full height of fill has been placed.

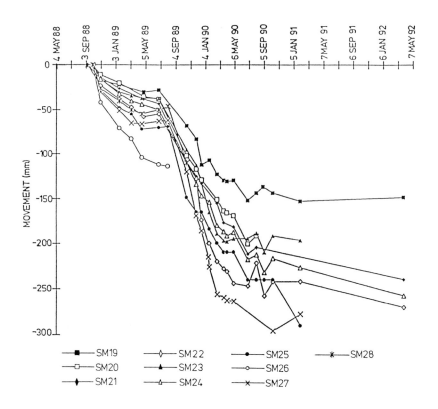

Figure 5 - Typical Surface Settlement - Time Profiles

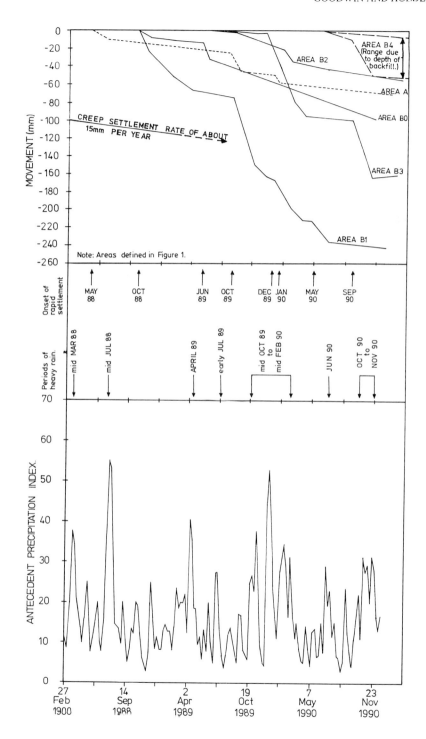

Figure 6 - Average Settlement Profiles and Rainfall

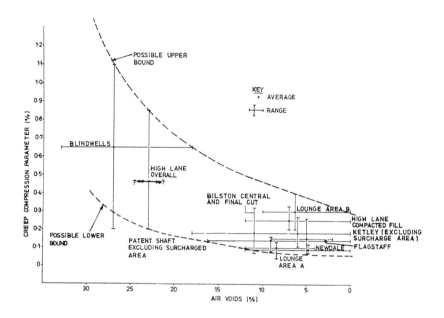

Figure 7 - Relationship Between Creep Compression Parameter
and Air Voids

19. The creep compression parameter for the fill varied
between 0.04% and 0.55% (Table 5), with values typically
between 0.1% and 0.3%. The range of results together with
those from other British Coal sites are plotted against air
voids in Figure 7. This shows that the results obtained are
typical of experience at other sites, if slightly larger than
average in Area B. A possible explanation for this is that
the controlled zone lies immediately adjacent to the side wall
and shear stresses between the rock wall and fill would have
developed . These would have slowed the settlement rate in
the early stages until material degradation progressively
allowed movements to occur. This effect would diminish as the
monitoring period increases, but would have resulted in
higher measured creep strains at Lounge because a smaller
proportion of the strains would have occurred before the
relatively early installation of the monitoring
instrumentation at this site compared with other sites.
20. To investigate the causes of the periods of rapid
movements, a comparison was made between the records and the
construction history. No significant influence of surcharging
on settlements was found, which was largely as expected from
a comparison of the embankment height (which was less than
7.5 m) and the fill depth (which was up to 40m). Some small
heaves were recorded where no surcharging was applied.
424

21. In the absence of significant groundwater recharge, the influence of rainfall upon the behaviour was investigated. The rainfall records for the site were interpreted using a modified form of the Antecedent Precipitation Index to obtain a weighted average for the rainfall over the preceding 4 weeks and hence highlight the peaks relative to "background average". The equation used was:

$$API = 0.044P_4 \quad 0.088P_3 + 0.177P_2 + 0.354P_1 + 0.707P_0 \quad \ldots \ldots (1)$$

where P_i = average weekly rainfall for the i weeks before the current week (week 0).

22. The resultant rainfall pattern is shown in Figure 6. It is apparent that the periods of heavy rainfall preceded the periods of rapid movement by about 2 to 3 months. Some time lag was expected as the majority of water induced movements are thought to be associated with softening of the inter-particle contacts and associated reduction in air voids as the backfill compresses under the weight of the overlying materials. Such softening occurs continually due to weathering and the "background" rainfall pattern, and contributes to the creep settlement. The heavy rainfall periods would have resulted in wetting fronts that penetrated deeper and more uniformly through the fill. A time lag would result from the time required for the following to occur:

(i)   for the wetting front to descend through the fill, particularly below the upper horizon regularly wetted by the background rainfall;
(ii)   for softening of particles;
(iii) for localised movements to propagate upwards through the fill to the surface.

23. Total inundation strains varied between 0.1% and 1.7% but were typically between 0.6% and 1.2%. These values are typical if slightly towards the upper limit found from past experience (see Figure 8). This similarity supports the view that strains due to repeated cycles of temporary saturation associated with wetting bands descending from the surface (as at Lounge), trend towards those strains that arise from permanent saturation associated with restoration of a high groundwater table (as at the other sites reported in Figure 8). However, strains arising from the former would be expected to occur less rapidly, as demonstrated by the Lounge data which shows that periods of rapid movement continued to occur for about two years. Explanations for the settlements being slightly higher than typical may be that early installation of the monitoring instrumentation could have led to the full extent of these movements being recorded, and that dry mudstone dominated the fill. Such materials are recognised to be more susceptible to inundation settlement.

425

24. The magnitude of each saturation induced settlement cycle does not appear to be directly related to the magnitude of the rainfall peak, where this peak is defined in relation both to the intensity and duration of the rainfall. It is likely that local variations in permeability of the backfill affected the flow paths of the water and that the preferential flow paths varied with time. Thus the inundation history of each part of the backfill would have varied and affected the behaviour, as would the depth of backfill and variations in its compacted state.

25. Despite these masking influences, it was observed that slightly larger creep settlements and more pronounced inundation strains occurred in box cuts B1 to B28, and in particular B11 to B18. This has led to problems with the ride quality of the A42 at this location, as although sufficient time was left before pavement construction for the majority of inundation strains to occur, the slightly larger than usual creep strain rate in this area has caused significant differential settlements relative to adjacent areas. Investigation of the possible causes showed that the

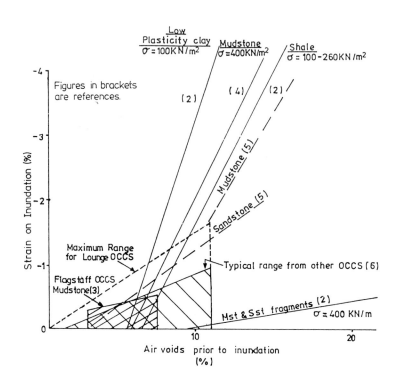

Figure 8 - Relationship Between Inundation Strains and Air Voids

compaction characteristics of the fill in this locality were not different from those elsewhere, but it was noted that the affected area coincided with a sag curve on the A42 in a generally low lying area. Ditch drainage from a long stretch of the highway outfalls into a stream which crosses the A42 at the low point of the sag curve. Furthermore, a seasonally marshy area lies to the west of the road at this point, and this is fed by flows from a pond on the opposite side of the highway. In combination, all these factors resulted in a greater potential for saturation of the backfill and associated settlement in this locality.

CONCLUSIONS AND RECOMMENDATIONS

26. The compaction monitoring data shows that the Specification for the site resulted in a high standard of compaction of a dry mudstone, with air voids typically less than 10%. The requisite air voids were achieved through the imposition of a minimum target dry density and moisture content limits, both of which are more quickly verified on site than air voids.

27. The settlement records show creep compression parameters typically between 0.1% and 0.3%, and inundation strains typically between 0.6% and 1.2%. These magnitudes accord with previous experience. Little difference in saturation strain magnitudes resulting from full groundwater recharge and repeated temporary saturation due to surface inundation is expected. The records show also that, in the absence of groundwater recharge, periods of rapid movement continue to occur over a period of several years, and are associated with heavy rainfall events. It is recommended for future sites that at least two wet seasons be allowed between completion of backfilling and construction to allow the majority of these movements to occur.

28. It is recommended that careful consideration be given to the topography and drainage to avoid inundation from the surface accelerating settlements in some areas, which would then cause large differential movements relative to areas which have not been inundated. This applies in particular to opencast sites with low groundwater tables.

29. Design should allow for differential settlements arising from settlements delayed as a result of shear stresses between backfill and excavation walls.

ACKNOWLEDGEMENTS

Grateful acknowledgement is given to British Coal and the Department of Transport for permission to utilize the compaction and settlement data, respectively. The input of colleagues within SWK was gratefully received also, particularly from Jon Saunders who was present on site during the restoration works.

REFERENCES

1 Department of Transport (1976) "Specification for Road and Bridge Works", HMSO

2 Charles, J A (1984) "Some geotechnical problems associated with the use of coal mining wastes as fill materials", Proc Symposium on the Reclamation Treatment and Utilisation of Coal Mining Wastes, Durham

3 Thompson J, Holden J M W and Yilmaz M (1990) "Compaction of opencast backfill beneath highways and associated developments", Proc 3rd Int Symposium on the Reclamation Treatment and Utilisation of Coal Mining Waste, Glasgow

4 Cox, D W (1978) "Volume change of compacted clay fills", Proc Conf on Clay Fills, ICE

5 Egretti I and Singh R N (1988) "A laboratory investigation into the effects of air void and water saturation on the collapse settlement of opencast mine backfill", Min. Sci. Technol., Vol 7, No 1, pp 87-97

6 S J Hodgetts, J M W Holden, C S Morgan and J N Adams (1993), "Specifications for and performance of compacted opencast backfills", Proc Conf. on Engineeered Fills 1993

34. Settlement of opencast coal mining backfill at Horsley 1973–1992

J. A. CHARLES and D. BURFORD, Building Research Establishment, UK, and D. B. HUGHES, British Coal Opencast, UK

SYNOPSIS. A 70 m deep mudstone and sandstone opencast coal mining backfill at Horsley near Newcastle has been monitored throughout a 20 year period from 1973 to 1992 during which the ground surface has settled by as much as 0.8 m. A 34 m rise in ground water level between 1974 and 1977 has been a principal cause of the settlement. Settlement at depth within the fill has been measured using magnet extensometers and surface movements have been measured by precise levelling. The implications of the ground movements for building developments on opencast backfills are briefly discussed.

INTRODUCTION
1. The effect of a rising ground water level on the settlement of an uncompacted opencast coal mining backfill has been monitored at Horsley in the north east of England. The investigation has been carried out jointly by the Building Research Establishment and British Coal Opencast. Interim reports were presented in 1977 and 1984 (Refs 1 and 2). Monitoring at the site was discontinued at the end of 1992 and this paper presents a final report on the investigation.
2. Backfilling took place between 1961 and 1970 with restoration completed in 1973. It was necessary to de-water the site during opencast mining and pumping continued for some time after the completion of backfilling. Instrumentation was installed in 1973 to monitor settlement at different depths within the backfill. When pumping stopped in April 1974, the ground water level rose 34 m reaching a new equilibrium level in 1977. The site was returned to agriculture after the completion of backfilling and, so far as is known, no building development is planned.
3. The opencast workings covered an area about 1500 m long by 600 m wide and the backfill has a maximum depth of almost 70 m. The excavated strata belong to the Middle and Lower Coal Measures of the Carboniferous system. In the upper part of the workings excavation of the overburden was carried out by face shovels and backfilling was by end tipping from dump trucks. In the lower part of the workings excavation was by dragline.

4. Information about the backfill was obtained in 1973 when boreholes for the installation of instrumentation were drilled using a rotary air flush rig. The heterogeneous fill was composed principally of mudstone and sandstone fragments, with mudstone predominating. Less than 10% of the backfill was boulders. Cavities up to 0.5 m deep were found. Open drive 100 mm diameter samples indicated that about 10% of particles were finer than 0.075 mm and about 50% were coarser than 2.36 mm. The average dry density was 1.70 Mg/m^3 and the average moisture content was 7%, but these properties showed great variations between samples. The degree of saturation of the fill as placed varied between 10% and 100%. The relative density (I_D) was estimated to be about 60% and the mean SPT N value of 29 confirmed that the fill was of medium relative density. A field test in a borehole at the site of the lagoon indicated a permeability greater that 10^{-4} m/s.

INSTRUMENTATION

5. The following types of field measurement have been made:

   (a) settlement at different depths within the fill has been monitored using 5 magnet extensometers installed in boreholes drilled through the full depth of fill,
   (b) surface settlement of the fill has been monitored by precise levelling of traverses of surface settlement stations,
   (c) ground water level has been monitored in standpipe piezometers.

6. Late in 1973 five magnet extensometers were installed in 0.15 m diameter boreholes drilled through the backfill. These are referred to as borehole settlement gauges and details of the installation have been given in ref 1. The bottom magnet marker of each gauge was installed in bedrock and formed a stable reference point. From measurements taken on successive occasions, settlement of the magnets relative to the reference magnet could be computed. In this way settlement has been monitored at different depths within the backfill.

7. The locations of the five borehole gauges, which were selected to provide as much information as possible about the behaviour of the backfill, are shown in Figure 1. Information about the gauges is summarised in Table 1. Gauge D1 was installed in fill which had been preloaded by an overburden heap with a maximum height above restored ground level of 30 m and gauge C11 was installed in an old lagoon area. Gauge A9 was installed in the oldest fill and gauge D15 in the most recently placed fill.

8. A traverse of surface settlement stations was established adjacent to each borehole gauge to supplement the information on surface settlement. Precise levelling has been carried out from bench marks established on undisturbed ground outside the limits of the opencast workings. Traverse E, in an area of intermediate age, had no borehole settlement gauge.

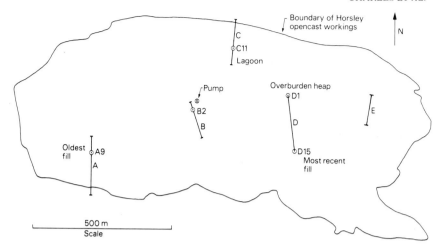

Figure 1. Plan of Horsley site showing location of magnet
extensometers and traverses of surface settlement stations

Table 1. Ground conditions at magnet extensometers

Gauge no	Ground level	Rock-head	Fill depth	Fill date	Inundated depth	Fill condition
	mAOD	mAOD	m		m	
A9	98.6	38.0	60.6	1961	46	oldest
B2	101.8	38.7	63.1	1964	45	deepest
C11	94.9	49.2	45.7	1965	35	lagoon
D1	108.1	52.6	55.5	1966	31	preloaded
D15	119.2	72.7	46.5	1970	11	recent
E12[1]	115.8	68	48	1966	17	intmdt age

[1] There was no magnet extensometer at E12, but it is listed
because the maximum settlement was recorded at this location.

MONITORED BEHAVIOUR OF FILL
    9. The surface settlements monitored by precise levelling at
the five borehole gauges are plotted in Figure 2. The maximum
movement measured anywhere within the site is at surface
settlement station E12 where the fill has settled 0.8m and this
is also plotted in Figure 2. The smallest movement measured at
a borehole settlement gauge has occurred at gauge D1 and this
can be attributed to the effect of preloading by a 30 m high
overburden heap. The settlement at gauge C11 has also been
small but it should be noted that, prior to the rise in water
level, the settlement rate at this gauge was greater than in
the other parts of the site. It may be that in the years
before monitoring commenced the settlement of the lagoon area,
which is composed of a wet and more cohesive fill, was large.
    10. Figure 3(a) shows the settlement measured at different
depths in the fill at gauge B2 where the fill is deepest and

Figure 3(b) shows the rise in ground water level plotted to the
same time scale. A surface settlement of 0.5 m has been
measured at this gauge and vertical compression has occurred
throughout almost the full depth of the gauge. Figure 4 shows
the total settlement measured during the 20 year period plotted
against depth for gauges B2, C11, D1 and D15.

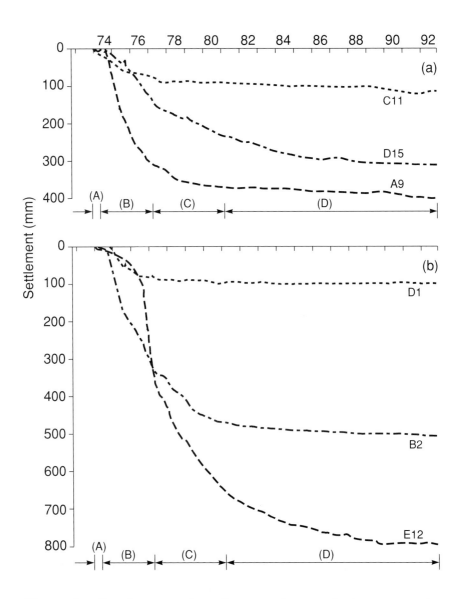

Figure 2. Development of surface settlement during successive
periods (A), (B), (C) and (D):
    (a) settlement at locations A9, C11, D15,
    (b) settlement at locations B2, D1, E12.

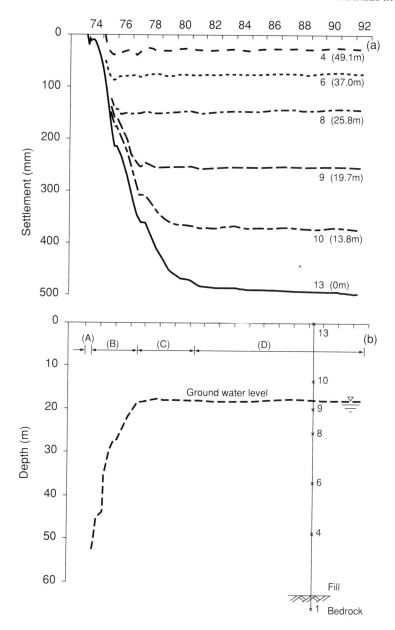

Figure 3. Relationship between settlement at different depths within the deepest fill (gauge B2) and the rise in ground water level:
- (a) settlement plotted against time for selected magnet markers, depth of magnet marker is shown in () after reference number of magnet eg 8 (25.8 m),
- (b) ground water level versus time with depths of magnet markers shown in relation to ground water level.

433

11. It is helpful to examine the settlement of the backfill during four periods:

Period (A) the few months prior to April 1974 when settlement was monitored while pumping kept the water level down in the bedrock below the backfill,

Period (B) the three years from April 1974 to April 1977 when the ground water level rose some 34 m through the backfill,

Period (C) the four years from April 1977 to April 1981 immediately following the rise in ground water level when ground movements were still affected by the rise in water level,

Period (D) the last 12 years during which the water level has shown only minor fluctuations.

These periods are shown on Figure 2 and Figure 3. The settlement measured during these four periods is summarised in Table 2.

Table 2. Surface settlement measured by precise levelling at Horsley 1973-1992

Gauge	Period (A) Dec 1973- Apr 1974		Period (B) Apr 1974- Apr 1977		Period (C) Apr 1977- Apr 1981		Period (D) Apr 1981- Dec 1992		Dec 1973- Dec 1992
	total mm	rate mm/y	total mm	rate mm/y	total mm	rate mm/y	total mm	rate mm/y	total mm
A9	002	005	310	103	057	014	028	002	397
B2	002	006	331	110	133	034	033	003	499
C11	017	043	061	020	015	004	016	001	109
D1	-010	-021	098	033	006	002	001	000	095
D15	001	002	152	051	081	021	073	006	307
E12	008	017	354	118	288	074	138	012	788

Period (A)
12. Between December 1973 and April 1974 ground movements were monitored prior to the rise in ground water level. However considerable settlement may have occurred prior to the commencement of monitoring in December 1973. The rate of settlement was greatest at gauge C11 which had been the site of a lagoon during opencast mining. In the four months of monitoring during this period the settlement at ground level measured at this gauge by precise levelling was 17 mm. The magnet extensometer indicated that compression was occurring over the full depth of the fill.
13. The backfill in the locality of gauge D1 previously had been loaded by a large overburden heap with a maximum height above restored ground level of 30 m. This had been removed two years before the measurements began. Four months of precise levelling prior to April 1974 showed a heave of 10 mm at ground level. The magnet extensometer indicated that this movement was caused by expansion at the base of the fill.

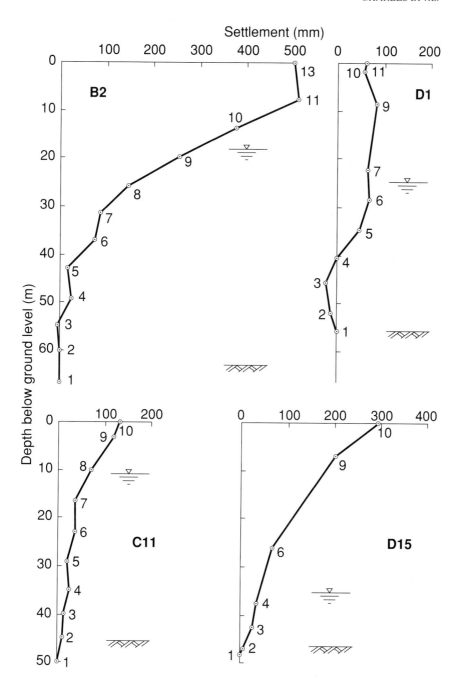

Figure 4.   Variation of settlement with depth within the fill
measured during period 1973-1992 at four locations B2, C11, D1,
D15; reference numbers of magnet markers are given for each
settlement measurement.

14. At gauge B2 close to the pump the water table varied between 5 m and 10 m above rockhead during this initial period in which pumping continued. The settlement measured by precise levelling was 2 mm during the four month period.

Period (B)

15. Ground movements were monitored from April 1974 to April 1977 as the ground water level rose. The submergence of a fill by a rising ground water level increases pore pressures thus reducing effective stresses and consequently heave of the ground might be expected. In practice most partially saturated fills undergo a reduction in volume, termed collapse compression, when their moisture content is increased. Partially saturated fills are susceptible to collapse compression under a wide range of applied stress when first inundated if they have been placed in a sufficiently loose and/or dry condition. Thus saturation often causes settlement due to collapse compression.

16. The ground water level at gauge B2 rose by 20 m between April 1974 and April 1975, 9 m in the following 12 months and 5 m in the 12 months after that. From June 1975 onwards the water level measured in the five borehole gauges has been virtually the same height above OD. Having reached a new equilibrium water level in April 1977 at about 83 mAOD subsequent fluctuations in water level have been small. A maximum water level of 84 mAOD was recorded in August 1978. The final equilibrium level of the ground water level in the opencast backfill appears to have been controlled largely by the topography of the site.

17. At gauge B2, where the fill was deepest, 0.33 m settlement occurred at the ground surface during this 3 year period as the water level rose 34 m. Figure 5 shows the relationship between the rise in ground water level and the vertical compression of the backfill. In Figure 5(a) the vertical compression between adjacent magnet markers in the borehole settlement gauge is plotted against time. Figure 5(b) indicates with arrows the dates at which the ground water reached the level of successive magnet markers. Figure 5(c) shows the positions of the magnet markers in the borehole settlement gauge. As the ground water rose from the level of magnet 5 to the level of magnet 6 a vertical compression of just over 1% occurred over the depth of fill between these two magnets. Then as the ground water continued to rise large compressions occurred successively between magnets 7 and 8, 8 and 9, and, 9 and 10. The effect of saturation in producing collapse compression within the backfill was clearly demonstrated. Locally compressions as large as 2% were recorded but the average compression over the 34 m depth of backfill saturated by the rising water table was smaller than 1%. It should be noted that at an early stage a small expansion occurred between magnet markers 4 and 5 and that little compression occurred between magnets 6 and 7.

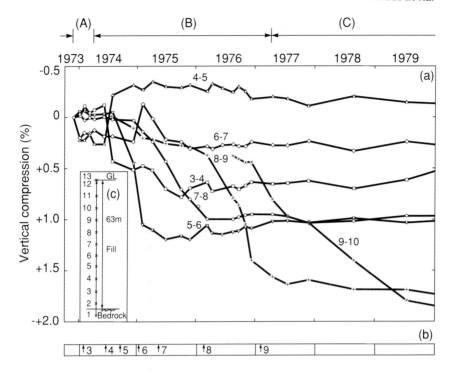

Figure 5. Relationship between vertical compression at
different depths within the deepest fill and the rise in ground
water level (measured at gauge B2):
  (a) vertical compression between adjacent pairs of magnet
      markers plotted against time,
  (b) time at which ground water level reached successive
      magr   markers,
  (c) dep   f magnet markers.

18.   The rise in water level saturated 31 m of the backfill
at the location of gauge D1 and a surface settlement of 0.1 m
was observed. As the ground water rose from the level of
magnet 2 to the level of magnet 3, a small vertical expansion
occurred in the backfill between these two levels. As the
water level continued to rise, vertical compressions occurred
successively between magnets 3 and 4, 4 and 5, 5 and 6. A
further rise in water level above magnet 6 caused no
compression between magnets 6 and 7 and most of the settlement
observed at gauge D1 was located more than 28 m below ground
level. At depths where compression was caused by the rising
water level the magnitude was much smaller than at gauge B2 and
this can be attributed to the effect of the preloading produced
by the overburden heap.
19.   The settlement at C11 gave little indication of being
affected by the rising water level. As this area was the site
of a lagoon during opencast working, it is probable that the
backfill was sufficiently wetted at that stage to prevent

further settlement occurring due to the rising ground water
level. Only when the water level rose from the level of magnet
5 to the level of magnet 6 did compression occur that was
clearly associated with the rise in water level. Precise
levelling recorded 61 mm settlement during this period. At
this gauge settlement occurred fairly uniformly through the
full depth of the backfill.

20. The rising ground water level had little effect on
settlement at gauge D15 because the gauge is situated on high
ground and only the bottom 11 m of the backfill have been
inundated (Table 1). Between April 1975 and April 1977 as the
ground water inundated the bottom 11 m of the backfill, there
was some increase in the rate of settlement.

Period (C)

21. Ground movements were monitored in the period
immediately following the rise in ground water level between
April 1977 and April 1981. At A9, B2, D15 and E12 the rate of
settlement was between three times and ten times as large as
the rate prior to the rise in ground water level. Large
movements continued at B2, but Figure 2 shows that the
continuing settlement was largely caused by compression of the
fill above the ground water level. Precise levelling has shown
that the greatest movement during this period occurred at E12
where 0.29 m settlement was observed. The reason for such
large movements at this location is not known.

Period (D)

22. Ground movement monitoring continued in the period from
April 1981 to November 1992 which commenced several years after
the rise in ground water level. At most locations the rate of
settlement during this period has been smaller than the rate
prior to the rise in ground water level. The only magnet
extensometer that provides an exception to this is gauge D15.
This is in the most recent fill and currently the settlement
rate is still greater than at the other gauges. At the other
borehole gauges the rate of movement is 3 mm per year or
smaller. The rate of movement at E12 is greater than at any of
the five borehole gauges.

DIFFERENTIAL SETTLEMENT

23. It is differential settlement rather than total
settlement which causes damage to buildings. The traverses of
settlement stations provide information on differential
settlement. In assessing this data it should be remembered
that if a structure had been built at the location where the
differential settlement has been measured, the differential
movement would have been modified and probably greatly reduced
by the stiffening effect of the structure.

24. The maximum differential settlement has been recorded on
traverse E. At a location where the total settlement is 737 mm
surface stations 5 m apart have shown a differential settlement
of 146 mm corresponding to a deflection ratio of 2.1×10^{-3}.

CONCLUSIONS

25. A rising ground water level caused significant settlement of an opencast mining backfill composed of mudstone and sandstone fragments and a correlation between settlement and the rise in water level has been clearly established. Vertical collapse compressions on inundation were locally as large as 2% but the average settlement measured over the full depth of inundated backfill at the borehole gauges was smaller than 1%.

26. Temporary preloading of an area with a 30 m high surcharge of fill during opencast coal mining greatly reduced subsequent settlement due to the rising ground water level and virtually eliminated any long term creep settlement. When the backfill was saturated collapse compression was significantly smaller than for similar fill which had not been preloaded and, during the last 12 years of monitoring, virtually no compression has occurred at this location. Prior to the rise in water level the ground surface was heaving.

27. Pre-wetting during opencast mining at the site of a lagoon greatly reduced the effect of the rising water level on settlement at this location. However the wet and more cohesive fill at this location probably suffered large settlements immediately following backfilling and it may be that in the years before monitoring commenced large settlements occurred in this area. As this settlement took place at an early stage, it might be of less importance for building development.

28. Although the investigation has demonstrated the effect of a rising ground water level in producing collapse compression and has shown the relationship between some of the features of the opencast mining and subsequent settlement behaviour, the pattern of settlement of a restored opencast mining site has been found to be complex. A number of observations are difficult to explain. For example, a significant proportion of the settlement monitored at the borehole gauges was due to compression in the upper 10 m of backfill which was not saturated by the rising ground water level. It must be concluded that even with a good knowledge of the opencast operations that had occurred on the site, it would have been difficult to predict the magnitude and rate of the settlement of the backfill.

29. The investigation has considerable significance for building developments on restored opencast mining sites. The complexity of the settlement pattern measured at Horsley points to the need for careful investigation. As much information as possible should be obtained about opencast operations, and settlement and water levels should be monitored over a realistic period. In addition to settlement due to self-weight and applied loads, the possibility of collapse settlement on inundation should be addressed. The results from the Horsley investigation have shown the effects that could be caused by inundation of a partially saturated opencast mining backfill subsequent to building on the site.

ACKNOWLEDGEMENT
30. The investigations described in this paper form part of
the research programme of the Building Research Establishment.
The authors wish to thank British Coal Opencast for their
permission to publish this paper. The views expressed herein
are entirely those of the authors.

REFERENCES
1. CHARLES J A, NAISMITH W A and BURFORD D. Settlement of
backfill at Horsley restored opencast coal mining site. Large
Ground Movements and Structures (ed J D Geddes). Proceedings
of Conference held at Cardiff, July 1977, pp 229-251. Pentech
Press, Plymouth, 1978.
2. CHARLES J A, HUGHES D B and BURFORD D. The effect of a rise
of water table on the settlement of backfill at Horsley
restored opencast coal mining site, 1973-1983. Ground
Movements and Structures (ed J D Geddes). Proceedings of 3rd
International Conference held in Cardiff, July 1984,
pp 423-442. Pentech Press, Plymouth, 1985.

35. Engineered fill for Sheffield and Rotherham City Airport

C. S. MORGAN and J. M. W. HOLDEN, Scott Wilson Kirkpatrick, UK,
P. S. O'BRIEN, R. J. Budge (Mining) Ltd, UK, and J. W. GARDNER,
British Coal Opencast, UK

SYNOPSIS   An opencast coal site at Sheffield has been
reinstated with engineered fill, placed for the most part in
800 mm thick layers, to a maximum depth of nearly 100 m.   In
situ testing has shown that the fill has relatively high air
voids and that much of the material through which the
groundwater will recover is a mudstone which will soften on
saturation. Settlements are predicted which are greater than
those normally expected from controlled compaction, and
these are considered with regard to the operational limits
on settlements for an airport runway which is to be
constructed at the site. Some comparisons are made with the
early results from the monitoring since completion of
backfilling.

INTRODUCTION
   1 Sheffield with its population of some 500,000 is the
largest city in the European Community not to have its own
regional airport.   The city authorities have considered many
proposals over a period extending back to between the two
World Wars (ref 1) but until very recently none has been
taken beyond a very preliminary feasibility stage.   One
major difficulty has been the local terrain, with the
location of Sheffield on its seven hills excluding much of
the city from consideration.   The most suitable land in
terms of uniformity of level is to the east along the valley
of the River Don but this was, until a few years ago, among
the most intensely industrialised areas in the world.
   2 However, the decline in the coal, steel and related
industries in the last two decades fundamentally altered the
appearance and vitality of the Lower Don Valley with the
result that it presented to Sheffield City Council, and to
its neighbour Rotherham Metropolitan Borough Council, a
broad landscape of industrial dereliction.   In the 1980s the
area, by then with its motorway access, was identified by
Sheffield City Council as a possible location for a future
airport although it was recognised that the valloy floor
presented considerable problems arising from the industrial
exploitation of many years.

THE SITE HISTORY

3   In the past the site had been extensively undermined
for coal both by shallow and deep workings, the former from
early bell-pits, adits, shafts and more recent open-casting,
and the latter from local collieries. As a result of these
various activities there were a number of mining spoil heaps
on the site as well as the old workings created beneath the
surface.

4   Additionally, part of the general area has been
occupied more recently by a large steelworks built in the
early 1960s and which operated until its closure some twenty
years later. The legacies from this operation were
substantial basements and buried foundations, slurry
lagoons, and spoil heaps of slag, flume dust and other
hazardous materials. Photograph 1 shows the site at that
time in an aerial view over which the proposed development
outline has been superimposed.

Photograph 1: Aerial view of proposed development

THE RECLAMATION AND DEVELOPMENT PROPOSAL

5   In reinstatement terms the area presented immense
problems for conversion to a state suitable for development
of any kind, but the presence of the remaining and
relatively shallow unworked coal seams offered the
opportunity for one particular type of reclamation scheme:
an opencast mining project which would remove the problems
as a matter of course in the coaling operation and which
could provide an engineered and/or landscaped fill on
completion.

442

6 The results of subsequent discussions and negotiation
between the principal parties, by then including British
Coal Opencast, was that an opencast mining operation was
decided upon and planning of the work commenced.

7 In February 1989, under a contract awarded by British
Coal Opencast, Scott Wilson Kirkpatrick began the
preparation of a specification for the controlled compaction
works as a precursor to their responsibilities for
supervising, in due course, that part of the overall
reinstatement. This specification was incorporated
subsequently in the main contract document prepared by
British Coal Opencast for the overall mining operation and,
following a tender process, the contract was awarded to
Budge Mining Ltd, later R J Budge (Mining) Ltd, and work
commenced in July 1989 at Tinsley Park OCCS.

8 In early 1990 Budge (Mining) Ltd entered into a
separate agreement with Sheffield Development Corporation,
which had taken responsibility in 1988 for the planning,
development and regeneration of the 800 ha of the Lower Don
Valley. The agreement was to build the airport and runway
and to operate them for a period of ten years; additional
commitments by the Contractor were to construct an access
highway through the restored site, to provide some
90,000 sq m of business park space, and to landscape the
remainder of the site to comprise about 35 ha of woodland
and a public amenity area; also, as part of the mining
contract there was a requirement to form a 1M cu m void
to meet future waste disposal needs of British Steel.
Shortly afterwards the scale of the operation was extended
considerably when negotiations led to the coaling operation
increasing by 50% to 1.5 m tonnes with a contract period
requiring completion of restoration by August 1993, an
extension from February 1993. Fig 1 provides the site plan
and Fig 2 a longitudinal section along part of the runway
indicating the size and shape of the main opencast void in
relation to the proposed development.

9 Thus a project was put into effect whereby over a 3.5
year period a 135 ha site of industrial despoliation was to
be further exploited but now with the purpose of providing a
reclaimed and improved area of major development potential.

THE SITE RESTORATION
10 The reinstatement of the site consisted of two
operations, often carried out concurrently: bulk placement
of fill by scraper and dumptruck in the landscaped and
amenity areas, and placement in layers with controlled
compaction in the airport and development areas. The former
areas were intended to accommodate the unacceptable
materials from the excavation, eg the contaminants such as
the flume dust, in clay cells within the backfill, and also
the larger blocks of sandstones and mudstones which
inevitably feature in a large scale opencast excavation.

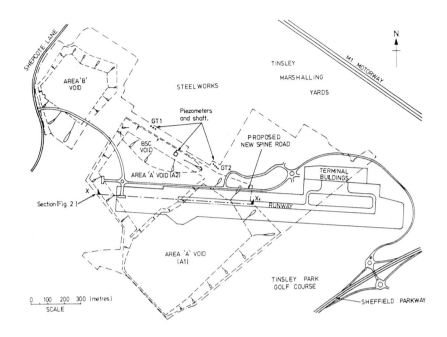

Fig 1: Plan of site and proposed development

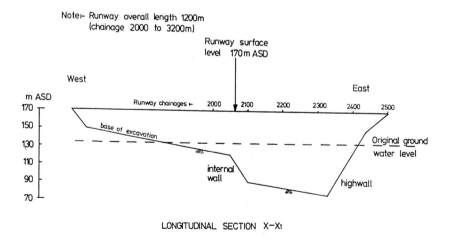

LONGITUDINAL SECTION X–X₁

Fig 2: Longitudinal section beneath proposed runway (part)

11 The mining contract statistics are given in Table 1
which, with the increase in both coal reserves and the
amount of controlled backfilling, required by Budge (Mining)
Ltd a re-appraisal early in the contract period of the
already demanding programme.

Table 1:   Tinsley Park - Site Statistics

Site Area/Life	135 ha/ 3.5 years
Coal Reserves	1,500,000 tonnes (6 seams)
Coal Output	12,000 tonnes per week
Maximum depth	100 m (overall ratio 27 cu m/t)
Overburden excavation	250,000 cu m/per week
Works commenced	July 1989
Coaling commenced	September 1989
Compaction commenced	June 1991
Coaling completed	April 1993
Works completed	August 1993

12 To meet these new requirements a greater output rate
was necessary in the earthmoving aspects of the opencasting:
excavation of overburden and reinstatement, but it was in
the latter phase that the difficulties of achieving higher
rates were judged to be the most marked. The increase in
the amount of controlled compaction was accompanied by a
reduced facility for the disposal of 'oversize' material,
which now would have to be dealt with either by a processing
operation or by incorporation in the controlled zone in
block sizes greater than permitted by the specification.
Budge (Mining) Ltd concluded that the former was impractical
because of the high volume throughput and devised a modified
compaction procedure which would enable sufficient of the
large material to be accommodated and also achieve the
necessary high placement rates.

THE SPECIFICATION AND THE MODIFIED PROCEDURE
13 The principal material types excavated were strong
sandstones and moderately weak to strong mudstones. The
as-dug sandstones generally could be classified as a coarse
granular material in accordance with the specification and
placement was permitted in layers up to 800 mm thick with
the appropriate compaction. The specification however
required placement of mudstones in layers of 275 mm maximum
thickness except below a depth of 40 m where a 50% increase
was permitted with a pro-rata increase in compaction. The
as-dug mudstones were often in large blocky form with arris
dimensions of over 500 mm. The excavation was carried out
by O&K hydraulic face shovels models RH200 (20 cu m bucket)
and RH 120 (12 cu m bucket). Although initially of very
hard and durable appearance the mudstones softened
significantly on saturation, as demonstrated by the 10%

445

fines test where the average values dropped from 115 kN in the standard 1-day test to 25 kN in the 5-day soaked test.

14   The modified procedure for compaction of the mudstones, as adopted by Budge (Mining) Ltd, consisted for the most part of spreading the materials in layers of approximately 800 mm loose thickness and compacting with five or more passes of vibratory rollers. These rollers were of 5000+ kg mass per metre width of vibrating roll and were operated at a speed of approximately 5-6 km/h rather than the 1.5-2.5 km/h specified.

15   The consequences of the modified procedure were that the mudstones, comprising the majority of the compacted material, did not receive the specified degree of compaction nor were they placed in the sizes and gradings envisaged. It was recognised that the resulting fill would have a lower density and higher air voids, with a greater potential for creep settlement and that, additionally, the high air voids in the moisture susceptible mudstones would result in significant strains on inundation.

16   This presented major implications for the proposed airport and other developments, but Budge (Mining) Ltd considered that only the modified procedure would enable the programme completion dates to be met whilst at the same time ensuring that the fill received the greatest amount of compaction that could be applied in the time available.

THE REINSTATEMENT - TESTING

17   The placement and compaction of fill in layers 800 mm thick provides a number of problems, not the least of which is obtaining a measure of the degree of compaction achieved. Conventional sand replacement or nuclear density gauge tests are impractical because of their depth limitations and recourse is made generally to large-scale water or gell replacement tests which are very time-consuming. All these tests were tried at Tinsley during the compaction trial stage but a new type of nuclear density gauge, used for the first time in the United Kingdom and known as a Strata Gauge, was found to be the most successful. This gauge has twin probes, Photograph 2, which permit testing of density and moisture contents at any number of depth increments to a maximum of 600 mm below ground surface level. On the recommendation of SWK this instrument was acquired by Budge (Mining) Ltd specially for the Tinsley Site and was found to be the most effective and efficient means of density testing for thick layers; moisture contents were determined in a more traditional manner because of the usual difficulties associated with NDG testing in Coal Measures materials.

THE REINSTATEMENT - SETTLEMENT

18   From the early compaction trials it was confirmed that high air voids were a feature of the compacted fill with average values greater than 20% being recorded. This order of result was also obtained during the initial stages of the

446

Photograph 2: Strata gauge

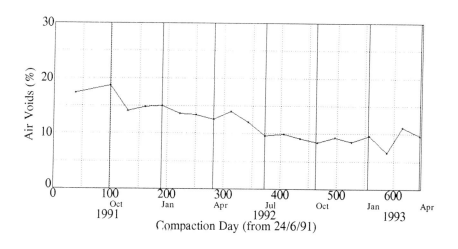

Fig 3: Monthly average air voids

compaction proper when, in the dry conditions of summer 1991, the monthly averages of air voids were in the range 17% - 19% with individual results from 5% - 31%. The compaction standard improved with the cooler, wetter weather of the 1991/92 winter when the monthly average range reduced to 13% - 15%, (Fig 3). All the test locations were surveyed and recorded in three dimensions and this information, together with the test results, was stored on the gINT computer database which permitted manipulation and analysis of the data and their plotting in plan or section form.

19   The results obtained from the routine in situ tests confirmed the original predictions that high air voids values would result from the compaction procedure adopted. The difficulty was in converting those values to estimates of amounts of settlement, whether through creep or inundation, so that an assessment could be made of the likely effects on the various proposed developments in due course.

20   The creep or self-weight settlement of this type of fill can be described by means of the logarithmic creep compression parameter (alpha). There are a number of records and case histories which present typical values of alpha (refs 2,3) and these range from as low as 0.15% to over 1.0%, for well-compacted and uncompacted fills respectively. Unfortunately, much of the data relates to the better compacted sites, where air voids are less than 10%, so that correlation between alpha values and high air voids remains uncertain.

21   Partially saturated soils, such as freshly placed opencast backfill, exhibit volume reductions on inundation as a result of softening of point contacts, whether from a recovery of groundwater levels or downward percolation of surface water. This is most marked in the less dense fills and, in the case of groundwater recovery, is a relatively rapid phenomenon whilst, for surface percolation, it can take two wet seasons or more. Strains on inundation can range from negligible amounts in very well compacted materials, eg those with air voids less than 5%, to over 6% in uncompacted colliery spoils (refs 2,3). Monitoring data which relate inundation settlement with higher air voids values, ie above 12%, mostly originate from laboratory tests, as rigorous in situ testing is not normally a feature of less well compacted sites.

22   From a consideration of available data, and experience generally of the performances of opencast backfills, estimates were made of likely creep compression parameters and inundation strains at Tinsley. This was done so that predictions could be made of the settlement behaviour of the backfill, thus enabling an assessment to be carried out of the possible effects of the modified compaction procedure on the proposed development.

23   It was determined that for preliminary calculations, and in the absence of any settlement monitoring data, the

alpha value should be taken as between 0.50% and 0.75%. Strains on inundation were assumed to be 1.5% for groundwater recovery and 0.75% for surface water percolation, with the groundwater level assumed restored some two years after substantial completion of backfilling. A further consideration arose with the groundwater recovery in that the excavation of void area B, Fig 1, would only take place to any real extent when most of void area A had been reinstated. The result would be that groundwater recovery in void area A would be arrested, and possibly reversed, until coaling was completed in void area B. This would have the effect of delaying the completion of inundation strains in area A, the deeper of the two voids.

24 The settlement predictions were made for various periods of up to 25 years and Fig 4 shows the generalised settlement profiles along the line of part of the proposed airport runway. In the area shown the runway alignment crosses the maximum 100 m depth of void and traverses both the high wall and an internal wall of the excavation.

25 The Civil Aviation Authority document, CAP168 Licensing of Aerodromes, detailed the alignment standards for the 1200 m long runway and the settlement predictions gave a preliminary indication of the nature and timing of reprofiling operations likely to be necessary to maintain the required alignment.

26 Not surprisingly, the most critical areas were shown to be over the buried high wall and internal wall where predictions showed that maintenance works would be required at an early date. Localised differential movements would also occur in other locations but their magnitude and distribution were not possible to predict and they could only be dealt with by local attention as necessary.

SETTLEMENT PERFORMANCE

27 Settlement monitoring has been taking place (at June 1993) for some seven months and whilst this is only a short time in respect of creep behaviour the initial data are enabling interpretations to be made of the performance of the fill material. Only the surface settlement markers and piezometers have been installed at present but it is hoped that it will be possible soon to install extensometers in order to provide a fuller understanding of the settlement behaviour with depth. The discussion of the settlement relates to void Area A which can be subdivided into a deep section A1 (max 95 m depth) and, above the internal wall, a shallower section A2 (max 60 m depth).

28 The groundwater records have been obtained from piezometers located in the undisturbed ground adjacent to the north-east high wall of void area A, Fig 1. Overall groundwater level movements are depicted in Fig 5, which shows an initial depression to approximately 106 mASD in October 1991, recovering to a temporary high of about 116 m

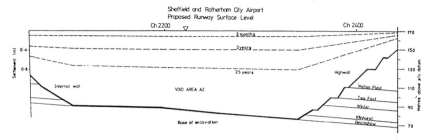

Fig 4: Predicted settlement profiles

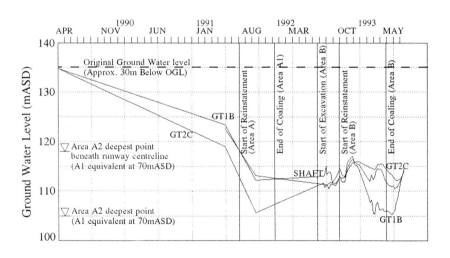

Fig 5: Groundwater levels

ASD in December 1992, followed by a further depression to
about 105 mASD by March - May 1993. [Note: Site Datum (SD)
is 100 m below Ordnance Datum]. Projections indicate that
the groundwater should recover its original level of some
135 mASD during the winter 1993/94. Whilst these
predictions are based on piezometers outside the void,
recently installed instrumentation within the fill has
provided comparable readings.

29 The settlement performance in the shallower part of
the void (A2) is given by the two survey markers shown in
Fig 6. The depths of fill beneath these markers are 47 m
for C7 and 56 m for C9 and, as such, the material will have
been subjected only slightly, if at all, to inundation by
the recovering groundwater. The general uniformity of the
graph suggests that any inundation by downward percolation
has not had a significant effect. Consequently, the slope
value given in Figure 6 is likely to represent the creep
compression parameter, ie alpha = 0.35% approximately.

30 For the deepest part of the void the settlement
monitoring data are given in Fig 7 from which it is evident
that the settlement performance has been affected by
groundwater inundation and also possibly by subsequent
groundwater drawdown with accompanying increases in
effective stresses. The lowest slope value is 0.51%, for
D11, and this has been recorded over the last six months or
so during a period when the groundwater recovery has been
reversed because of the excavation of adjacent void area B;
consequently, it remains to be seen whether this slope value
of 0.51% is indicative of the creep compression parameter
value. It is significantly greater than that recorded in
the adjoining and shallower section A2, although it should
be noted that this deeper section A1 is where the earliest
filling took place and where the greatest air voids were
measured, para 18.

31 Over the last four months in void area A1 the
settlement markers C15, D16 and E17 have provided
corresponding slope values of about 0.8%, 0.85% and 1.0%,
and this period would appear to be beyond the main effects
of the first-phase inundation as indicated by the initial
steeper portions of the curves, Fig 7. Again, it is possible
that the groundwater drawdown has increased the settlement;
the greatest slope of 1.0% approximately is given at marker
E17, which represents the deepest and earliest fill area
which had the highest air voids. Any influence of the
groundwater drawdown will now have ended in that the second
phase of groundwater recovery has recently begun and the
level is approaching once again the 116 m ASD mark
temporarily attained in autumn 1992. A new period of
groundwater inundation will follow as the groundwater rises
through some 19 m of previously unaffected fill to reach its
original level of about 135 m ASD. It is after this phase
has been completed that the various ground settlement
parameters will become more clearly defined.

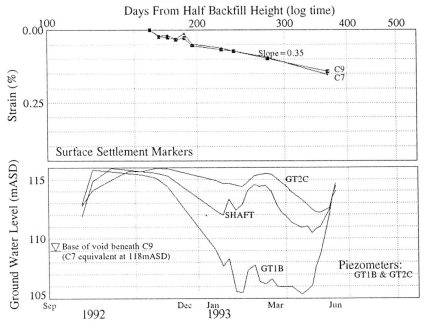

Fig 6: Void area A2 - Surface settlement and groundwater

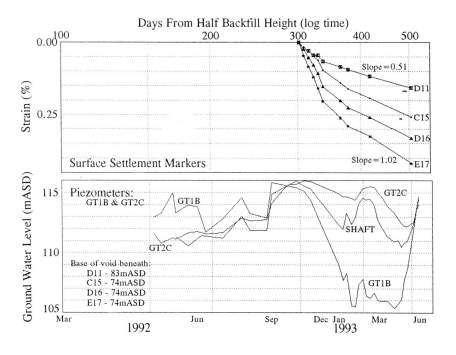

Fig 7: Void area A1 - Surface settlement and groundwater

452

CONCLUSIONS

32   In order to review the original ground settlement predictions it is necessary to distinguish between the void sub areas A1 and A2.   In the latter and shallower area, whilst the inundation compression has yet to take place, the indicated alpha value of 0.35% suggests that the overall settlement will be less than that predicted, ie on the basis of a lower bound alpha value of 0.5%.   In the former and deeper area there is insufficient information available at present to enable a complete review of the original predictions.   On the basis of surface settlement markers alone, it is not yet possible to determine with any confidence either the alpha value or total strains on inundation in area A1.   The situation has been further complicated by the temporary drawdown in the groundwater table, although this may be beneficial subsequently if significantly increased effective stresses have been generated temporarily such that the potential for creep compression is reduced in the longer term.

33   The maximum rate of groundwater recovery is that presently recorded by piezometer GT1B, at some 6 m/month. The latest prediction of complete recovery, ie during the winter of 1993/94 (para 28) will be about 15 months after substantial completion of backfilling; this compares with the originally predicted period of 24 months.

34   An estimation of inundation strain can only be made at present using a presumed alpha value.   For the settlement markers in Fig 7, if a creep strain equivalent to an alpha value of 0.35% is deducted the remaining total strains vary from 0.2% (D11) to 0.75% (E17) in relation to the estimated depth of inundation.   These values are lower than the predictions for inundation compression but do not take account of any such settlement which had already taken place prior to the start of monitoring, Fig 7.

35   In overall terms it would appear that the general performance of the fill is likely to be within the bounds of the original predictions made to take account of the modified compaction procedure.

REFERENCES

1   HUNTER D AND STOTT.   Tinsley Park Lives Again.
    Third Int Conf on Land Reclamation: An End to Dereliction,
    Univ of Wales, July 1991, Cardiff 411-420

2   CHARLES J.   Building on fill: geotechnical aspects.
    Building Research Establishment Report, 1993

3   SCOTT WILSON KIRKPATRICK.   State-of-the-Art-Review of the
    Compaction of Opencast Backfill.   British Coal Opencast,
    1993

36. Compaction, monitoring and performance of engineered fill at the Dixon opencast site, Chesterfield, UK

N. A. TRENTER, Sir William Halcrow & Partners Ltd, UK

SYNOPSIS. Two areas within the major Dixon opencast site at Brimington, near Chesterfield, England, required eng-ineered compaction in order to permit the construction of an industrial development and By-Pass. The paper describes the approach to compaction trials, monitoring of compaction and instrumentation, which comprised the installation of surface monuments, magnetic extensometers, piezometers and a United States Bureau of Reclamation (USBR) gauge. The results of settlement monitoring for a period of up to 1700 days are described.

INTRODUCTION

1. Quite properly, attention has turned to the use of marginal land for building purposes. Such land may be the derelict and contaminated product of nineteenth century in-dustrialisation; alternatively it may be the consequence of exploitation by the extractive industries. Whatever the reason, a method must be found for its rehabilitation and its return to productive use. This is all the more imp-ortant when, as is often the case, marginal land is at or close to major industrial centres.

2. This paper describes the approach to the engineered compaction of the Dixon opencast site at Brimington, near Chesterfield, England, in order to render the site suit-able for later industrial development and a By Pass. The site (Fig 1) is some 119 ha in extent and was worked by British Coal Opencast Executive between February 1987 and August 1992. About one million tonne of coal was mined, involving the removal of 16 million m^3 of spoil.

3. Two parts of the site required engineered compaction: Area A of 20ha extent, excavated up to about 74m depth and

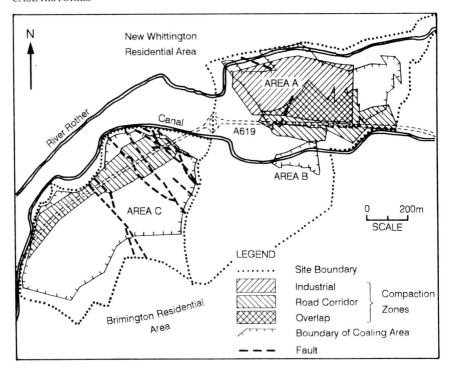

Fig. 1 The Site

Area C, a 8ha road corridor for the proposed A619 Brimington By Pass, which involved excavation to an average of
about 15m depth.

GEOLOGY OF SITE

4.   Eighteen coal seams were worked: four in Area C and
the remainder in Area A. All coal seams were of Lower Coal
Measures age, ranging from the Brampton Low (oldest) to the
Chavery Upper. Strata between these coal seams comprised
mudstones with siltstones and occasional sandstones, and
very occasional ironstones.   Three north-west trending
normal faults, known as the Inkersall Fault Zone cross Area
C and the north-west trending Brimington anticline crosses
the site between Areas A and C. The solid rocks were overlain by gravel and alluvium from the River Rother flood
plane which forms the north and west site boundaries.

5.   Information provided by British Coal, derived from
cored boreholes, indicated that the backfill material
(mudstones, siltstones and sandstones) would be in the
proportions 74:16:9, respectively.   The river gravel and

456

alluvial deposits were removed from all of Areas A and C and were stored for later top-soiling

6. In parts of the site, Made Ground, mainly foundry sand and slag, was present on top of the alluvium, and there were large tips east of Area A. This material was removed and incorporated with other spoil in the compacted areas.

7. There were no active underground workings in the seams below the compacted areas in Area A and Area C, with none planned for the future. Old workings were encountered in the north west corner of the compaction zone of Area A. The old adits were backfilled with sandstone blocks and then sealed with clay at the positions where they entered the high wall.

8. Groundwater was observed at about 155m OD in Area A during excavation. In Area C, groundwater varied, being 155m OD in the northern third of the road corridor; 151m OD in the middle third; and absent in the southern third. These variations are attributed to the Inkersall Fault Zone which, as already remarked, crosses Area C. (NB : all levels quoted are 100m above the true Ordnance Datum level).

COMPACTION TRIALS

9. Proper engineered compaction requires preparation of a specification; and the following approach was adopted:

- trials to determine the most suitable equipment, layer thickness, moisture content and number of passes
- in situ and laboratory testing to measure density and moisture content of the fill during compaction
- monitoring using appropriate instrumentation of both settlements and groundwater rise and the relationship between them.

10. A trial area 35m x 20m was selected on a prepared surface with the top and sub-soil removed. Base levels were taken on the trial area, prior to the test material being deposited. Further levels were taken before and after each trial strip had been compacted, to establish layer thickness and the compaction achieved.

11 On the basis of the compaction trial results, the compaction method adopted was to spread the material in loose layers 300mm thick by a CAT 825 Tamping Roller,

followed by six passes of a BOMAG BW217 PD Vibrating Roller. Bomag BW6 and Stothert and Pitt T208 vibrating equipment were also employed.

MONITORING OF COMPACTION

12. Supervision of the compaction works and the comp- action trials involved regular testing of the compacted layers, using the in situ density sand replacement test (BS 1377 : 1975 Test No. 15B) and moisture content test (BS 1377 : 1975 Test No. 1). Standard Proctor compaction tests (2.5kg rammer method) and sieve analyses were also performed.

13. The mean level of compaction achieved for the whole site was 96 per cent of standard Proctor maximum dry den- sity (see Fig 2) with a moisture content average of 7.5 per cent and a dry density average of 1.94 Mg/m^3. This corr- esponds to an average air voids content of about 12 per cent. Particle size distribution analyses indicated that the spoil varied substantially but was normally silty sandy fine to coarse gravel size with some boulders, roughly in the proportions clay and silt (25 per cent), sand (25 per cent), gravel and coarser particle sizes (50 per cent).

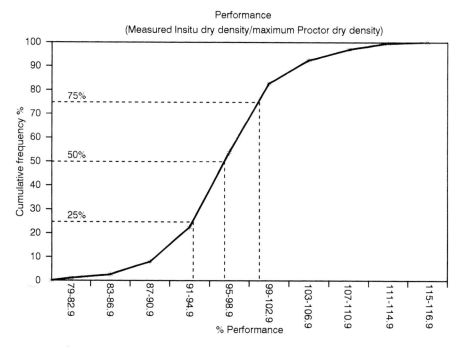

Fig. 2 Frequency distribution of in-situ test results

INSTRUMENTATION

14. Settlement, particularly any residual settlement which might occur during and after later building operations, is one of the most important criteria for the engineered compaction of fills and three methods were selected for its measurement:

- surface monuments
- magnetic extensometers
- USBR gauge (see below)

Unlike the USBR gauge, surface monuments and magnetic extensometers can only determine settlements after the fill has been placed to its final level, since it is only at that stage that the instruments are installed, i.e. these devices measure settlement "post placement". However, by its construction, the USBR gauge which is installed at the commencement of fill placement, can monitor settlement of the fill as it is placed. Likewise, standpipe piezometers can only monitor settlement changes after the fill has been placed to its finished level, whereas the USBR gauge can provide an indication of groundwater level changes during fill placement.

15. Surface monuments installed were concrete cubes (1m x 1m x 1m) incorporating a fixed button on the upper surface for surveying purposes. Nineteen were installed into the fill about 300mm below restored level. Using standard optical surveying methods, settlement of the fill (post placement) was monitored and recorded on a regular basis.

16. At selected points in Area C three boreholes of 150mm diameter were sunk to allow the installation of magnetic extensometers to monitor settlement and were also read (post placement) on a regular basis. At each of the above mentioned borehole locations sunk to install magnetic extensometers, parallel 150mm diameter boreholes were sunk to allow the installation of standpipe piezometers to provide complementary ground water data.

17. At one of the deepest parts of Area A, a USBR settlement gauge was installed (US Department of the Interior, 1974). The USBR gauge comprises a series of cross-arms mounted on 50mm diameter steel pipe sections, each being separated by standard length spacers. The gauge is installed from the pavement (base of fill) upwards at a rate consistent with the rate of fill being deposited and settlements of the various cross-arms are read by a torpedo

459

unit lowered down the pipe. For this installation, cross-arm width was 3m; the average spacing between cross-arms being about 6m. The gauge was damaged after 58.9m of fill had been placed; however, sufficient data had been collected to estimate the settlement behaviour of the recently comp-acted opencast backfill.

RESULTS OF MONITORING
Monuments (Area A)

18. Typical graphs of settlement as a percentage of fill thickness v time presented in Fig 3 demonstrate initial heave followed in most cases by small settlements. After about 350 to 400 days from installation, there was a period of relatively rapid movement which reached up to about 0.7 per cent of fill thickness by about 600 days from install-ation.

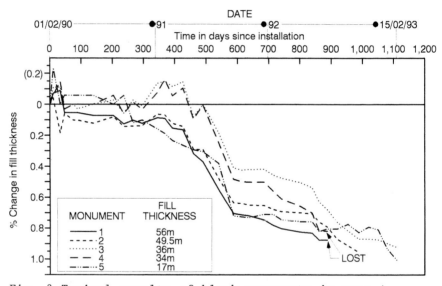

Fig. 3 Typical results of block monuments in area A

19. This settlement corresponded to a rise in groundwater table from about elevation 113 to 136m OD. Further settle-ments commenced thereafter as the pumps were moved around the excavation and, at the time of the most recent readings, when the groundwater level was at approximately 154m OD, the maximum settlements were up to 0.9 per cent of fill thick-ness.

Monuments (Area C)

20. Graphical presentation of typical settlement meas-

460

urements is given in Fig 4. In this area, fill thickness was generally less than in Area A, as was the saturated fill thickness beneath the monuments. Only in the case of monuments J and M was a water table established in the fill; beneath the other monuments, groundwater table was below pavement level.

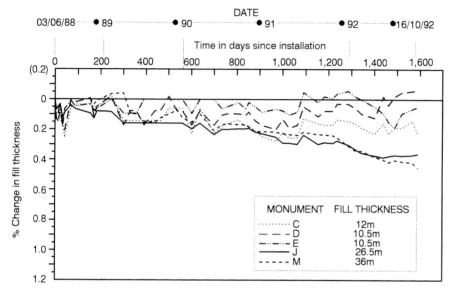

Fig. 4 Typical results for block monuments in area C

21. Groundwater began to rise near both monuments J and M some 1100 days after installation and settlement corresponding to 0.4 per cent of fill thickness was observed. Such large amounts of resulting settlement as occurred in Area A cannot be expected, however, because the thickness of fill inundated was much smaller.

Magnetic extensometer R2 (Area C)

22. Groundwater table conditions at this extensometer location are shown in Fig 5, and typical extensometer results in Fig 6. The deepest of the two arms of R2 (arm 6 being one of them in Fig 6) were installed at or below the groundwater table in the fill and percentage settlements were small. The remaining arms were installed above the groundwater table and demonstrate relatively uniform movement until about 1150 days from installation.

23. Pumps were switched off permanently at this time; thereupon groundwater rose rapidly to about 153m OD, and thereafter more slowly to about 157m OD. During this period

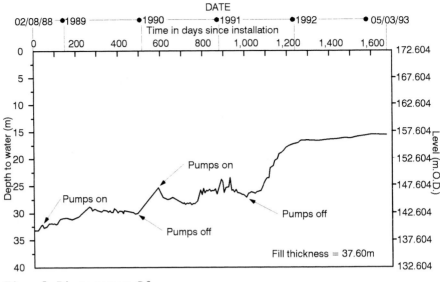

Fig. 5 Piezometer R2

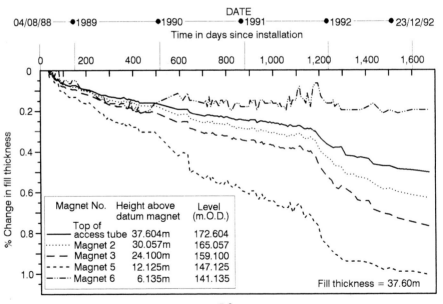

Fig. 6 Magnetic Extensometer R2

of rising groundwater table, all extensometer arms except those which were below groundwater at the time of installation show marked settlement, although at the time of the most recent readings the settlement had substantially reduced. For the inundated extensometer arms, maximum

462

movements vary between about 0.5 and 1.0 per cent of fill thickness.

Magnetic Extensometer R3 (Area C)

24. Reference to Fig 7, where groundwater conditions for this extensometer are illustrated, will indicate that the

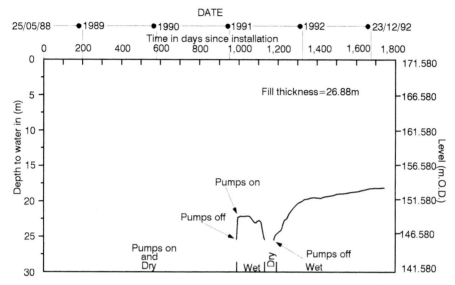

Fig. 7 Piezometer R3

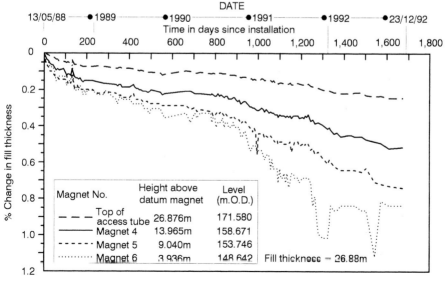

Fig. 8 Typical results for magnetic extensometer R3

fill at the adjoining piezometer location was generally dry
until about 1200 days from installation when pumps were
switched off and groundwater rose to about 152.8m OD. Up
to that time, settlements varied between about 0.1 and 0.5
per cent of fill thickness (Fig 8). Settlements increased
as groundwater rose in the fill and the recent readings
indicate settlements varying from about 0.25 to 0.8 per cent
of fill thickness.

Magnetic Extensometer R4 (Area C)

25. The fill was dry at this location. Movements were
very small with maximum settlements corresponding to only
about 0.2 per cent of fill thickness (maximum). This
demonstrates that for soundly compacted material, post
placement settlements are unlikely to be significant unless
groundwater rise is involved.

USBR Gauge.

26. The results of the settlement monitoring undertaken
in Area A with the USBR gauge are given in Fig 9 (a and b)
for the period 20 March 1990 to 9 January 1992 (660 days),
when the apparatus was unfortunately lost due to site
operations. During this period, ten cross-arms had been
installed and about 58.9m of fill placed and compacted.

27. Installation of the cross arms will have coincided
with the rise in groundwater table discussed in relation to

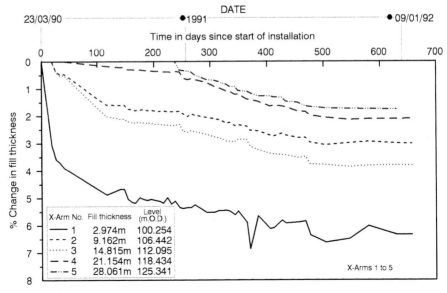

Fig. 9(a) U.S.B.R. gauge

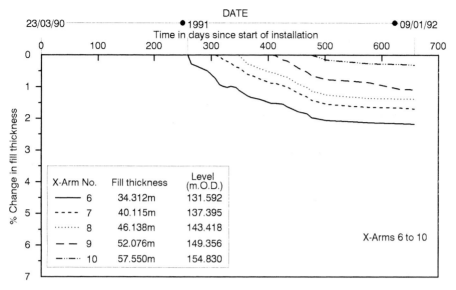

Fig. 9(b) U.S.B.R. gauge

the settlement monuments for Area A.   However, because of
the size of the movements recorded by the USBR equipment,
"groundwater induced" settlements of relatively small magnitude are not readily identifiable from the plots, although
their magnitude may be estimated (see below)

DISCUSSION

28. Settlement in opencast backfill as a consequence of
groundwater rise has been measured by several workers
including Charles, Naismith and Burford (1978) and Charles,
Hughes and Burford (1989) at Horsley; together with Smyth-
Osbourne and Mizon (1984) at West Auckland. Measurements
made at the Dixon site, for a period in excess of 1700 days
in some cases, demonstrate that a relationship exists
between settlement (expressed as a percentage of original
fill thickness) and saturated fill thickness (also expressed
as a percentage of original fill thickness).

29. In Fig 10, such a plot is prepared for the individual
magnets installed (post placement) as part of magnetic extensometers R2 and R3.   It will be seen that the data lie on
a straight line, with relatively small scatter.   It is
interesting to note that the magnets which were installed
below groundwater level in the fill, demonstrated small
settlement of about 0.2 per cent of fill thickness, very

465

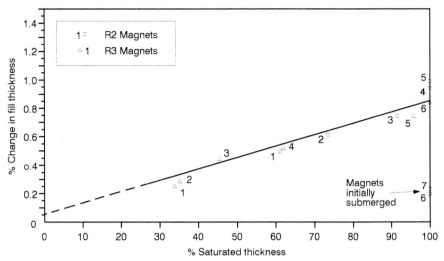

FIG. 10 Relationship between percentage settlement of fill
and saturated thickness (Extensometers R2 and R3)

similar to the post placement settlements of dry fill. This
confirms the point that significant post compaction settlement is caused by inundation, and does not occur to anything
like the same extent in dry or initially wet fills.

30. A similar plot of percentage change in fill thickness v percentage saturated thickness is prepared in Fig 11,

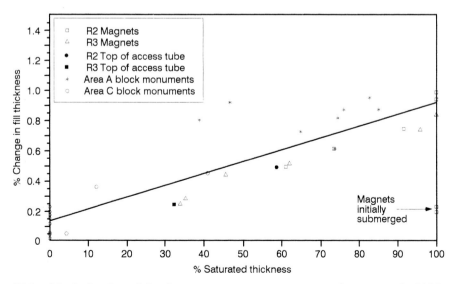

FIG. 11 Relationship between percentage settlement of fill
and thickness (Extensometers and Surface Monuments)

for both the individual magnets of extensometers R2 and R3 and also the various surface monuments established in Areas A and C. The scatter is considerably larger than for extensmeters R2 and R3 alone, as might be expected, given the various combination of fill types monitored by the surface monuments, and the various times they were installed relative to completion of compaction. Nevertheless, a clear trend of increasing percentage settlement with increasing percentage saturated thickness emerges.

31. It is of interest to estimate the proportions of "dry" and "wet" settlements experienced by the fill surrounding the USBR gauge when it was in operation. The total settlement between the pavement and the uppermost cross-arm of the gauge whilst it was in use corresponded to 2.55 per cent of the fill placed. During this period, monuments which had been positioned elsewhere in Area A settled by an amount corresponding to about 0.6 per cent of fill thickness due to groundwater rise. Since this rise will have affected the USBR gauge during its installation, it is reasonable to assume that 0.6 per cent of the 2.55 per cent total settlement was due to inundation. The remaining 1.95 per cent will be due to self weight compaction. It should be noted that recorded settlement may be

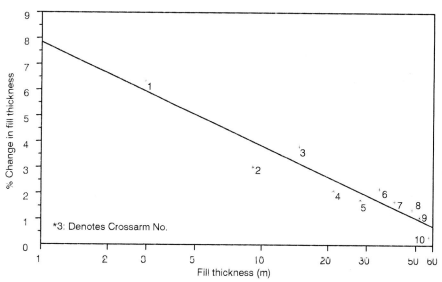

Fig. 12 Relationship between percentage settlement of U.S.B.R. gauge and fill thickness

somewhat on the high side because compaction plant were instructed to keep away from the instrument in order to reduce the possibility of accidental damage, which did ultimately unfortunately occur. Compaction in the immediate vicinity of the gauge was not therefore to such a high specification as elsewhere.

32. A plot of maximum change in fill thickness against original fill thickness (to a log scale) is given in Fig 12 demonstrating a linear relationship.

CONCLUSIONS

33. Two areas, totalling approximately 28 ha and extending in one case up to about 74m deep were restored using engineered compaction techniques at the Dixon opencast site, near Chesterfield, England. The spoil comprised mudstones, siltstones and sandstones in the proportion 74:16:9, respectively. Compaction trials demonstrated that spreading in 300mm thick loose layers by a CAT 825 tamping roller and compaction using six passes of a Heavy Vibrating Roller achieved an average of 96 per cent of standard Proctor (2.5kg rammer method) maximum dry density.

34. Instrumentation consisted of surface monuments, magnetic extensometers, piezometers and a USBR gauge. Settlements of dry and initially wet fill (measured after all fill had been placed) averaged about 0.20 per cent of original fill thickness. The expected accelerated settlement rate was observed when groundwater rose up in the fill and both the surface monuments and the magnetic extensometers indicated that there was a simple linear relationship between change in fill thickness (expressed as a percentage of original thickness) and the degree of saturation of the fill (also expressed as a percentage of the original fill thickness). It is suggested that quoting settlement figures for a fill without clearly stating its degree of saturation consequent upon inundation is not particularly helpful, and indeed could be most misleading. Maximum settlements (measured after all fill had been placed) were about 0.9 per cent of original fill thickness, where groundwater rose to achieve full fill saturation.

35. A USBR gauge installed at a location where 58.9m of fill was placed demonstrated total settlements corresponding to 2.55 per cent of fill thickness, of which about 0.6 per cent was due to groundwater rise in the surrounding

fill. This suggests that roughly 1.95 per cent was self weight settlement occurring during and shortly after fill placement. However, this may not accurately represent the actual strains in the majority of the fill since compaction plant were kept well clear of the instrument in order to avoid accidental damage. A linear relation was observed between percentage change in fill thickness and the logarithm of fill thickness.

ACKNOWLEDGEMENTS

36. The author is grateful to British Coal Opencast Executive and to Sir William Halcrow & Partners for permission to publish this paper. The author acknowledges that by giving permission to publish the paper, British Coal Opencast Executive do not necessarily agree with all statements made and conclusions drawn. He is also grateful to his colleagues D M Park and I Bamborough for their help in its preparation.

REFERENCES

1. US Department of the Interior, Earth Manual. US Government Printing Office, Washington, 1974.
2. CHARLES J A, NAISMITH W A and BURFORD D. Settlement of backfill at Horsley restored opencast coal mining site. In 'Large ground movements and structures. Proceedings of a conference held in Cardiff, July 1977'. London, Pentech Press, 1978. pp 229-251.
3. CHARLES J A, HUGHES D B and BURFORD D. The effect of a rise of water table on the settlement of backfill at Horsley restored opencast coal mining site, 1973-1983. In 'Ground movements and structures. Proceedings of the 3rd international conference held in Cardiff, July 1984'. London, Pentech Press, 1985. Volume 3. pp 423-442.
4. SMYTH-OSBOURNE K R and MIZON D H. Settlement of a factory on opencast backfill. In 'Ground movements and structures. Proceedings of the 3rd international conference held in Cardiff, July 1984'. London, Pentech Press, 1985. Volume 3. pp 463-479.

37. Observed fill pressures on a concrete culvert in a steep-sided valley

A. K. HUGHES, Rofe Kennard & Lapworth, and C. G. HOSKINS, RKL Geotechnical, UK

SYNOPSIS. Loadings from engineering fills onto buried culverts have traditionally been assessed using a factor applied to the nominal total overburden pressure. The factor has frequently been taken as two based on experimental work earlier in this century and monitoring work by the Building Research Establishment. This value was used for the design of the RC culvert beneath a 35m high rockfill dam in a steep sided valley in Cyprus. The loadings on the culvert were measured to verify the design assumptions in view of the valley profile and the variable nature of the soft rockfill. Observations indicated that measured pressures on the culvert reached only just above 80% of the nominal overburden. The paper will present the results of the monitoring and discuss the conclusions from the observed behaviour and observations. Results are compared and contrasted with those for a recently completed dam in the UK.

INTRODUCTION

1. Symvoulos Dam is a dam built within the Western Sovereign Base Area in Cyprus which provides 1.1 million cubic metres of storage. This enables a proportion of the winter yield of water from boreholes and springs to be stored for use in the following summers as well as the run-off from the small catchment and used to balance a reduction in summer borehole yields from depleted underground aquifers.

2. The dam is formed above the concrete drawoff culvert, of 270,000 cubic metres of low grade rockfill obtained from a quarry within the reservoir area. The dam is straight in plan and has an upstream 3.5mm thick HDPE membrane and associated filters, which acts as the waterproofing membrane. It is some 40m high and 130m long, with a downstream slope of 1 (vertical) to 2.3 (horizontal) and an upstream slope of 1 (vertical) to 2.7 (horizontal). A typical cross section and plan of the dam is shown on Fig. 1.

3. A reinforced concrete culvert was constructed along the line of the existing river channel to carry the flow of the river during construction and the 500mm ductile iron draw-off and scour pipes upon completion. A concrete plug at the

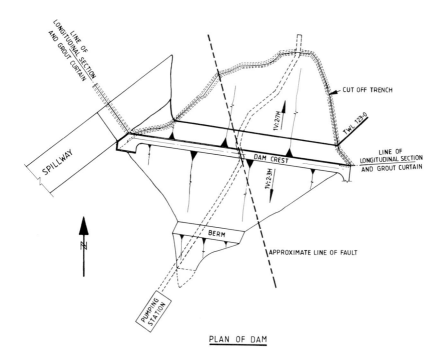

PLAN OF DAM

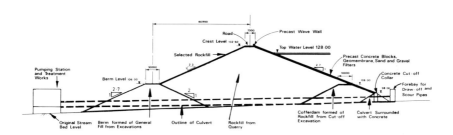

CROSS SECTION OF EMBANKMENT

Fig. 1. Symvoulos Dam

upstream end of the culvert was placed to allow impounding to commence, whilst the downstream end of the culvert connects with a pumphouse and treatment works.

PREVIOUS PRACTICE

4. Means of assessing the loadings on buried pipes have been formulated for various loading conditions by Young and O'Reilly (ref. 1). These allow for varying conditions and loadings on the pipes, depending on whether the pipes are in trenches or projecting, and factors including the trench dimensions and shape, the compressibility of the bedding and backfill and amount of fill above (Spangler (ref. 2)).

5. Works for conveying diversion or flood flows beneath an embankment dam are generally substantial and are normally constructed of cast insitu concrete to form a culvert. They are usually located on low compressibility or near rigid foundations and normally constructed in the open without any significant trenching into the general foundation surface. Consequently the culvert loading is normally higher, with little or no reduction for the effects of a trench construction. The opposite phenomena occurs, in fact, as the culvert is more rigid than the surrounding material and thus compression of the adjacent fill results in an enhanced loading on the structure. Penman et al (ref. 3) presented several case histories of earlier dams which showed a vertical pressure from the fill on the culvert of between 174 and 192 percent of the nominal overburden pressure. Information on pressure measurement at Winscar Dam was also given which showed a measured vertical pressure of 176 percent of the theoretical overburden pressure at the culvert centreline. Additional information was also available from cells which had been placed at the same elevation at increasing distances from the centreline, both above the culvert and into the fill beyond. These showed that the maximum pressure decreased rapidly, falling to about 80 to 90 percent of the theoretical value at the edge of the culvert and just beyond, and to approximately 55 percent at four to five culvert diameters from the centreline. They concluded that the simple criterion initially proposed by Trollope et al (ref. 4) of assuming a vertical pressure of twice the overburden pressure was reasonable.

6. The culvert for Roadford Dam is described in greater detail in a later section, but the vertical design loading was taken as 200 percent of the overburden pressure at the crown, decreasing to 133 percent at springing level.

DESIGN

7. The Symvoulos dam is located on the Symvoulos River across a deeply incised valley in the Pakhna Formation. At the damsite, this comprises a series of weak limestones and stiff calcareous clays dipping essentially downstream at approximately three degrees. The valley has been formed by two periods of incision by the river and consequently

comprises a narrow ravine-type feature in the valley bottom
as a result of rapid erosion by the river. A fault runs
NNW-SSE across the damsite and thus the generally trending
NE-SW valley is crossed by the fault near the dam centreline
as show on Fig. 1. This has controlled the local development
of the valley and resulted in a length of valley following
the line of the fault. Thus the bottom of the valley is not
straightforward and varies both in size, direction and slope
of the lower valley sides. The valley profile at the
location of the dam is shown on Fig. 2.

8. Unless substantial excavation was to be undertaken, the
topography dictated that the line of the culvert should
follow the valley floor as shown on Fig. 1. Some excavation
of weathered material was envisaged on the slopes of the
valley, but the lower levels and floor were expected to be
sound rock with little need for excavation beyond a
superficial depth. The exact profile and shape of the
excavation in the valley bottom was not known with any
certainty at the design stage and thus a conservative
assessment of loadings from the embankment fill was
considered to be appropriate. The small size of the culvert
and the potential difficulties of access dictated that the
arrangement should be straightforward and easy to construct
and thus the cross section on Fig. 3 was adopted. This
comprised gently sloping sides extending to a sufficiently
high level to allow the crown of the culvert to be cast
without sloping shutters. The crown was sloped at
approximately 20 degrees therefore, which also suggested that
the vertical loads could be substantial across the full width
with little reduction towards the springing. A factor of two
applied to the nominal overburden was adopted with varying
depths of overburden pressure assessed on the line of the
culvert.

9. Most of the Middle East is known from historic records
to be active seismically and Cyprus is no exception. The
main activity has been along the southern coast and
particularly in the south western section. The dam site is
thus located in the area of greatest risk and thus the
effects of seismic action were included in the design.

CONSTRUCTION AND MONITORING
10. Evidence from the investigation of possible quarry
sites led to the conclusion that the variable nature of the
rock must be carefully considered as the use of material
approaching the Marsal lower bound limit for rockfill, whilst
adequate for stability considerations, may cause difficulties
during construction. Such materials could be associated with
substantial mechanical breakdown during placement resulting
in the formation of layers of crushed material, which would
prove troublesome to compact, especially in wet weather and
could lead to excess constructional pore pressures.

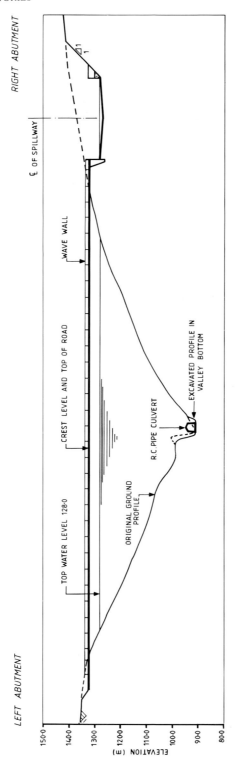

Fig. 2. Longitudinal section along dam axis

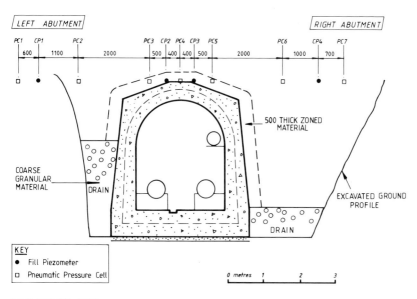

SYMVOULOS DAM

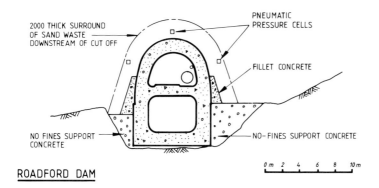

ROADFORD DAM

Fig. 3.   Sections through culverts showing positions
of instrumentation

11.   In order to achieve high dry densities, the
specification called for the rock to be placed in 500mm thick
layers so that less compactive effort was required. This
would reduce the breakdown of the rockfill caused by the
compaction process. The specification also precluded double
handling of the rockfill and stated that compaction would be
by not less than 4 passes of a 10 tonne vibratory roller.

12.   Blasting, excavation and loading of materials in the
quarry were by methods designed to produce a well graded fill
material without excessive fragmentation and degradation.
Site trials were used to determine the optimum means of
compaction. These trials showed that it was possible to
adopt a layer thickness of 1000mm and use a maximum size of
rock fragments of 600mm compacted by four passes of a 15
tonne dead weight vibratory roller without detriment.

13.   The moisture content of the rockfill excavated from
the quarry was of the order of 8-10% which was then raised to
the optimum moisture content by spraying/sluicing.
Particular care was taken to ensure segregation did not take
place at the edge of each layer by tipping the rockfill on
top of layer and then pushing the material forward with the
blade of a bulldozer. The upstream and downstream profile of
the dam was achieved by overbuilding the embankment and then
cutting back to the profile so that the face was formed of
well-graded, well-compacted rockfill. Special attention was
paid to the placing and compaction of material close to
structures, the foundation rock surface where the standard of
spreading and compaction was done with care to avoid
segregation and the collection of nests of boulders. A 500mm
cushion of finer material was placed adjacent to the culvert.

14.   An area of two square metres of drainage material and
an associated filter was placed adjacent to the culvert to
act as a collector drain down either side of the culvert.

15.   Pressure cells were installed at seven positions on a
section under the centreline of the dam and perpendicular to
the culvert as shown on Fig. 3. Piezometric pressures were
measured by four piezometers installed on the same section.

16.   The results of the monitoring are presented on Figs. 4
and 5. Fig. 4 shows the development of pressures with time
over a period including construction and impounding. The
apparent changes in readings in April 1991 occurred because
pressure cells were disconnected to allow the instrumentation
station to be moved and pipes to be installed within the
culvert. However, with time the readings have generally
returned to the level prior to disconnection.

17.   Fig. 5 shows the vertical pressures on the culvert at
the end of filling, on impounding, and the situation in
February 1993. The pressures are shown, in a manner similar
to that for Roadford, as a load factor relative to the
nominal overburden pressure.

18.   The conclusions that may be drawn suggest the vertical
load factor relative to the nominal overburden pressure
varied from a maximum of about 0.75 at the end of filling

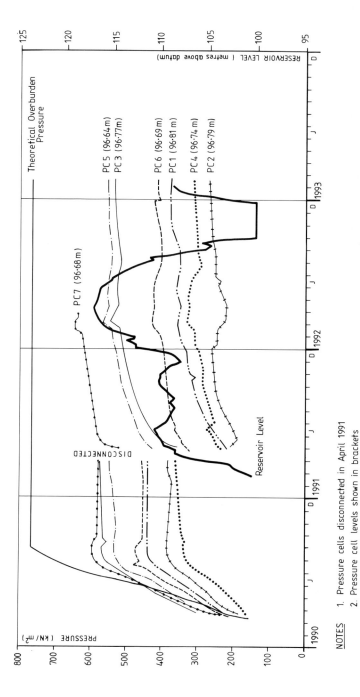

Fig. 4. Development of pressures on culverts - Symvoulos Dam

NOTES 1. Pressure cells disconnected in April 1991
2. Pressure cell levels shown in brackets

just off centre of the culvert on the sloping portion of the crown, to 0.44 on the centreline of the culvert. The load factor was generally higher on the right hand side of the culvert where the valley side was more gently sloping.

ROADFORD DAM

19. A section through the culvert at Roadford Dam is shown on Fig. 3, together with the details of the hydraulic pressure cells installed on the crown and at the springing level on both sides. The culvert is incised into the sloping east side of the valley as shown on the figure. A wedge of no fines concrete was placed to each side of the lower portion to improve the stability of the excavation during construction and to provide a longitudinal drain. A concrete fillet was also provided to aid compaction and to provide a transition between the fill and foundation, whilst a two metre thick cushion of sand waste was placed around all the exposed portions of the structure.

20. Pressure cells were installed at five positions along the crown and three positions along the springing level at both sides, corresponding with the three central crown locations. Provision was made to measure stresses in the three orthogonal directions at each location, but the results for vertical pressure only are compared in this paper.

21. The results of the monitoring have been presented by Evans and Wilson (ref. 5) and by Babtie Shaw & Morton (ref. 6). The detailed results are not presented in this paper, but Fig. 6 has been plotted from the available information to indicate the variations in measured pressure beneath the crest. Pressures are presented at the end of filling, after a further increase prior to impounding and the latest published values in September 1992: again they are shown as a load factor relative to the nominal overburden pressure.

22. The conclusions drawn from the latest observations (ref. 6) suggest the vertical load factor on the nominal overburden pressure varied from 0.72 to 1.18 with the highest value occurring beneath the dam crest. The load factors at the springing positions beneath the crest varied from 0.63 on the west side to 0.67, although higher factors of 1.15 and 0.98 respectively were found further upstream of the crest. Vertical pressures approximately 10m upstream of the dam centreline were broadly equal or just above the nominal overburden pressure at that position, whilst at a similar distance downstream, the load factors were found to increase from just below 0.7 to the west side to almost unity on the east side.

COMPARISON OF RESULTS - SYMVOULOS V ROADFORD

23. The results obtained from the pressure cells at Roadford show that the vertical pressures measured are considerably less than those assumed for design, being a maximum of 1.28 times the overburden pressure whilst the pressures either side of the culvert were only of the order

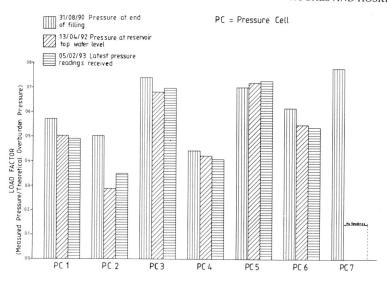

Fig. 5. Vertical pressures on culvert -
Symvoulos Dam

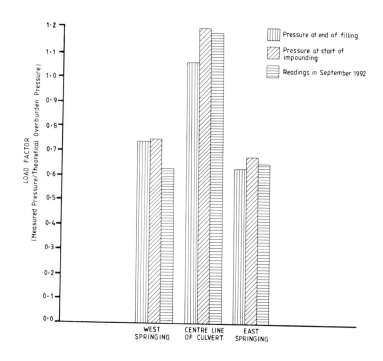

Fig. 6. Vertical pressures on culvert -
Roadford Dam

479

of 0.65 times the overburden pressure. However, the pressures to the left and the right of the culvert were sensibly equal. The valley slope appears to have had little effect on the value or the distribution of the measured pressures.

24. At Symvoulos, the vertical load factors were just in excess of 0.70 between the centreline and springing on both sides, but the maximum value of about 0.8 occurred at the end of the instrumentation array towards the right abutment. The value on the centreline was 0.44, whilst locally reduced pressures were evident above the drain to both sides of the culvert.

25. The apparent variations in readings at Symvoulos are thought to be as a result of valley shape which differs from one abutment to the other; being quite steep and close to the culvert on the left hand, and more gently sloping and distant on the right hand side as indicated on Fig. 3. The influence and proximity of the abutments to the culvert are thought to contribute to an arching effect which reduces the measured vertical pressure on the culvert in the manner shown.

26. The steeply sided valley, in conjunction with the relatively uniformly graded drain material also appears to result in localised arching to either side of the culvert, leading to the reduced pressures measured above the drain and enhancing the pressures recorded between the springing and the centreline.

CONCLUSIONS

27. Measurements of vertical pressures on culverts at two recently completed dams have shown that these are less than the commonly accepted value of twice the nominal total overburden pressure. The culvert shape and lower valley profile are considered likely to have influenced the recorded pressures.

28. Insufficient evidence is available to comment further on the effects of the culvert profile, excepting that the flatter shape of the Symvoulos culvert might have influenced the pressure distribution across the crown. Alternatively, the low pressures recorded on the centreline may have resulted from a localised zone of lower density fill associated with the pressure cell immediately above the culvert.

29. The influence of the lower valley profile appears to be more evident with little or no arching across the valley at Roadford. The reduced vertical stresses at Symvoulos appear to result from a significant arching effect due to the steep sided valley, which may possibly be influenced further by the relatively flat profile of the top of the culvert.

30. It is clear that the valley profile has a major influence on culvert loading, which may be further affected by the culvert profile. Additional reporting of case

histories appears necessary to gain further insight into this problem.

REFERENCES
1. YOUNG O.C. and O'REILLY M.P. A Guide to design loadings for buried rigid pipes. Second Impression. HMSO, 1989.
2. SPANGLER M.G. Discussion on the modification of the pressures on rigid culverts with fill procedures. (Pawsey S. and Brown G.B.) Highway Res Record, no. 249, 1968, 41-43.
3. PENMAN A.D.M., CHARLES J.A., NASH J.K.T.L. and HUMPHREYS J.D. Performance of culvert under Winscar dam. Geotechnique, vol. 25, no. 4, 1975, 713-730.
4. TROLLOPE D.H., SPEEDIE M.C.E. and LEE I.K. Pressure measurements on Tullaroop dam culvert. Proceedings of the 4th Australia - New Zealand Conference on Soil Mechanics and Foundation Engineering, 1963, 81-92.
5. EVANS J.D. and WILSON A.C. The instrumentation, monitoring and performance of Roadford Dam during construction and first filling. Proceedings of the Seventh Conference of the British Dam Society, Stirling, 1992, Thomas Telford, 157-165.
6. BABTIE SHAW & MORTON. Roadford Reservoir - Report on results from pressure cells adjacent to access/overflow culvert. Report to South West Water, 1992.

38. Temporary works bund design and construction for the Tees Barrage

Q. J. LEIPER and C. T. F. CAPPS, Tarmac Construction Engineering Services, UK

SYNOPSIS. The Tees Barrage, currently under construction, provides a focalpoint for the Teesside redevelopment area. A temporary works scheme to provide an open cofferdam (150 m x 70 m) within which to construct the barrage, by constructing two bunds across the Tees and diverting the river, was adopted by the design/build contractor. Design of the bunds, including dewatering requirements, instrumentation and the monitoring programme is described in the paper together with comments on the construction process to date.

INTRODUCTION

1. The Tees Barrage will provide a centre piece to a massive 101 hectare regeneration project at Thornaby between Stockton-on-Tees and Middlesborough in the north-east of England, UK. It will establish a relatively static water level upstream of the Barrage, thus enhancing the new river frontage of marina, housing and office developments.

2. The construction of the Barrage was awarded to Tarmac Construction as a lump sum design and construct package in October 1991. The contract package included the Barrage itself, together with recreational facilities, international standard canoe slalom, fish pass, navigation lock, link road to the A66 and a tubular steel arch bridge over the Barrage, Fig. 1. The Barrage, comprising cill beam, five concrete piers and four fish belly gates, is founded just below the existing river bed level on stiff glacial clays.

3. During the tender period for the works, it was evident that construction of the cill and piers in the river using individual sheet pile cofferdams would present problems of programming and accessibility for both the construction and river traffic. These problems also had considerable cost implications. In order to overcome these and also to provide both an engineering and cost effective solution, the construction team conceived and developed a scheme to create a large single cofferdam encompassing the Barrage site. The 70 m x 150 m cofferdam was created by initially diverting the river to the south of its existing course and then by creating cut-offs using fill bunds across the river channel. The permanent works piling for the "south island" between the Barrage and navigation channel was used as part of the diversion channel created, Fig. 2.

4. The paper describes the design process and comments on the implementation of the temporary works scheme.

Engineered fills. Thomas Telford, London, 1993

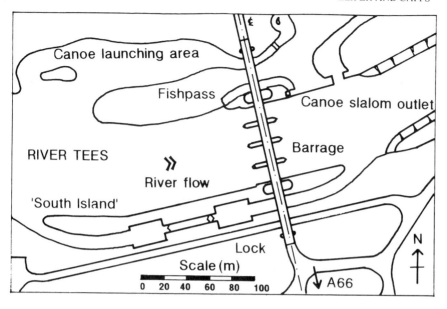

Figure 1: Plan of completed Tees Barrage

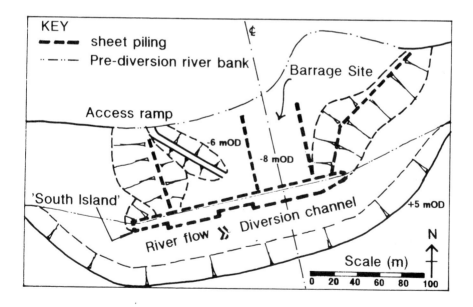

Figure 2: Concept of construction method utilising
permanent works piling, bunds and river diversion

SITE CONDITIONS

5.   The geology of the site is a variety of glacial and post glacial drift deposits, overlying Sherwood Sandstone bedrock at a level of approximately -20 m OD.   The drift deposits comprise 15 to 20 m of stiff to very stiff Glacial Till and Laminated lacustrine clays overlying about 5 m of Fluvio Glacial sand and gravels.   Subsequent action by the River Tees has cut channels through these glacial soils and deposited alluvial soft clays, silts and sands down to levels of -18 m OD.   Industry has also had its effect on the present day river course. Channels were cut in order to remove the long meanders, to make the river more navigable and to allow land to be reclaimed.

6.   It is on such a cut called "old cut" that it was decided to site the Tees Barrage, thus providing the benefit of founding on stiff Glacial soils rather than on the considerable thicknesses of alluvial soils in the former river channel, Fig. 3.   Although the Barrage was ensured a firm foundation, the space required for the temporary works meant that the bunds for the cofferdam scheme were pushed to the boundaries of the old river course, Fig. 4.

ADDITIONAL SITE INVESTIGATION

7.   The site investigations carried out for the Barrage provided a considerable volume of data at the Barrage location itself, but not at the proposed bund locations, where the interface between the stiff boulder clays and alluvial soils was uncertain.   Additional investigations in the river channel were carried out following contract award by means of shallow boreholes and Mackintosh probes to define the extent of soft alluvium beneath the proposed bund areas and to enable appropriate design parameters to be determined.

8.   The additional investigation more precisely located the old channel margin beneath the east bund where it was found that the depth of alluvium increased from 0.5 to 10 m beneath the proposed bund location.

9.   Boreholes carried out at the proposed west bund location revealed that the bund would be predominantly on the stiff boulder clays, but they identified a zone of very soft alluvium at the south end of the bund in the river bank.

BUND LOCATION

10. The bund locations were constrained by the Barrage and adjacent structure locations.   The east bund location was critical because the northern end had to be sandwiched between the canoe slalom structure and the poor ground at the margin of the old river channel while the southern end of the east bund had to lie between the temporary sheet pile wall for the Barrage construction (enabling construction to -8 m OD) and the eastern end of the river diversion channel (see Fig. 2).

11. As a result of the constraints, the northern part of the east bund had to be designed to overlie poor ground at the old channel margin where there was known to be a considerable depth of very soft alluvial clay which could not be economically removed.   A further complication was that temporary tied back sheet pile cofferdams were required to allow the Barrage, fish pass and canoe slalom structures to be built.

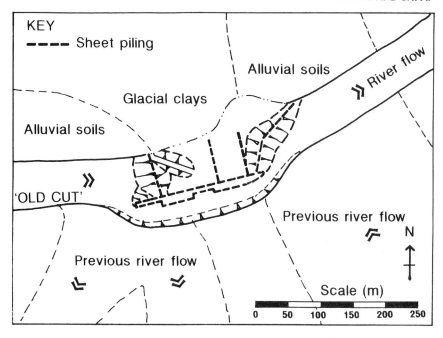

Figure 3: Plan of old river course and
construction site location

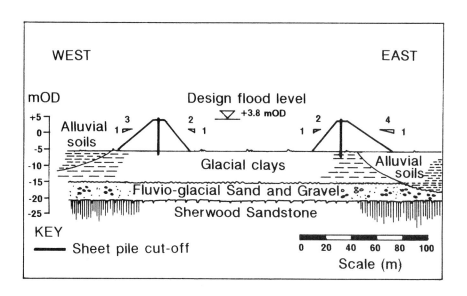

Figure 4: Long section through construction
site showing ground conditions and bund locations

485

12. The west bund location was dictated by access ramp requirements into the main excavation area, and the benefit gained by constructing the bund on the boulder clay in preference to the weaker alluvial soils of the former river cut to the west of the bund (see Fig. 3). The bund location chosen did, however, overlie the margin of the old channel. Following assessment of the stability of the bund and its method of construction, it was considered necessary to remove approximately 4 m depth of soft alluvial soils in the river bank at the south end of the bund.

FILL MATERIALS AVAILABLE

13. Fill materials available on site were predominantly natural clays and highly variable fills. These were considered unsuitable for placement under water to form what were to be essentially two small earth dams. A cheap source of granular material was therefore sought which resulted in a processed slag material being located.

14. Teesside Slag treats slag from Redcar Steelworks to remove heavy metal contaminants to provide a clean coarse granular material. Use of the material was discussed with and approved by both the National Rivers Authority and the Tees and Hartlepool Port Authority. This material was also suitable as granular fill elsewhere on the site, and in fact, all the 70,000 m^3 required for the bunds and access ramp was allocated for re-use as drainage and general fill.

15. Checks were made on leaching potential and gradings available. The grading of the selected material is shown in Fig. 5. A number of angle of repose tests were carried out to provide a guide as to how the material would be placed under water. In combination with these tests and available published information, parameters of ϕ = 35° to 45° were adopted for design.

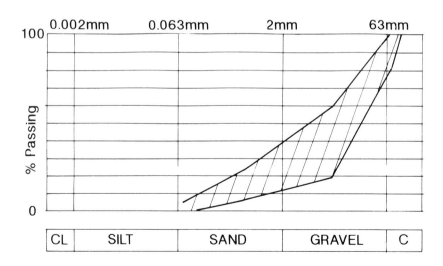

Fig. 5: Grading Envelope for Processed Slag

BUND DESIGN

16. The construction team's objectives in selecting the river diversion and bunding option were simple, namely to build the Barrage more quickly and more economically.

17. The design team's brief was equally straightforward, to provide a safe working environment and to create a workable scheme to construct the permanent works. The risks were appreciated by all concerned and it was these two points which dictated the design and construction process. Very careful thought was given to the natural soil and fill material parameters and the adoption of an appropriate factor of safety for the bunds. The tidal range at the Barrage location is 4.0 m. The bunds would, therefore, be subjected to considerable water pressures which were cyclical. Design flood level was +3.8 m OD, with an excavation level of -8 m OD. At the design stage it was recognised that instrumentation and monitoring of the bunds would be important elements of the design, together with the need for engineering control during the construction process.

18. Slope stability analyses were carried out using Bishop's Simplified Method and a factor of safety of 1.3 was adopted.

19. In areas where the bunds rested on stiff boulder clays, design side slopes were 1V:3H for the external (tidal) slopes and 1V:2H for the internal slopes. The north end of the east bund was a particular problem due to the soft alluvial clays underlying it. In order to achieve the design factor of safety required, slopes were flattened out to 1V:4H locally and temporary restrictions placed on access to this area of the bund until monitoring instrumentation was in place and confidence had been gained in the performance of the bund.

20. Consideration was given to alternative cut-off wall options, but due to the abundance of permanent works sheet piling on the contract, sheet piles were used for the cut-off walls through the bunds. Care was taken in handling and driving in order to maximise re-use of the sheet piles in the permanent works.

21. Where the sheet pile cut-offs abutted the south island permanent works piles, the junction was grouted using tube à manchette. At the north end of each bund, the sheet pile walls were extended into the bankside to extend any possible water flow path. The sheet piles were designed to penetrate a minimum of 2 m into the boulder clay beneath the bunds to ensure an appropriate cut-off.

22. The piezometric pressures within the sand and gravel horizon beneath the boulder clays and the possible presence of sand and gravel inclusions within the glacial clays indicated that groundwater control would be required to prevent heave at formation level.

23. To maintain engineering control throughout the proposed 11 months life of the bunds an instrumentation and monitoring programme was established as part of the design. A combination of inclinometers, extensometers, surface movement markers and standpipe piezometers was designed to monitor movement of the bunds and the groundwater control systems, Fig. 6.

24. A positive management system was established for the monitoring, with a proper inspection record kept by an approved and suitably experienced and qualified individual.

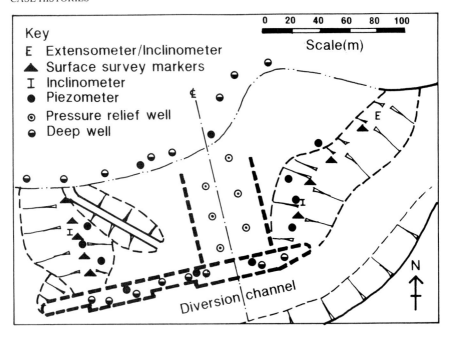

Figure 6: Layout of Instrumentation, Deep Wells
and Pressure Relief Wells

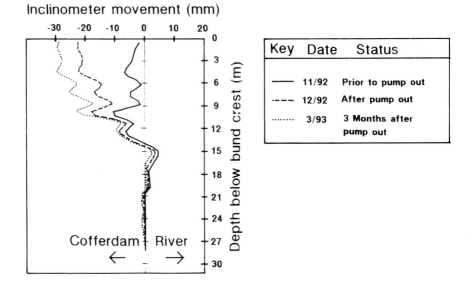

Figure 7: Typical Inclinometer Reading

CONSTRUCTION

25. Bund construction took place during November 1992. A stockpile of slag was established on the site to enable the bunds to be constructed quickly and, in particular, to enable closure of each bund to be accomplished without the tidal effects of the river destabilising the bunds or removing placed material.

26. The bunds were constructed by pushing out a spit of land directly into the river, on the bund centre line, allowing the advancing bund to push away any loose and soft silt as it progressed across the river. Material placed underwater, below low tide level, was end tipped and compacted only by tidal action. Above low tide level, the fill was compacted in approximately 0.5 m layers by a D6 bulldozer towing a 2 tonne roller.

27. The West Bund was constructed first. The slag was placed from the north side of the river, excavation to remove the soft alluvium adjacent to the south island having been completed prior to the diversion works (see Fig. 3). The 25,000 m^3 of slag was placed in a two week period and the slopes graded above low tide level to the design profiles. It was appreciated at design stage that the tidal action would considerably affect (and improve) the compaction of the bunds. The sheet pile cut-off was commenced one week after completion of the bund profile and took two weeks to install, Plate 1.

28. The East Bund and access ramp (approximately 35,000 m^3 and 10,000 m^3 respectively) were similarly placed in a two week period. Closure was delayed slightly due to the requirement to remove a quantity of clay fill placed in the river to facilitate driving the sheet piles at the east end of the south island. Following a re-assessment of the construction programme and temporary works requirements, a tied retaining wall at the south end of the bund and one at the canoe outfall works, combined with installation of the permanent training wall sheet piles meant that piling works for the bund took a period of four weeks.

29. Following completion of the instrumentation installation and the sheet piling cut-off works, the Barrage was pumped out. 17 No. pumps took ten days to pump the excavation dry, Plate 2.

30. It was evident on completion of pumping that there was a considerable ingress of water into the cofferdam area. Checks were made to the cut-off/pile connections to the south island and to the north of the bunds the cut-offs were extended. The cause of the problem was identified to be leakage through sheet pile clutches.

31. It was possible that some piles had declutched, but inspection of driving records through the boulder clay and the fact that piles had been vibrated through the bund material led to this theory being discounted.

32. At the time of writing, further attempts to limit water flow by allowing Pulverised Fuel Ash to penetrate the clutches were planned.

INSTRUMENTATION MONITORING

33. Prior to pumping out the cofferdam the inclinometers showed very little movement, apart from some induced by adjacent piling works. During pump-out in November, movement

Plate 1: Bunded cofferdam prior to pump out
November 1992

Plate 2: Bunded Cofferdam after pump out
December 1992

of up to 10 mm was measured towards the cofferdam. Three months after completion of the pump-out, the inclinometers have stabilised with a maximum recorded movement of 30 mm on the East Bund, Fig. 7. Small fluctuations are evident in current readings, possibly due to tidal effects.

34. The deep well system was switched on in mid-November 1992. Standpipe piezometers in the fluvioglacial sand and gravel stabilised at levels of between -9 and -12 m OD within three weeks with a tidal fluctuation of between 0.5 and 1 m.

35. Since completion of the bunds there has been approximately 10 mm of settlement at the crest. The majority of this settlement occurred during the November pump-out.

36. Water levels in the pressure relief wells in the base of the Barrage stilling basin area (which was excavated to -8 m OD) remained constant to within 0.5 m. Fluctuations in readings were generally associated with maintenance of the dewatering system where, on occasions, two, but more generally one, pump was being serviced or replaced.

37. Due to the inflow of water into the cofferdam area, water flows and its fines content were monitored rigorously and included within the positive monitoring system.

38. After an initial period of daily instrumentation readings, which was extended to twice weekly or weekly for some instruments, the monitoring programme was reduced, whilst still maintaining tight control by daily readings of the key piezometers and relief wells.

39. The monitoring programme also provided for:

   (a) Daily visual inspections by the instrumentation engineer and specified senior managers.
   (b) A positive recording and checking system, incorporated into the site quality management system.
   (c) Regular liaison with and checks by the design team.

CONCLUSIONS

40. The river diversion and bunded cofferdam provided an appropriate engineering option for construction and enabled programme savings to be made.

41. A processed slag material was successfully used to form the bunds which provided further savings and the potential for reuse of the material elsewhere within the works.

ACKNOWLEDGEMENTS

42. The authors are indebted to Teesside Development Corporation for permits to publish the paper, John Franklin, (Project Manager) Sam Robinson (Barrage Construction Manager) and Bob White (Design Liaison Manager) of Tarmac Construction Ltd., Major Projects Division, for their help in its preparation and also for their constructive comments during the design phase.

REFERENCES
1. BISHOP A. W. The use of the slip circle in the stability analysis of earth slopes. Geotechnique Vol.5, No.1, 1955, 7-17.

39. Engineering landfill by dynamic compaction to support highways and buildings

N. J. MAPPLEBECK and N. A. FRASER, W. S. Atkins Consultants Ltd, UK

SYNOPSIS. Two contrasting case studies of ground improvement by dynamic compaction are presented. The first case study concerns the construction of a 6.5 m high road embankment over fill containing domestic refuse. The second study describes a scheme involving the construction of a variety of lightly loaded buildings over a former limestone quarry backfilled with quarry spoil and overburden. Performance testing is described in detail and $\propto$ values for various types of fill materials are summarised. This paper demonstrates that by careful considerations of the geotechnical and environmental aspects it is possible to engineer difficult ground and enable the economic development of otherwise derelict land.

ROYAL QUAYS OFF-SITE HIGHWAYS, NORTH SHIELDS, TYNE & WEAR
Background
1.   The Tyne and Wear Development Corporation (TWDC) are presently developing an extensive area of former riverside industrial land known as Royal Quays. The site is situated on the northern bank of the River Tyne and is being developed for housing, business, retail, industry and leisure.
2.   To provide improved access to the site the existing A187 Howdon Road is being upgraded to dual carriageway standard and a new link is being constructed between the Tyne Tunnel and the A193 Wallsend Road (Ref Fig 1).
3.   The chosen route for the Off-Site Highways makes use of several "second-hand" sites. In fact the alignment is almost entirely over disused landfill sites containing domestic refuse.
4.   Ground investigations confirmed the presence of weak and compressible domestic refuse of varying content and ages. Furthermore gas probes installed around the perimeter of the tip by Gateshead MBC in May 1990 indicated that the landfill was still gassing.
5.   Ground improvement of the landfill was clearly necessary prior to road construction. Several methods of ground treatment were considered including removal and replacement, vibro compaction, dynamic compaction and pre-loading. The use of lightweight fills was precluded on economic grounds and the client's requirements to use surplus suitable fill materials from the Royal Quays site.
6.   A combination of dynamic compaction (DC) followed by pre-loading was chosen as the best option for the southern spur of the Tyne Tunnel/A193 link on the basis of suitability, cost and programme.

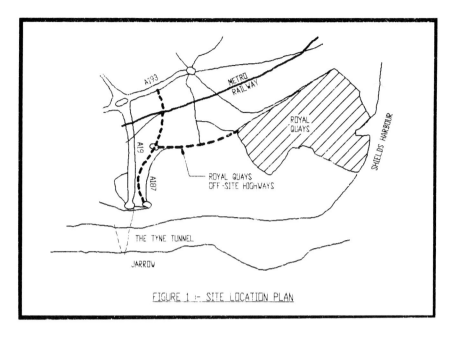

FIGURE 1 :- SITE LOCATION PLAN

7.   It was accepted by TWDC and the local highway authority (North Tyneside Council) that there would be some long term maintenance costs because of settlement arising from biodegradation of the organic materials within the landfill. These settlements will be most apparent at the embankment junction with the proposed Metro Bridge which is to be supported on piled foundations.

8.   An advanced earthworks contract was prepared for a 300 m section of embankment which ranges in height between 1.0 m and 6.5 m.

9.   This case history describes both the engineering and environmental aspects of the ground treatment. Also presented are the results of the performance testing and some post-treatment data.

Site description

10.   The southern section of the Tyne Tunnel/A193 link runs across open scrubland which caps a former refuse tip known as Low Flatworth Tip. The tip was closed in 1974 and contains approximately 0.8 million cubic metres of domestic refuse. The site is bounded to the north by the Tyne and Wear Metro railway and to the south by terraced housing. There are also a number of major services crossing the site including a 400kV O/H powerline, 33kV buried cables, medium and low pressure gas mains and the Tyne and Wear Interceptor Sewer (2.9 m dia).

Ground conditions

11.   Ground investigation works for the new road comprised boreholes and trial pits which proved up to 5 m of refuse. Standard resistance testing in the fill gave 'N' values of less than 10. Additional dynamic penetration testing was carried out at the beginning of the advanced earthworks contract on a 20 m grid which further confirmed the poor load bearing and settlement characteristics of fill. The fill material was

found to be dry and consisted of varying amounts of glass bottles, paper, plastic, clinker, timber and building rubble in a sandy or clayey matrix. In order to monitor gases perforated polypropylene pipes (60 mm dia, 4 m length) were installed in each of the trial pits. The top 1 m length of each probe was unperforated and fitted with a gas-tight cap. Follow up monitoring for methane, carbon dioxide and oxygen was carried out after allowing a two week stabilisation period. Representatives from ground treatment contractors attended the trial pitting exercise and later submitted their recommendations on the most appropriate form of treatment.

12. The gas monitoring revealed that significant concentrations of landfill gas were accumulating within the probes. Fill generating concentrations of up to 24% methane and 20% carbon dioxide were recorded although, generally, the gas composition was dominated by carbon dioxide rather than by methane and had a low but measurable concentration of oxygen. This suggested that the period of exclusively anaerobic gas activity, during which methane production within the waste mass is at its peak, had passed and the waste mass had entered a partially aerobic phase of gas activity. This was not unexpected given that the deposited refuse was at least 15 to 25 years old at the time of the study.

13. Below the landfill the superficial geology consists of Glacial Till comprising Pelaw (or Upper) Clay, Tyne-Wear Complex and Durham Lower Boulder Clay. The underlying solid geology comprises productive Coal Measures.

Advanced earthworks contract

14. The advanced earthworks contract documents were prepared in accordance with the Department of Transport Specification for Highway Works (6th Edition). The works included ground treatment by DC, the provision of a basal "no-fines" gas blanket and associated gas venting trenches and the construction of a 300 m long section of road embankment ranging in height from 1.0 m to 6.5 m with side slopes at 1 (vert) to 2 (hori).

15. An end product specification was adopted for the DC with performance testing to measure and check the effectiveness of the treatment.

16. From north to south the embankment reduces in height from 6.5 m to 1.0 m and therefore the area for treatment was divided into two parts with a transition zone between the two zones (Ref Fig 2). In Zone A, the contract specified that the treatment must provide a factor of safety of 2.5 against bearing capacity failure at a working load of 130 kN/m^2. Similarly in Zone B, a factor of safety of 2.5 was required for a working load of 70 kN/m^2 A 10 m by 10 m preliminary trial area was included within zone A to allow the successful contractor to test, monitor and refine his preliminary treatment design if necessary.

17. The performance testing consisted of three 2.5 m square zone load tests taken to 2.5 times working load and dynamic cone testing on a 20 m square grid.

18. The ground treatment specification contained a clause placing the responsibility on the Contractor to design his treatment to ensure that there was no risk of damage to nearby structures or services. Of particular concern was the effect of ground transmitted vibrations on the stability of the Metro railway embankment and associated signalling equipment.

Ground improvement works

19. The advanced earthworks contract commenced in January 1991. The first operation was the placement of a 0.6 m thick granular working platform. Initially, the DC was carried out using a single 60 tonne crawler crane with an 8 ton tamper, although later a second rig was mobilised to site. The treatment design submitted by the Contractor for the preliminary trial area was as follows:

Pass 1 - 8 tons from 16 m, 2 blows at a 4 m spacing
Ironing pass - 8 tons from 5 m, 1 blow at a 3.5 m spacing
Total energy input 28 Tm/m^2

The proposed total energy input was later increased to 35 Tm/m^2 following discussions with the Contractor.

20. On completion of the trial compaction, the first zone load test was set up. Zone tests enabled the performance of the treated ground to be tested over a wider and deeper zone than is normally possible when using conventional plate bearing tests. The tests consisted of a 2.5 m by 2.5 m square rigid steel test slab loaded by jacking against a reaction load comprising concrete blocks. The kentledge was carried by steel girders which were supported on mass concrete piers. Load was applied using a calibrated hydraulic jack with a load cell. Settlements were monitored using dial gauges mounted on a remotely supported rigid reference frame to ensure the readings were unaffected by ground movement. Precise levels were also taken to monitor the settlements of the test arrangement. The tests were of long duration typically 4 days and involved several cycles of loading.

21. The load applied was taken up to 2.5 times the working load required for each zone. The acceptance criteria were two fold;

a) The total recorded settlement at the working load must not be greater than 30 mm at a stage when the settlement rate had reduced to less than 0.5 mm/hour and

b) bearing capacity failure must not occur at 2.5 times working load.

22. The zone test in the preliminary trial area failed as settlements were in excess of 30 mm at 100% working load. At 250% working load, the ground carried the load in bearing. Hence the test showed that the required bearing capacity could be achieved however, adjustments were necessary to ensure that the settlements were reduced to the required level. Dynamic cone testing carried out on a two metre grid before and after the preliminary trials showed that little improvement had been effected in the upper 1 m of the ground.

23. During the compaction trial a detailed vibration study was carried out. The main objective of the study was to measure the levels of vibration on and around the Metro railway embankment for varying impact energies and distances from the impact centre. Background vibrations from the Metro trains were also measured. The main parameters measured were the resultant peak particle velocity (PPV) and frequency.

24. On completion of the trial exercise, PPV limits were set for the Metroline and the various services. The limits ranged between 4 mm/s for the Metroline signalling equipment and 25 mm/s for the utilities. As a precaution, a number of condition

surveys were carried out on nearby houses and close circuit television surveys were undertaken on several sewers. In order to minimise vibrations adjacent to Howdon Road, the Contractor opted to excavate a 1.5 m deep cut-off trench. Vibration monitoring continued throughout the contract whenever the compaction approached within 30 to 50 m of any sensitive areas.

25. On the basis of the preliminary testing the Contractor modified his treatment design as follows:

Zone A, Total Energy Input 92 Tm/m^2
Pass 1 - 15 tons from 10 m, 3 blows at a 4 m spacing
Ironing pass - 15 tons from 5 m, 3 blows contiguous

Zone B, Total Energy Input 50 Tm/m^2
Pass 1 - 8 tons from 16 m, 1 blow at a 4 m spacing
Pass 2 - 8 tons from 5 m contiguous
Additional treatment following results of 1st zone load test
Pass 3 - 15 tons from 5 m, 1 blow at a 4 m spacing
Ironing pass - 9 tons from 3 m, 2 blows contiguous

Performance testing
26. The second and third zone tests in zones A and B met the performance criteria with 25 mm and 15 mm settlement at 100% working load respectively. The load settlement curves are shown on Figure 2.
27. The dynamic cone testing in zone A indicated an average improvement ratio of 1.6 in the upper 1 to 3 m of fill, whilst negligible improvement was shown in the remaining depth of fill.
28. In zone B the dynamic cone testing showed an average improvement ratio of 1.4 over the full depth of fill and some improvement of the underlying natural ground.
29. Ground levels taken before and after the dynamic compaction indicate an average enforced settlement of 0.46 m in both zones.
Post treatment monitoring
30. In order to obtain some post treatment monitoring steel pins were driven 1 m into the top of the embankment at 20 m intervals along the centre line. Records collected over a 3 month period, following the completion of the earthworks, showed settlements of between 37 mm and 11 mm at the Northern and Southern ends respectively.

INVICTA BARRACKS, MAIDSTONE

Background
31. Maidstone Barracks is presently being relocated to Invicta Park (which contains the existing Invicta Barracks) because of the proposed improvements to the A229 Maidstone Spine Road. The cost of the works is approximately £15 million (excluding improvements to A229) of which 50% is for civil engineering works and is split between the three distinct sites at Invicta park. The Quarry site area is approximately 50,000 m^2 and is located 1.5 kilometres to the north of Maidstone town centre (Ref Fig 3). The client is Kent County Council.
32. The Quarry scheme comprises a miltary transport zone with parking, workshops and office facilities. The buildings consist of steel framed workshops with composite insulated panels and brick cladding together with two all brick

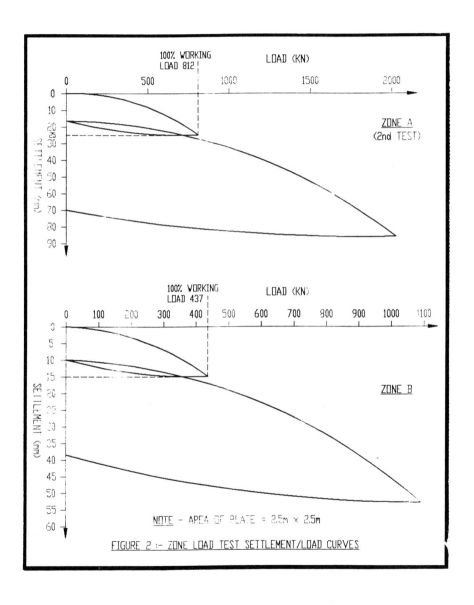

FIGURE 2 :- ZONE LOAD TEST SETTLEMENT/LOAD CURVES

buildings. A major component of the scheme included the regrading of a former limestone quarry that had been partly backfilled with quarry spoil and overburden.

Ground Conditions

33. The original geology consists of Hythe Beds of Lower Greensand (Cretaceous) age. In the undisturbed and unweathered state, the Hythe Beds comprises alternating layers (less than 1 m) of weak, calcareous sandstone (termed 'Hassock') and strong sandy limestone (termed 'Ragstone'). Local structural disturbance and subsequent weathering of the material has resulted in the creation of large cracks ('Gulls') that were subsequently infilled with weathered Hythe Beds consisting of fragments of rock in a silty clayey sand matrix.

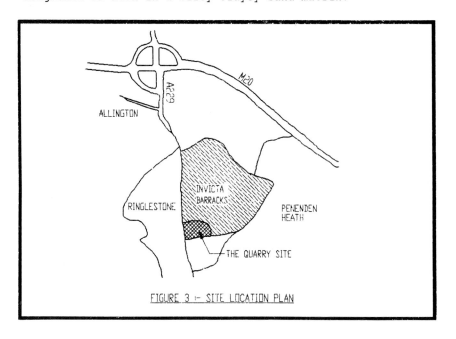

FIGURE 3 :- SITE LOCATION PLAN

34. Since the mid 1800's the site had been extensively quarried and subsequently loosely infilled with principally quarry spoil material. The SPT 'N' values of the spoil were typically less than 10 blows. The resultant topography consisted of hollows up to 8 m deep with numerous partly vegetated spoil heaps, together with a relatively flatter area to the southern half of the site.

Earthworks

35. For the construction of the proposed barracks it was necessary to undertake substantial earthworks prior to the construction of the proposed buildings. In order to reduce the settlements to an acceptable level the loose quarry spoil (made ground) had to be compacted. At the design stage it was considered that there were two principle alternatives, namely:

a) excavate all the made ground and compact as a structural fill;

b) regrade the site and use dynamic compaction in order to adequately compact the material.

498

36.   A major problem with option a) was that it was calculated that a substantial quantity (approx 50%) of the made ground would be unacceptable as a fill material, and this would have required either an extensive amount of treatment on site (such as lime stabilisation) or the removal off site coupled with the importation of large quantities of suitable fill to a site on an urban environment. Furthermore the cost of the excavation of the made ground was very large, there being an average of 4 metres of uncompacted made ground covering the site.

37.   It was therefore decided to use dynamic compaction. This was calculated to be substantially cheaper, but it was anticipated that careful attention was required to the design of the DC, principally due to the noise/vibration effect of the process in an urban environment; together with the problem of using DC on a relatively wet and fine grained material.

38.   It was decided to use a performance specification for the proposed works. The performance specification inserted into the contract documents was that the compacted fill should undergo not more than 10 mm settlement for a load of 100 kN/m^2. An additional clause was inserted to ensure that all of the made ground was adequately compacted, namely that the improved fill should have an equivalent Standard Penetration Test (SPT) 'N' value of 20.

39.   In order to check the performance of the compacted made ground fifteen zone load tests were proposed. In addition dynamic probes were specified on a 20 m by 20 m grid both before and after DC treatment.

Ground Treatment

40.   The contract was awarded by the client (Kent County Council) to Balfour Beatty Building Limited in April 1991. The bulk earth handling subcontractor was L&B Contactors Limited and the ground improvement sub-contractor was Vibroflotation Limited. Dynamic Compaction trials commenced in August 1991 and the works was completed in May 1992, using 1-2 rigs (35 rig weeks). The bulk earthworks included the excavation and placement of 70,000 m^3 of made ground and imported material (including granular blanket) prior to DC. The maximum layer thickness treated was 5 m, which in some areas the contractor used the DC on two separate layers. The compaction settlement as an overall average was in the region of 300 mm depth.

41.   The original method statement submitted by the sub-contractor was as follows:

Pass 1 - 8 tonne from 15 m, 10 blows at a 5 m spacing
Pass 2 - 8 tonne from 10 m, blows at a 5 m spacing
         (offset)
Pass 3 - Surface roll with a 6 tonne smooth wheel
         vibrating roller.
Total Energy Input = 64 Tm/m^2

42.   During the works the sub-contractor decided to vary his method as a result of zone load tests during trials together with the variable thickness and density of the made ground. The principal amendments were to use a 10 tonne weight, and the use of a 6 tonne weight for Pass 3 (ironing pass). The buildings were subsequently constructed using typically 1600 mm square pad foundation and 800 mm width strip footings.

43.   At the design stage it was felt that a fines value (<63 μm sieve) of 20-30% was the upper fines limit which could be compacted by DC. The use of DC on fine grained material is

liable to produce excess pore pressures that would render DC ineffective and make the treated made ground unsuitable for supporting fill (Ref Fig 4). During the works it became apparent that a 30% fines content was the highest acceptable. In one case where the fill was rapidly treated by two phase DC, excess pore water pressures resulted in the zone load test failing at only 75kN/m². The made ground, was found however to slowly recover its strength over a period of 2-3 months although some fill was partially replaced by imported granular material.

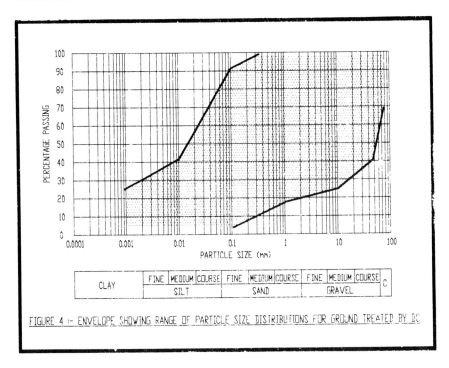

FIGURE 4 :- ENVELOPE SHOWING RANGE OF PARTICLE SIZE DISTRIBUTIONS FOR GROUND TREATED BY DC

44. At the design stage there was the concern that the DC would be discontinued in some areas as a result of vibration problems. Condition surveys were recommended to the adjacent housing. During the works the vibration was monitored by the contractor by using a CEL 221 Digital Vibration Meter and a CEL 223 filter unit. The maximum allowable peak particle velocity was 2.5 mm/sec.

45. During the works it was found that DC could be undertaken up to 20 m from existing buildings. However at one location approximately 150 m from the DC rig there was an unacceptable level of vibration attributed to the unique presence of a continuous layer of rock extending from the DC location to the house in question.

Performance testing

46. The performance specification of 10 mm settlement at a load of 100 kN/m² was measured by use of 14 zone load tests across the site. The zone load test was undertaken by applying a load on a rigid 1.5 m x 1.5 m steel plate located up to 1 m below ground level. The zone load tests were therefore

approximately equivalent to the proposed foundation and the stressed area. The kentledge consisted of 3 blocks of concrete, each of 4.2 T supported by a purpose made structure. Settlement measurements were recorded by 4 dial gauges at each corner of the plate. The results of these tests are given in table 1.

Table 1

Test No	Settlement (mm) at 100 kN/m² WL	Settlement (mm) at 200 kN/m² 2 x WL
1	8.6	
2	17	
3	5.2	
4	5.2	
5	19.6 (At 75 kN/m² Fail)	
6	10.8	
7	10.7	
8	12.7	
9	15	
10	6	15
11	7.2	18
12	8.7	22.7
13	13.2	24
14	12.3	22.3

Where the zone tests indicated settlements that were greater than the specification, it was necessary to undertake more DC.

47.   The requirement of the improved fill to reach an equivalent SPT 'N' value of 20 was assessed by means of dynamic probe testing.   There are published correlations (Card 1991) between the dynamic cone (N100) value and the SPT 'N' value with values N/N100 of between 1.5 and 2 (silty sands) and 3-4 for cohesive soils. An extensive amount of work was undertaken during the works in order to obtain a site specific correlation. It was tentatively proposed that over an averaged depth of 0.5 m, the N/N100 ratio was frequently 3.1-3.5, however this ratio was found to vary from 1-6. The accuracy of the dynamic cone to produce a N/N100 correlation or to define or indicate an improvement of the ground as a result of DC, was found to be poor. It was concluded that the dynamic cones sole use were to indicate areas where the DC may have inadequately improved the made ground, although this could never be quantified or proved without the use of the relatively more expensive zone load test.

48.   In one part of the site, the materials present were partly cohesive, having clay contents of up to 5-10%. As a result of the principally granular nature of the soils it was anticipated that the consolidation settlement would be relatively rapid. In order to confirm this assumption, a zone load test was extended for a period of nearly two months (1200 hours). Figure 5 depicts the results indicating that the consolidation settlement was approximately 6 mm.

CREEP SETTLEMENT RATE PARAMETER ∝

49. The long term creep component of the settlement of fills can be estimated using ∝ values. The parameter ∝ is defined as the percentage vertical compression of the fill that occurs during a log cycle of time, generally between one and ten years (Sowers et al 1965).

50. The value of ∝ will depend on several factors and is only valid where conditions in the fill remain unaltered. The main factors to be considered are as follows:

. Fill type, organic content
. Depth of the fill
. Initial method of placement and degree of compaction of fill
. Groundwater conditions
. Method of ground improvement and subsequent construction

51. An unavoidable feature of the Royal Quays site is that there will be long term settlement of the road embankment, as a result of the biodegradation of the landfill. There is a relatively small amount of published information available on the ∝ values of landfill. A summary of ∝ values for different soil types is shown on Table 2.

Table 2

Fill Type	Typical Values of ∝	Reference
Domestic Refuse	3% - 13%	Oweis and Khera (1986)
Domestic Refuse	2% - 10%	Charles (1982)
Domestic Refuse	7% - 20%	Perelberg (1986)
Well Compacted Sandstone Rockfill	0.2%	Charles (1982)
Uncompacted Opencast Mining Backfill	0.5% - 1%	Kilkenny (1968)

52. More work is required to determine ∝ values for fill materials, especially landfill. This can only be done by the continual monitoring of existing schemes. It is anticipated that correlations will eventually be produced by future authors between ∝ and the bulk biodegradable content of material. It is suggested that ground investigations should always carefully determine the biodegradable content of the material under consideration.

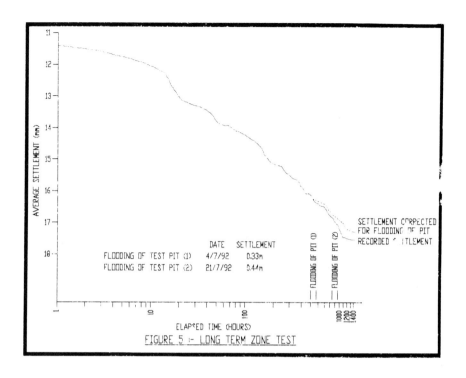

FIGURE 5 :- LONG TERM ZONE TEST

CONCLUSIONS

53. DC is a cost effective means of densifying landfill materials to enable redevelopment.

54. Although zone tests are relatively expensive they are in combination with penetration testing the most reliable means of performance monitoring.

55. The dynamic cone results for both case studies had a poor level of reproductability, correlations between N/N100 were frequently between 1 and 6, and this sometimes led to uncertainties on the relevance of the comparison of dynamic cone results before and after the DC. Consequently it is suggested that dynamic cone tests are best used for indicative purposes where it is suspected that the ground has been inadequately improved by DC, and proof could only be obtained by the use of zone load tests.

56. Creep settlement values depend on several factors and are best assessed using long term loading tests (ie skip loading tests).

57. 25% to 30% fines content is a maximum for effective DC and even this can lead to problems if the soil is wet.

58. Ground transmitted vibrations are a very important consideration when assessing the feasibility of DC.

ACKNOWLEDGEMENTS
The authors wish to express their thanks to the Tyne and Wear Development Corporation and Kent County Council for allowing the information in this paper to be published. Special thanks also to Corinne Bainbridge and Mark Docherty for preparing the text and figures.

REFERENCES
Charles, JA (1982) 'Settlement of Fill' Ground Movements and their effects on structures Ed. Attewell and Taylor. pp 26-45.

Kilkenny WM (1968) 'Settlement of Restored Opencast Sites and their Suitability for Building Development'. Departmental Bulletin No 38 Department of Civil Engineering, University of Newcastle upon Tyne.

Oweis, IS and Khera R (1986) 'Criteria for Geotechnical Construction of Sanitary Landfills. International Symposium of Environmental Geotechnology pp 205-222.

Perelberg, S, Boyd, PJH, Montague, KN and Greenwood, JR (1986) 'M25 Bell Lane Pit : Ground Improvement by Dynamic Compaction' Building on marginal and derelict land ICE Glasgow pp 267-280.

Sowers, G F, Williams, R C and Wallace T S (1965) compressibility of broken rock and the settlement of rockfills. Proc 6th International Conference on Soil Mechanics and Foundation Engineering, vol 2, pp 561-565.

40. Investigation and reconstruction of a failed crib retaining wall

A. McKENZIE, Suffolk County Laboratory, UK

SYNOPSIS. Following the collapse of the 3.5 m. high precast concrete crib wall retaining an 8.0 m. deep cutting slope, an investigation was carried out into the cause and mechanism of failure. The investigation included dismantling sections of wall both in the failed and unfailed sections, together with field and laboratory testing of the wall units, infill, backfill and natural ground. This information was used to develop a design for a replacement crib wall. The subsequent reconstruction was monitored using both load cells between the crib units and earth pressure cells in infill and backfill to the wall, to validate the design method.

INTRODUCTION

1. As part of the works for the construction of the grade separated interchange on the A12 trunk road at Capel St. Mary, Suffolk, a retaining wall was required to support the cutting slope to the underpass slip road. The maximum height of the cutting was some 8.0 m. with a total length of 300 m. To minimise the visual impact of the works and to maintain the semi-rural aspect of the area, a crib retaining wall was specified.

2. The geology of the area comprises an approximately 4.0 m. thick capping of chalky boulder clay; a grey, weathering to brown, silty or sandy clay of intermediate plasticity with chalk fragments and flint gravel (Boulder Clay), overlying glacial sands and gravels followed by Pleistocene Red Crag and London Clay at depth. Between the boulder clay and glacial sands and gravels a transition zone of very clayey, gravelly sands is sometimes present. Thus, the upper part of the cutting was within the Boulder Clay with the lower parts penetrating the glacial sands and gravels. Groundwater is present within the Crag some 2 m. below the base of the cutting.

3. The wall was constructed using a precast concrete system of crib walling to the contractors design as shown in fig. 1. At its maximum height the wall comprised a total of twelve cells with a triple cell width at its base on a reinforced concrete strip foundation to provide a sloping wall face. The system is built of two basic precast concrete units; header units lain perpendicular to the wall face with location nibs on either end, and stretcher units lain parallel to the wall. Each cell comprised a set of headers and stretchers laid upon each other, producing a cell height of 280 mm. Infill and backfill are placed and compacted in lifts as the wall is constructed.

4. Backfill was specified as Class 6N (selected well graded granular fill) in accordance with the Department of Transport's Specification for Highway Works, but with a restricted fines content of 5% passing the 63μm. sieve to meet a minimum permeability requirement of 10^{-6} m/s. This specification was

considered appropriate to achieve the system manufacturer's suggestion that backfill be a "free draining" material. A perforated pipe drain surrounded by Type B filter material was included at the base of the wall.

5. Wall construction took place during the late summer of 1990, and generally followed the planned sequence of excavation and foundation casting in bays, followed by wall erection in lifts, with simultaneous infilling and backfilling. The back slope of the excavation was restricted to a maximum slope of 2 vertical to 1 horizontal to safeguard the works during construction. Both infill and backfill were obtained from a local pit of glacial sand and gravel as an 'as dug' material with compaction of the infill between units by hand ramming and the backfill behind the wall using a small vibrating plate compactor.

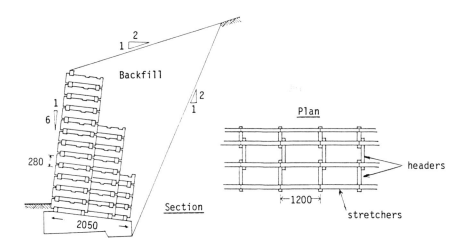

Fig.1 Original Crib Wall

WALL FAILURE AND INVESTIGATION

6. Following completion of the wall the slip road to the underpass was flooded during late October by rainwater to approximately 0.5 m. above the foundation along the deepest section of wall. In December the spoil was stripped close to the wall toe to allow slip road construction and a lateral displacement in the wall was apparent at the second crib cell above foundation level over the central highest one third of the wall. This was followed, after a period of heavy rainfall, on the 2nd January, 1991 by total collapse of the highest 20 m. length of the wall. Within this collapsed section the upper two thirds of the crib wall was completely destroyed and the infill and retained backfill slumped forward to form a slope of some 45°. No water flow was apparent within the collapsed material. A buttress was immediately placed to provide support to the slope throughout the collapsed length and adjacent distorted sections of the wall and an investigation into the failure was instigated.

7. A preliminary investigation was carried out during January, 1991 which included the installation of a series of monitoring points on the foundation and wall to check for further movement, dynamic probing, sampling, a reassessment of the wall's stability and literature search. The dynamic probing indicated that no softened zones were present beneath the foundation and the survey indicated that no noticeable movement of the foundation had

occurred. Overall stability appeared to have been adequate and
it was considered that failure may have occurred due to internal
failure of the wall itself.

8. The design of crib walls is included in the Code of
Practice for Earth Retaining Structures, CP2 (ref. 1) but the
recommendations for internal stability and design of crib members
are not unequivocal. A literature search indicated that an
extensive programme of both laboratory and field investigations
of crib walls had been carried out at the Technical University of
Vienna by H. Brandl (ref. 2, 3, 4). A summary of some of this
research was included in Brandl (1985) (ref. 5). A controlled
dismantling of sections of the wall was proposed to determine the
precise mode of failure and, together with an examination of the
published research, to derive a methodology for the design of the
replacement wall.

MAIN INVESTIGATION

9. Three sections of wall were excavated and dismantled during
March, 1991; one section within the collapsed zone, an adjacent
bulged but uncollapsed section and an unfailed section of wall at
the limit of the zone of deformation. These excavations took
place by dismantling each successive layer of crib walling with
manual excavation within a drag box to prevent collapse of the
adjacent sections of wall and backfill. The results of these
investigations are summarised in fig. 2 as discussed below:-

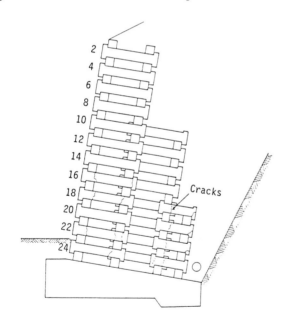

Fig.2 Typical section of failed wall

(a) Both the infill and backfill materials generally fell
within the grading envelope for modified 6N, although
a tendency of excess fines (5% to 10%) was apparent.
However, the material appeared to behave in a free
draining manner and when tested in a 1m x 0.3m x 0.3m
permeability box (ref. 6) gave permeability values of
approximately 4×10^{-4} m/s at compacted in-situ density.

(b) Infill and backfill in-situ densities were typically greater than 95% of proctor maximum within the centre of the cells. Conversely tests in the infill immediately beneath the header units, indicated low densities (87% of proctor maximum) equivalent to densities achieved in collapse settlement tests following inundation.

(c) Voids were apparent beneath the header and stretcher units towards the base of the wall with a thin silt deposit on the voids below level 18. Some siltation of the drain surround had occurred.

(d) Misalignment of some of the headers and stretchers was apparent due to imperfect laying (although it was difficult to see how this could be perfectly achieved without the use of compressible packing between the members).

(e) Extensive cracking was apparent within the bottom four cells of crib wall. The predominant crack pattern was sub-vertical cracking through the headers at the edge of their bearing on the underlying or overlying stretcher. Secondary cracks were occasionally present through the header nibs. Cracks within stretchers were rare. The headers were found to be unreinforced, whereas the stretchers generally contained a single polypropylene rope as handling reinforcement.

(f) Crushing tests carried out on cores cut from the precast units indicated an equivalent mean cube strength of 72 N/mm². Offset loading tests were carried out to define the moment of resistance of the unreinforced headers to eccentric loading at the nodes of the wall. These tests indicated a mean moment of resistance of 2.29 KNm. Some of these header units were found to contain dispersed fibre reinforcement but no enhancement of strength was apparent.

ANALYSIS AND REDESIGN

10. Following a review of the available literature it was considered that the internal stresses within the crib wall structure should be analysed by applying silo theory as described by Brandl (ref. 5) and briefly summarised below.

11. The crib wall is considered to act as a series of silo cells loaded by the earth pressure on the rear of the wall. Stresses in the infill within the cells are calculated using Janssen's silo theory modified for the rectangular crib cells (fig. 3). For equilibrium of a thin section of fill within a vertical silo, the mean vertical silo pressure at depth 'z' is given by:

$$P_{\overline{vz}} = \left[\gamma_v - c\frac{u}{A}\right] z_o \left[1 - e^{-z/zo}\right] \qquad \text{eqn. (1)}$$

$$z_o = \frac{A}{u} \frac{1}{K \ \tan \delta s} \qquad \text{eqn. (2)}$$

where γ_v = infill density

c = fill cohesion

u = perimeter length of infill cross section ($= 2(a + b)$)

A = area of infill cross section ($= a \times b$)

z_o = hypothetical depth of wall when geostatic overburden pressure, γ_{vz} is equal to the theoretical maximum silo pressure (ie P_{vz} max. for $z \rightarrow$ infinity)

k = earth pressure coefficient

δ_s = angle of internal friction between the silo wall and soil.

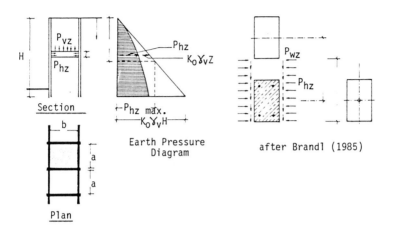

Section

Earth Pressure Diagram

after Brandl (1985)

Plan

Fig.3 Silo pressure in crib walls.   Fig.4 Stresses in crib units.

12. For practical design using typical infill materials, the following parameters are appropriate:-

c = 0

K = K_o = 1 - sin ϕ (for light compaction within cell)

δ_s = ⅔ ϕ

where ϕ = angle internal friction of infill.

13. As shown in fig. 3, the result of the silo effect is to reduce the vertical stress from that expected due to the weight of the overlying fill.   The resulting stress difference is progressively transferred into the surrounding crib structure via shear stress on the perimeter of the cell as shown in fig. 4

14. Thus, the vertical pressure acting along the cell boundary (P_{wz}) is given by:

$$P_{wz} = P_{hz} \tan \delta_s \qquad \qquad \text{eqn. (3)}$$

$$\text{where } P_{hz} = K \, P_{vz}$$
$$= (1 - \sin \phi) \, P_{vz}$$

NB: The vertical stress near the cell boundary (P_{vz}) will be less than the mean value ($P_{\overline{vz}}$) due to a non-linear vertical stress distribution across the cell. However, for simplicity the mean value $P_{\overline{vz}}$ is assumed to act at the cell boundary.

15. These silo pressures are combined with the active pressures acting on the rear of the wall together with the self-weight of the crib structure and surcharge fill above the wall.

16. The crib structure was then analysed with these applied loads to determine design forces at the nodes. For this analysis of the wall two modes of behaviour could be considered as shown below:-

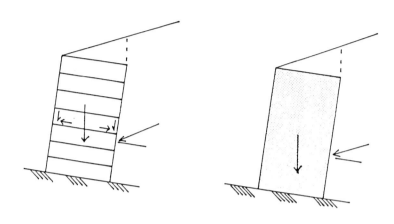

(a)   Virandeel truss.   (b)   Monolithic wall.

Fig.5 Analysis of crib structure

17. In case (a) the nodes are assumed to be effectively fixed due to the moment induced by the applied loads being less than the self clamping moment of the system. If the induced moments exceed the clamping moment, a pinned truss will develop as the wall deflects and produces a compressive stress across the cell infill. For case (b) the crib wall is assumed to act as a monolith, hence the base stress distribution can be calculated by moment equilibrium; as for a gravity wall. The results of these analyses confirmed that significant loads are transferred to the crib units producing maximum tensile stresses within the header units at the nodes.

18. To prevent failure of the members, redesign was specified as follows:-

(a) Design wall to allow for silo pressures within the cells, together with applied loads due to surcharge and active pressure from backfill.

(b) Allow for possible lack of fit at the nodes inducing moments in the header units (fig. 6 (a)). This was considered an upper bound case of the anticipated loading at the nodes due to inclination of the resultant nodal force (fig. 6(b)).

(c) Design members to BS5400, for ultimate limit state.

(d) Design nibs to resist possible shear loading from the overlying stretchers due to movement of the wall. Although calculations indicated there should be sufficient friction at the nodes to prevent lateral movement, it was considered prudent to ensure that the nibs could take the shear loading as this was observed as a secondary mode of failure in the failed wall.

510

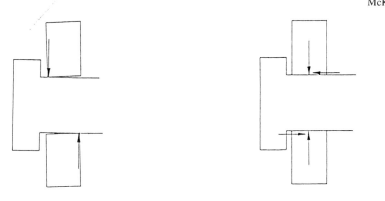

(a) Lack of fit   (b) Perfect fit

Fig.6 Nodal load conditions

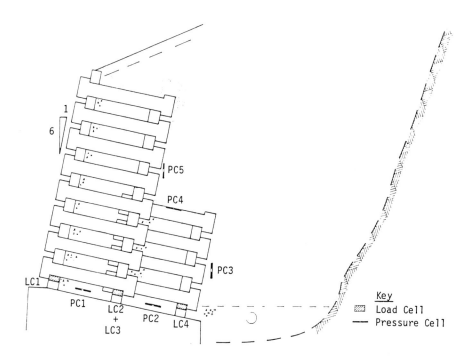

NB LC3 mounted beneath Header behind LC2

Fig.7 Replacement crib wall

19. The approved design for the replacement wall is shown in fig. 7. Key amendments from the original design were:-
   (i) Replacement crib wall to be built on existing foundation.
   (ii) Crib members of increased thickness, providing an increased cell height of 0.35 m.

    (iii)   Spacing of headers along wall reduced from 1.2 m. to 0.837 m.
    (iv)   Length of header units increased from 0.95 m. to 1.2 m. Thus only two cell widths were required at the wall base to maintain overall wall width.
    (v)   A reinforcing cage of 4 No. Y12 bars with shear links included in both header and stretcher units.
    (vi)   Header nib thickness increased from 75 mm. to 150 mm. to maintain cover to nib reinforcement.

RECONSTRUCTION AND MONITORING

20. The crib wall was reconstructed using the new RC units throughout its length during April/May, 1992. Modified Class 6N material was used for both infill and backfill, although the backfill specification was relaxed to allow up to 10% fines for maximum re-use of material on site. A drain was included at the rear of the wall as before. Compaction of infill was specified as 95% of proctor maximum and backfill compaction was to be to the specification for Highway Works ie 95% of vibrating hammer optimum. As before compaction was achieved by hand ramming between the units, although particular care was taken to ensure compaction beneath the members, and a vibrating plate compactor was used within the backfill. Control tests demonstrated compliance with the specification with mean dry densities of 1.98 Mg/m^3 and 2.12 Mg/m^3 for the infill and backfill respectively. Due to the different size of the new units, foundation levels were adjusted by over slabbing and consequently the maximum height of the crib wall was reduced from 3.5 m. to 3.0 m.

21. To validate the design method the highest section of wall was instrumented during reconstruction as shown in fig. 7. Five Glötzl earth pressure cells (PC1 to PC5) were installed to measure lateral pressures acting behind the wall and vertical pressures within both front and rear silo cells and surcharge pressure above the rear cell. To measure the loads between the precast units, four hydraulic load cells (LC1 to LC4) were installed at the bottom nodes. To avoid stress concentrations from misalignment due to installation, the load cells were recessed into the bottom stretcher by coring prior to wall assembly. These instruments were read frequently during wall construction and subsequently at monthly intervals; simultaneously the location of each header nib was recorded to detect any lateral movement of the wall. The earth pressure cell and load cell readings up to February, 1993 are plotted against time in figs. 8 and 9 together with the height of fill during construction.

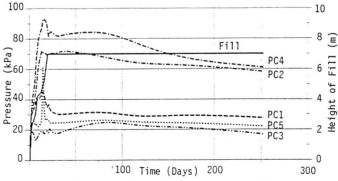

Fig.8 Earth pressure cell readings

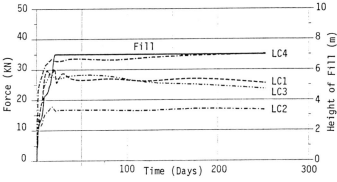

Fig.9   Load cell readings

22. The earth pressure readings generally show a sharp peak
shortly after installation due to compaction of the adjacent
fill.   After completion of the wall PC1, PC3 and PC5 generally
gave consistent readings, whereas the vertical earth pressure
above and within the rear crib cell (PC4 and PC2) showed a marked
decrease in pressure, particularly at PC4.   The maximum and
minimum post construction earth pressures are plotted against the
theoretical pressure distributions in figs. 10 (a), (b) and (c).
The lateral earth pressure at depth (fig. 10 (a) lies close to
the active value whereas at shallow depth, presumably due to
compaction pressures, PC5 exceeded the "at rest" pressure.   The
actual lateral pressure distribution and the deflected shape of
the wall are dependent upon the construction sequence.   The
observed distribution is consistent with completion of the wall
infilling, prior to completion of backfilling and slope
reinstatement, leading to higher pressures towards the top of the
wall.

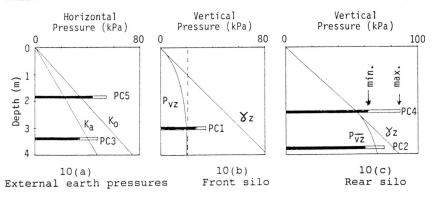

| 10(a) | 10(b) | 10(c) |
| External earth pressures | Front silo | Rear silo |

Fig.10 Predicted and measured earth pressures.

23. Figs. 10 (b) and (c) show the vertical silo pressures
within the front and rear cells respectively.   PC1 indicates a
pressure some 30% greater at the centre of cell than the
predicted mean value ($P_{\overline{vz}}$) as would have been anticipated given
the non-linear distribution of silo pressure across the cell
cross-section.   PC4 initially indicated a much higher vertical
pressure at the top of the rear cell than expected, although this
subsequently reduced to close to the overburden pressure.   PC2
suggested that the silo action is present within the rear cell
with a pressure close to that predicted.

513

24. The load cell readings in fig. 9 appear to show less response to temporary construction loads. The front cell nodes (LC1 and LC2) give fairly constant post construction readings. In comparison, within the rear cell, LC3 and LC4 indicate a decrease in load at the front of the rear cell and an increase at the back, consistent with rotation of the rear cell into the slope and the observed reduction in vertical pressure in PC4. Conversely, displacement monitoring has not shown any significant movement of the front cell. The nature and significance of these load variations at present remain uncertain.

25. The immediate post construction load readings are compared with the predicted values in Table 1 below. The nodal loads for the monolithic approach were derived from the base stress distribution assuming a three pin support. Thus, a single load is obtained for the central node, although within the wall this load is shared between two adjacent nodes (LC2 and LC3).

Table 1   Nodal Loads (kN)

Method	Front Node LC1	Central Node	Rear Node LC4	Total
Virandeel Analysis	43	57	31	131
Monolithic Analysis	41	63	21	125
Measured	30	46*	34	110

* Sum of LC2 and LC3 load.

26. Both analyses overestimate the front and centre nodal loads and underestimate the rear node load, and the effect of the variation of the load cell readings (para. 24) is to increase the difference from the predicted values. Assuming "lack of fit" due to misalignment of the crib members, the moment applied to the bottom header due to the measured nodal load is 3.1 kNm, which is greater than the failure moment measured on the original header units (para. 9) and is consistent with the postulated cause of failure.

CONCLUSION

27. The investigations suggested that failure of the crib wall could have occurred due to fracture of the unreinforced precast concrete units at the nodes. Collapse may have been triggered by settlement of infill immediately beneath crib members following saturation.

28. Significant stresses occur at the nodes of crib walls and concrete units require appropriate reinforcement. It is advisable to assume lack of fit in determining reinforcement schedules.

29. Internal stability should be considered in crib wall design. Silo theory may be used to estimate internal stresses which, together with self weight and external applied loads due to surcharge and backfill, can be combined to predict stresses within crib members. Alternatively, a simple monolithic approach can provide a guide to likely loadings.

30. Load cell readings indicate increasing loads on the rear of the wall, slightly above predicted values. Further monitoring is required to determine the long term equilibrium load distribution.

Acknowledgements
The instrumentation of the replacement wall was financed by the Department of Transport to whom thanks are given for permission to publish this paper. The permission to publish the paper does not imply that the statements made or opinions expressed reflect the views of the Department of Transport. The author would also like to thank P.J.Cearns and M.P.Young both of Suffolk County Council for many useful discussions during the project and his colleagues at the County Laboratory who carried out the investigations.

REFERENCES
1. INSTITUTE OF STRUCTURAL ENGINEERS, Civil Engineering Code of Practice No.2 (1951),"Earth Retaining Structures".
2. BRANDL, H.(1982) "Crib walls: Large scale tests, in-situ measurements, case histories, design, construction", Vol. 208.
3. BRANDL, H.(1984) "Crib wall systems" and "Crib walls with open back faces", Vol 251/1.
4. BRANDL, H.(1984) "Cases of damage to crib walls", Vol. 251/2
5. BRANDL, H.(1985) "Slope Stabilization and Support by crib walls and prestressed anchors",Third Int. Geotechnical Seminar, Soil Improvement Methods, Singapore, Nov. 1985.
6. DEPARTMENT OF TRANSPORT (1990),"A permeameter for road drainage layers", Highways and Traffic Departmental Advice Note No. 41/90.

References 2, 3 & 4 were published in German by the Federal Ministry of Construction and Engineering, Vienna, Austria and translated to English by the Transport and Road Research Laboratory.

41. Design and construction of spoil mounds

D. B. BLYTHE and D. B. HUGHES, British Coal Opencast, UK and
B. G. CLARKE, University of Newcastle upon Tyne, UK

SYNOPSIS. British Coal Opencast have undertaken a study of the
performance of a subsoil mound to facilitate the design of
future mounds. Subsoil mounds are constructed rapidly to
guidelines which cover the geometry of the mound. In this
study the mound was constructed in layers and observations were
made of the pore pressure response in the foundation soils and
the settlement of the mound. Samples of the foundation soils
and subsoil mound have been taken and tested. These data have
been used to establish the reasons for the failure of an
earlier subsoil mound and to produce preliminary design charts
for future development.

INTRODUCTION
   1. The extraction of coal by opencast mining is an
established economical technique which involves moving a
substantial amount of material in order to win the coal. This
material can include superficial deposits such as topsoil and
subsoil, plus glacial deposits and Coal Measures strata which
are known as overburden. In Northumberland, the extraction of
one tonne of coal may involve the removal and storage of up to
twenty cubic metres of overburden. Individual opencast mining
projects may yield from half a million to twelve million tonnes
of recoverable coal.
   2. It is a requirement on the operator of a mine to restore
the site so that other activities, for example agricultural,
can take place. Therefore, the topsoil, subsoil and overburden
have to be stored separately during the working life of the
mine and then used to restore the land surface.
   3. Each site will have, in order of increasing size,
topsoil, subsoil and overburden mounds. These are constructed
rapidly so that coal production can be achieved as quickly as
possible. They remain in place throughout the working life of
the mine which, in the Northumberland coalfield, can be between
two and fifteen years.
   4. Requirements for the ground investigation, design and
construction of the mounds are described in a Code of Practice
(ref. 1). Since these mounds are usually near the site

516

perimeter and/or adjacent to excavated faces it is necessary to undertake stability analyses to ensure safety of services, adjacent properties and excavations. These analyses are usually based on total stress parameters using data from site investigations undertaken at the time of exploratory work for the mine. Where failures have occurred, it has been usually due to failure in the foundation soils.

5. As part of an on going research and development study following the failure of a subsoil mound, the Northern Region of British Coal Opencast undertook to monitor the performance of a more recently formed subsoil mound in order to determine the changes in the foundation soils and subsoil materials during its construction. This paper introduces this study and gives recommendations for the design of future mounds constructed of glacial clay subsoils founded on glacial deposits.

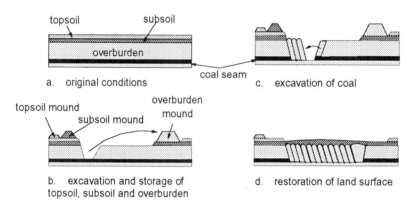

Fig. 1. The sequence of working for opencast coal mining

SEQUENCE OF OPERATIONS OF OPENCAST COAL MINING

6. The operations at an opencast coal site involve excavating through superficial deposits and Coal Measures strata in order to win coal from one or more seams. The simplified sequence of operations is as shown in Fig. 1. Firstly, topsoil and subsoil are excavated mainly by motor scrapers and transported for storage in mounds and dumps near to the site perimeter (1b). This is followed by excavation of the overburden to the coal seam(s) in the initial cut by face shovel or backactor. This overburden is transported by dump truck to the main overburden mound for storage. Successive cuts are then excavated by dragline and/or face shovel and/or backactor with spoils going into the previous cuts (1c). The final void is backfilled using the spoil from the overburden mound (i.e. from the initial cut), and the subsoil and topsoil are replaced (1d). Restoration to agriculture is the most usual end result, but other uses such as recreation, industrial

development or forestry are also common. Fig. 1 shows that there is usually a net volume increase in the backfill materials, (bulkage), and this is accommodated in the final restoration contours.

METHOD OF CONSTRUCTION OF SUBSOIL MOUNDS

7. Typically subsoil mounds are ten to fifteen metres high, have base widths of 50 to 150m, and side slope gradients normally between 1 in 1.5 and 1 in 2. Subsoil mound lengths can be several hundreds of metres. Thus larger subsoil mounds can include up to one million cubic metres of material.

8. Generally, in the Northumberland coalfield, the subsoils consist of glacial materials which are predominantly till. This till, described by Robertson et al (1993), is generally stiff sandy silty clay with gravel, the stone content increasing with depth. The foundation soils of a storage mound usually comprise the same glacial materials. The till is weathered near the surface and contains lenses of laminated clays, sands and gravels. After the topsoil has been stripped the top one to two metres of till is generally used to construct a subsoil mound. It therefore mainly consists of weathered material, though other Units of till may be present.

9. Subsoil materials are normally excavated by motor-scraper and transported to the mounding area where they are deposited in loose layers of 0.5 to 1m thickness. Occasionally, the subsoil may be excavated by face shovel or backactor loading onto dump trucks. Some compaction takes place due to the successive passage of the motor scrapers or dump trucks over the previously deposited subsoil. However, compaction is not considered desirable since increased porosity improves drainage and preserves the soil structure for agricultural restoration. The subsoil is deposited in layers though not necessarily in sequence. For example, one end of a mound may be constructed first. The final shape of a mound is achieved using a dozer vehicle with a blade.

THE CONSTRUCTION OF THE COLLIERSDEAN SUBSOIL MOUND

10. During this study it was decided to monitor the construction of a mound and the effect that this had upon the foundation soils in order to produce parameters for design. The Contractor, RJB Mining Ltd, was required to construct the mound in sequence in a controlled manner, that is completing each layer before placing the next.

11. Fig. 2 shows the two methods of construction using motor-scrapers (Caterpillar 631) and dump trucks (Caterpillar 777). Table 1 gives details of the earth moving vehicles. In addition to these vehicles a Komatsu D155 and/or a Caterpillar D8 dozer was used to spread end-dumped subsoil, assist the unloading or discharging of motor scrapers and form the slopes of the mound.

518

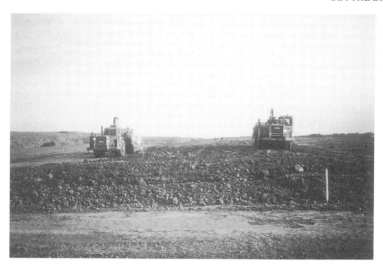

Fig. 2(a). Caterpillar 631 motor scraper

Fig. 2(b). Caterpillar 777 dump truck

Table 1 Earth moving equipment involved in the
construction of Colliersdean subsoil mound

Vehicle	Unladen Weight	Laden Weight	Ground Contact Pressure/axle
	tonnes	tonnes	kN/m^2
Cat 631	39.24	73.26	376
Cat 777	55.6	132.7	386

631 - Caterpillar 631D motor scraper
777 - Caterpillar 777 dump truck

519

12. Table 2 gives a summary of the construction which includes, for each day, the number and type of vehicles in operation and the source of the material. It took approximately 37 working days (a day was for the most part a twelve hour shift) to construct the mound over a 50 day period. Approximately 370,000m^3 of subsoil was moved to form a mound 12m high with a base width of 140m, a base length of 230m and a side slope of 1 in 2. This is in excess of 0.60M tonnes of subsoil or about 20000 tonnes per day on average which is equivalent to 260 movements of a Cat 777 per day.

Table 2   The construction of Colliersdean subsoil mound

Date	Vehicle Type	No of Vehicles	Source of Subsoil	Height above gl
4/9/91 - 7/9/91	631	7 or 8	stripping topsoil	
9/9/91 - 14/9/91	631	7 or 8	Boxcut A	5
16/9/91 - 18/9/91	631	5, 7 or 8	Boxcut B	5
	777	2 or 3		
19/9/91 - 20/9/91	631	6	dragline corridor	7.5
	777	2 or 3	Boxcut B	
23/9/91	631	6 or 8	dragline refurbishment	7.5
24/9/91 - 25/9/91	631	8	dragline corridor	8
	777	2 or 3	Boxcut B	
26/9/91	631	8	N-S haul road	9
	777	2 or 3	Boxcut B	
27/9/91	777	2 or 3	N-S haul road	9
30/9/91	777	2 or 3	unknown	9
1/10/91	777	2 or 3	general	9
2/10/91 - 3/10/91	777	2 or 3	N-S haul road	9
4/10/91	777	2 or 3	office area	9
7/10/91 - 12/10/91	777	2 or 3	Boxcut B	10
14/10/91 - 16/10/91	777	2 or 3	Boxcut B	10
17/10/91 - 18/10/91	777	2 or 3	Boxcut A	10
21/10/91 - 24/10/91	777	2 or 3	Boxcut A	11
25/10/91 - 26/10/91	777	2 or 3	general	11
28/10/91 - 29/10/91	777	2 or 3	general	12

13. Typically the subsoil was placed in approximately 0.5m thick layers by motor-scrapers or up to 1m thick layers with dump trucks. There was no planned compaction. Generally each layer was subject to more than one pass due to the movement of the earth moving vehicles which included a Komatsu D155 and/or a Caterpillar D8 dozers though the wheeled vehicles would tend to follow each others tracks. The subsoil tends to retain its original density because of the manner by which it was excavated and placed.

DESIGN OF SUBSOIL MOUNDS

14. Subsoil mounds are normally built within guidelines such as those given above. This is no different from techniques used in the design of highway embankments in which side slopes are specified for typical materials. However, there are instances in which it may be necessary to undertake

further studies and in those cases there is a need to select
parameters.

i. If there is a risk to adjacent property or services
it is necessary at the design stage to undertake a
stability analysis to ensure that in the event of a
failure there will be no damage to these structures
and that the side slopes give an adequate factor of
safety.

ii. It is also necessary to determine the ground profile
following construction of a mound since the
settlement could affect adjacent services.

iii. If the mound is adjacent to a proposed excavation
area it is necessary to undertake stability analyses
to determine not only the safety of the mound but
also the safety of the excavation face.

iv. Mounds sometimes fail because the foundation soils
contain layers of laminated clay. Failures may be
associated with periods of heavy rainfall.

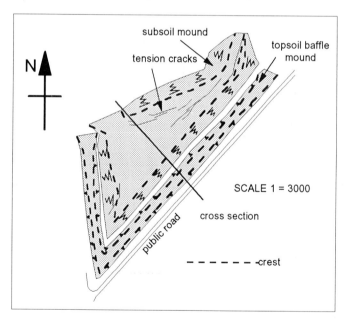

Fig. 3. Plan view of the Acklington subsoil mound
showing areas of slips

15. In order to undertake these analyses it is necessary to
make some assumptions about the parameters for the subsoil and
the effect the mound has upon the foundation soils. There is a

difficulty as is demonstrated by the back analysis of a failed mound.

ACKLINGTON SUBSOIL MOUND
The site and soil properties

16. Fig. 3 shows a plan view of the Acklington subsoil mound, which was located in the south west corner of the site. The rectangular shape mound was bounded on two sides by a topsoil mound and on the third side by the excavation.

17. The 13m high subsoil mound was completed by July 1981. Failures occurred during construction but these were contained within the site. Damage to the edge of the road occurred about four weeks after completion. Movements between the subsoil and topsoil mounds continued but these were not catastrophic. However, there was a need to check the overall stability both in the short term and the long term to ensure that no further damage occurred to the road.

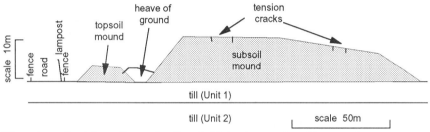

Fig. 4. Section through the Acklington subsoil mound

18. A typical section through the two mounds is shown in Fig. 4. A site investigation was carried out to determine the ground profile and parameters, the properties of the fill and to estimate the pore pressure regime.

19. Fig. 5 shows the typical soil profile and results of classification tests. Robertson et al (ref. 2) identified four Units of glacial materials in the Northumberland coalfield. All four Units, which together form a lodgement till, occur on this site. Unit 1 is a weathered ablation till, Unit 2 is a red brown ablation till, Unit 3 is a grey basal till and Unit 4 is a laminated clay. The laminated clay can occur at any depth but on this site it lies below the weathered till. Table 3 summarises the data for these soils. The figures in the brackets refer to typical values for those soils taken from an extensive database from the region.

20. The classification data for the subsoil were similar to the natural soils. The subsoil can contain all four Units of glacial soils therefore the classification data represent average values. The undrained shear strength of the subsoil was similar to the underlying soils confirming that the strength of the fill is unaffected by remoulding during placing.

Table 3 Classification and strength data for soils at
         Acklington mound

Soil	Water Content %	Liquid Limit %	Plastic Limit %	Undrained Shear Strength kPa	Angle of Friction
subsoil	19	40	17	90	(30)
mottled clay (Unit 1)	18(17)	39(33)	17(10)	124(150)	(30)
laminated clay (Unit 4)	25(20)	45(46)	21(26)	80(100)	(29)
red brown till (Unit 2)	16(14)	37(34)	17(14)	150(180)	(28)
grey till (Unit 3)	12(12)	30(32)	14(15)	305(200)	(31)

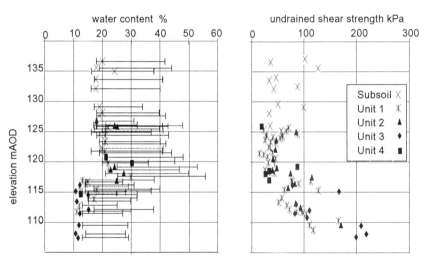

Fig. 5. Typical classification profile at Acklington

Assessment of the failure

21. The mound failed within one month of construction. An
undrained analysis, using the data given above and constraining
the slip surface to pass through the location of the mapped
tension cracks, gave a factor of safety in excess of 2. The
lowest factor of safety was for a circular slip wholly within
the subsoil mound. There was little evidence to show that the
slip surface was confined within the mound.

22. Inspection of the records of construction suggests that
toe heave between the topsoil and subsoil mounds occurred
during construction, perhaps due to the foundation soils being
squeezed out. Further heave between the mounds occurred once
construction was complete. These were accompanied by tension
cracks forming in the surface of the subsoil mound. Failure
did not occur as a result of increased precipitation since, at
the time, rainfall levels were relatively low.

23. About one month after construction heave was noticed outside the site boundary adjacent to the topsoil mound. There was no evidence of failure of the topsoil mound which suggests that the failure of the subsoil mound and the associated heave between the mounds caused the topsoil mound to be moved bodily. This mechanism would have resulted in heave beyond the topsoil mound.

24. Further movements of the subsoil mound on the other two sides occurred some months later. While these did not give cause for immediate concern there was the fear that they could affect the proposed excavation face.

25. A fully drained analysis would show that the subsoil mound was unsafe since the side slopes were greater than the angle of friction. Any failure surface passing through the foundation soils would give a factor of safety in excess of 1 in the longer term.

Pore pressure development beneath a subsoil mound

26. Piezometers were installed beneath the mound and in boreholes adjacent to the mound. In Northumberland there is widespread evidence of under drainage to the underlying Coal Measures with the phreatic surface being approximately at one metre below ground level but it tends to fluctuate seasonally. One piezometer beneath the mound indicated a piezometric head above the level of the mound suggesting an r_u value in excess of 0.5. At the time of the investigation this was attributed to instrument error because the values were much in excess of those anticipated based on published data.

27. It is generally accepted that the maximum pore pressure generated is about 40% of the increase in total vertical stress (ref. 3). However, if this is applied to the foundation soils and the subsoil mound, it produces factors of safety significantly less than 1 in the short term.

COLLIERSDEAN SUBSOIL MOUND

28. The Colliersdean subsoil mound was constructed as described above. It is about two miles from Acklington and founded on similar glacial soils. Given the speed with which these mounds are built and the size of equipment used, it was considered unwise to attempt to instrument the mound itself.

29. The depth to the Coal Measures is 14m, and all four Units of glacial soils are present. Piezometers, settlement cells and load cells were placed within the foundation soils at locations shown on Fig. 6. Additional piezometers and settlement cells were placed some distance from the mound to establish equilibrium conditions. This was necessary because of the limited time between installation and construction of the mound.

30. The instruments were recorded automatically, the interval between readings ranging from two hours to twelve

hours. During construction surface profiles were taken daily
using levels.

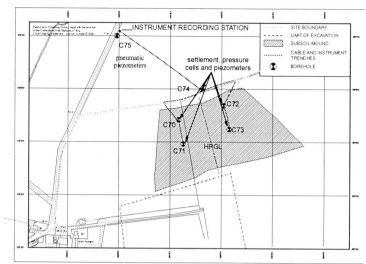

Fig. 6. Location of the instruments at Colliersdean

31. Details of the performance of this mound are to be
published. The data obtained, which are applicable to the
design of a mound, and relevant to the Acklington mound, are
the increase in pore pressure during construction and its
subsequent dissipation following construction.

32. Fig. 7 shows the changes in pore pressures at 1.7, 4
and 6m below the centreline of the mound. Piezometers remote
from the mound showed that equilibrium conditions were not
reached by the end of construction thus the changes shown in
Fig. 7 are likely to include some increase in pore pressure due
to the ambient conditions. The maximum increase in pore
pressure above the ambient conditions after correcting for the
time to reach equilibrium was 174, 97 and 32 kN/m^2 for depths
6, 4 and 1.7 m respectively. This is equivalent to r_u factors
of 0.60, 0.40 and 0.16.

33. It shows that the greatest increase in pore pressure
occurs at depth. There may be two reasons for this. The
shallow piezometer was placed in a trench which had to be 1.5m
deep because of possible damage to the cables during the mound
construction. The backfill material is likely to be more
permeable than the surrounding soil therefore the trench could
act as a drain. The upper till (Unit 1) is weathered and it is
likely that the permeability is greater than the unweathered
till below.

34. After 1.3 years the excess pore pressures were 115, 68
and 27 kN/m^2 at the same locations (that is r_u factors of 0.46,
0.30 and 0.14).

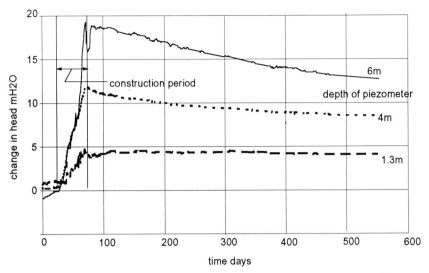

Fig. 7. The change in head of water during and after
construction beneath the centre of the mound

THE INSTABILITY OF THE ACKLINGTON MOUND

35. The pore pressure data from Colliersdean suggests that
it is possible to develop large pore pressures within the
foundation soils. It was not possible to determine the pore
pressure changes within the subsoil because of the difficulties
of ensuring that instruments would be unaffected by
construction.

36. Further analyses of the Acklington mound were
undertaken assuming undrained conditions in the subsoil using
the measured shear strengths and partially drained conditions
in the foundations soils with r_u factors of up to 0.7. This
combination of total and effective stress analysis may not be
totally satisfactory but the alternative is to develop a
peizometric distribution based upon an assumed model for the
development of excess pore pressure and its dissipation. It is
considered that the simple assumptions given above are
adequate.

37. A circular stability analysis showed that the critical
circle passed through the position of the mapped tension cracks
and the toe of the topsoil mound which conforms with the
observations. An r_u of 0.7 gives a factor of safety of 1.13.
The formation of a tension crack and infilling with water both
act to reduce this factor. The critical circle passes through
the weathered till (Unit 1).

38. It is proposed that the failure at Acklington occurred
in two stages. The first stage was the failure of the subsoil
mound which produced heave between the mounds. The second
resulted from the horizontal force of the failed subsoil mound

526

acting upon and causing movement of an otherwise stable topsoil mound. This, in turn, resulted in damage to the public road.

RECOMMENDATIONS FOR THE DESIGN AND CONSTRUCTION OF SUBSOIL MOUNDS

39. The failure at Acklington highlighted some of the problems associated with the design of the subsoil mounds, that is the selection of parameters and the pore pressure profile.

40. The undrained shear strength of glacial till is usually unaffected by remoulding. Therefore data from exploratory site investigations can be used. The subsoil can consist of a mixture of glacial soils but, for design, the strength parameters should be taken as those of Unit 1 in most cases, that is the top 2m.

41. Undrained analyses suggest that the Acklington subsoil mound was very stable. Fully drained analyses indicate that the subsoil mound would fail but because of the steepness of the side slopes the failure would be within the mound. Neither of these situations occurred in practice. Failures can occur during construction or sometime after construction.

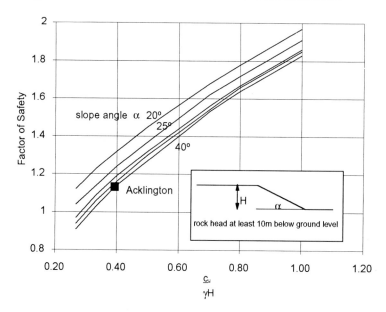

Fig. 8. A chart to assess the side slopes of a subsoil mound give the height of mound and the undrained shear strength of the subsoil

42. Evidence from an instrumented mound at Colliersdean shows that substantial pore pressures develop within the foundation soils, perhaps up to 0.7 of the overburden pressure. A simple design approach is suggested with the assumption that the strength of the subsoil is the same as the in situ

strength. A combined total and effective stress analysis was used in which the subsoil was assumed to have an undrained shear strength, whereas the foundations soils have a drained strength with an r_u factor of 0.7.

43. Fig. 8 shows a plot that could be used to give a preliminary assessment of the safe side slope angle for a given height of mound of differing strengths. It is assumed that the foundation soils are homogenous with an angle of friction of 30°, which is typical of tills in the region. It is recommended that if laminated clay layers (as opposed to lenses) are present, then a more rigorous non circular analysis should be undertaken. Further, if the mound could affect adjacent property, services or excavation high walls, then similarly a more rigorous analysis is recommended.

44. This design is based on the final shape. Typically subsoil mounds are not built uniformly, which can lead to instability during construction. It is recommended that mounds should be built in uniform layers while maintaining the final side slopes. This will reduce the possibility of instability.

CONCLUSIONS

45. It has been established that excess pore pressures developed in glacial tills following the rapid construction of a subsoil mound, are in excess of those expected from conventional construction of engineered fills.

46. It has been confirmed that the undrained shear strength of a subsoil mound is similar to the in situ strength of the source glacial materials.

47. A simple design procedure has been suggested that will allow preliminary estimates to be made of suitable side slopes for a given height of mound.

ACKNOWLEDGEMENTS

The authors would like to thank British Coal Opencast for permission to publish this paper and RJB Mining Ltd for providing invaluable assistance at Colliersdean. The views expressed in this paper are those of the authors and not necessarily those of British Coal Opencast or RJB Mining Ltd.

REFERENCES

1. Code of Practice (1982)'The siting and construction of temporary spoil mounds at opencast coal sites' Federation of Civil Engineering Contractors

2. ROBERTSTON T, CLARKE B G and HUGHES, D B (1993) 'Geotechnical Properties of Northumberland Till' to be published

3. VAUGHAN, P R, HIGHT, D W, SODHA, V G and WALBANCKE, H J (1978) Factors controlling the stability of clay fills in Britain, Proc. Conf on Clay Fills, Inst Civ Engnrs, pp 205-218

42. The ability of in situ testing to assess ground treatment

A. P. BUTCHER and K. McELMEEL, Building Research Establishment, UK

SUMMARY. The ability of in situ tests to assess the effects of ground treatment has been investigated in a programme of work which has used six different in situ test techniques on a range of fills, each fill treated by at least one of four ground treatment methods. The in situ test results are compared either with measurements of ground compressions due to the treatment or with the behaviour of test pads or foundations. Guidance is given on the suitability of each of the in situ testing techniques to measure the effect of ground treatment on the different fills.

INTRODUCTION

1. In situ testing of treated fills has, in the past, been restricted to plate loading tests, which are expensive and only test the surface layers, and the SPT which cannot give continuous profiles. Other in situ testing techniques, such as penetration type devices and geophysics, need investigating as to their suitability for the detection of ground treatment effects.

2. Six different in situ testing techniques have been assessed on a range of fills, including clay, brick rubble, ash fill and miscellaneous fills, each of which was treated by one of four treatment methods. Comparison of the pre and post treatment results from each of the in situ tests has allowed assessments of their ability to detect the effects of ground treatment.

IN SITU TEST METHODS

3. Brief descriptions of each in situ test are given.

Dynamic probing

4. Dynamic probing was carried out using procedures and equipment which conformed with the BS1377:1990, ISSMFE (1989) and DIN 4094. Dynamic probing is a simple in situ test carried out by a falling weight striking an anvil which is attached, via extension rods, to a conical tip. The number of blows needed to drive the tip into the ground 0.1m is recorded as

N_{10}. After each metre of penetration, when an extension rod is added, the torque needed to rotate the rods already in the ground is measured. The torque reading is then used to correct the blow counts for the friction on the rods. In most granular fills the torque readings have been found to be insignificant. The corrected blow counts are used to calculate the dynamic point resistance (q_d), using the following expression derived from that given in ISSMFE (1989):

$$q_d = \frac{Mgh}{A} \left(\frac{M}{M+M'} \right) \frac{N_{10}}{0.1}$$

where: M is the mass of the hammer in kg
  g is the acceleration due to gravity in m/s^2
  h is the drop height of the hammer in m
  A is the projected area of the cone tip in m^2
  N_{10} is the (corrected) number of blows/0.1m settlement
    of the cone tip
and M' is the mass of the extension rods and anvil in kg.

Static cone penetrometer (CPT)
5. The static cone penetrometer was used in accordance with Meigh (1987) and the ISSMFE (1989). The cone resistance (q_c) and the friction stresses (f_s) are monitored continuously as the cone is pushed into the ground at 1.2m/min.

Nuclear densimeter
6. The Nuclear densimeter outwardly resembles the static cone but, when pushed into the ground, an assessment of the density of the surrounding soil can be made by monitoring the backscatter from a Caesium 137 radioactive source, set in the tip of the instrument, using a photomultiplier. The device was built by Oxford University (Sills 1989) and calibrated at the Atomic Energy Research Establishment, Harwell to give densities to $0.01 Mg/m^3$.

Pressuremeter (PM)
7. A pressuremeter of the type originally devised by Menard was used according to the procedures recommended in Mair and Wood (1987). The pressuremeter test gives a pressure versus displacement curve; the slope of the initial elastic portion is used to calculate the pressuremeter modulus and the final portion is extrapolated to estimate infinite expansion to determine the limit pressure (Powell & Uglow (1985)).

Marchetti dilatometer (DMT)
8. The Marchetti dilatometer was used following the procedures recommended by Marchetti and Crapps (1981). The DMT is a flat blade shaped device with a thin steel diaphragm on one face. The device is pushed to the test level. A gas pressure behind the diaphragm is increased and readings taken of the pressure when the diaphragm just begins to move (pressure p_o), and when it has displaced 1mm (pressure p_1). The

530

tests are normally performed at 0.2m depth intervals.

Rayleigh wave measurements

9. Rayleigh wave measurements were made using techniques developed by Abbiss (1981) and Nazarian & Stokoe (1984). Rayleigh waves travel at a velocity about 5% slower than shear waves and at a depth between one third and one half a wavelength below the ground surface. By varying the frequency of the source, shear waves of different wavelengths are produced thereby travelling at different depths. This allows a depth - velocity profile to be obtained.

GROUND TREATMENT TECHNIQUES

10. The in situ tests were carried out on ground treated by one of four ground treatment techniques: dynamic compaction (DC), vibro, the rapid impact compactor (RIC) and surcharging.

Dynamic compaction

11. Dynamic compaction (DC) is one of the readily available ground improvement methods. It consists of dropping a heavy mass on to the surface of the ground from a height of several metres. Usually the treatment is carried out in two phases: a primary compaction using large weights from relatively high drop heights, followed by a secondary compaction using both a reduced weight and drop height. This secondary compaction is to re-compact the near surface layers which the heavier weights tend to loosen in compacting the deeper layers.

Vibro replacement with stone columns

12. Vibro replacement with stone columns uses a heavy vibrating poker to form a hole in the ground into which selected stone fill is placed and compacted by the poker in stages until the hole is full of compacted stone. The poker can be assisted in its penetration by either water or air jetting.

Rapid Impact Compactor

13. The Rapid Impact Compactor (RIC) (Watts & Charles, 1993) uses a heavy piling hammer to drive a large steel plate into the ground using successive blows. Although the energy per blow is lower than dynamic compaction the RIC can continue to drive the plate until a predetermined settlement per blow is reached, effectively imparting much more energy into the ground.

Surcharging

14. Surcharging is the simplest of the ground treatment methods used; existing or imported fill is placed to a specified height to produce the required vertical loading. Once this has been achieved the fill can be removed to another location on the same site using a continuous earthmoving process, thereby treating the whole site.

IN SITU TEST RESULTS

15. Testing was carried out on six fill material types: ash

531

fill, brick rubble fill, miscellaneous fill, clay fill, loose sand and in naturally infilled swallow holes. A range of in situ tests was used on each type of fill. Space does not allow inclusion of all the data so examples to illustrate the application of particular tests are given.

16. The ability of in situ tests to measure the effects of ground treatment has been assessed by carrying out tests, at the same location, both before and after the treatment. When the ground treatment is vibro the stone columns displace the fill and so post treatment tests were carried out at set distances from the centre of the tops of columns. This will detect changes in the fill due to the installation of the stone column and not necessarily reflect the behaviour of the column-fill combination.

Clay fill

17. The clay fill site was a backfilled open cast mining site where the clay had been end tipped after mining leaving a 12m deep layer of uncompacted clay which included air voids between the various sized lumps of clay. The clay fill was treated using a 7m high, 50m square surcharge to reduce the air voids. The surcharge produced substantial ground compressions in the fill down to about 8m depth which were measured by a magnet extensometer (Burford 1991) as shown in Figure 1(a).

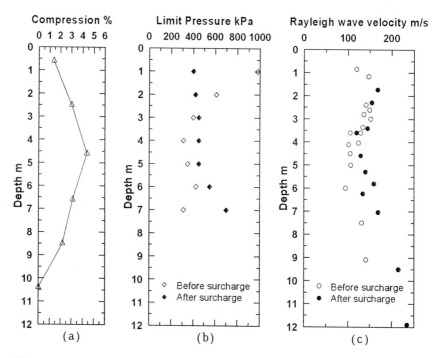

Figure 1: Clay fill: a) Magnet settlement gauge results
b) PM limit pressure profiles
c) Rayleigh wave velocity profile

532

18. All six of the in situ tests mentioned in an earlier
section were used but only dynamic probing, PM, Rayleigh waves,
and the DMT detected the effect of the reduction in air voids
in the fill induced by the surcharge. The dynamic probing
showed an increase in blows/0.1m after surcharging but this was
mainly due to an increase in extension rod torque rather than
tip resistance.
19. The PM profile is shown in Figure 1(b) and clearly shows
an improvement in measured limit pressures for tests between 3
and 7m depth. The Rayleigh wave data also showed an increase in
wave velocity particularly in the 3m to 9m depth range (Figure
1(c)). The pre treatment DMT profiles of p_0 and p_1 are given as
envelopes in Figure 2 with the post treatment envelopes
superimposed. The post treatment data suggests that only below
about 3.5m is there any improvement.
20. Although the magnet extensometer measured significant
vertical compressions from the surface the in situ testing did
not detect changes from the surface but from about 3m depth.
The magnet extensometer showed a compression of just over 3% at
3m depth increasing to almost 4.5% at 4.6m depth then reducing
back to about 3% at almost 7m depth. It appears that the in
situ tests only detected changes where the compressions were
greater than about 3%. The clay lumps were, in some cases,
larger than the in situ test devices and this would have
affected the results.

Ash fill
21. Two ash fill sites were investigated, one treated by
Vibro and the other treated in different areas by DC, the RIC
and Vibro.

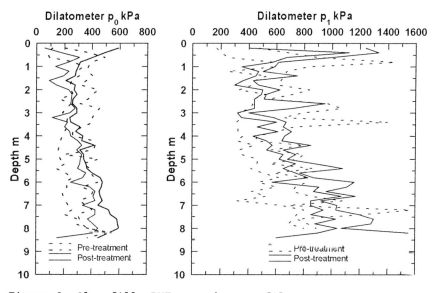

Figure 2: Clay fill: DMT p_0 and p_1 profiles

22. The first ash fill site had between 2.5 and 4.5m of ash and clinker fill underlain by 1.5m of soft clay resting on stiff boulder clay. The opportunity was taken to use the dynamic probing test to assess the fill treated by a standard stone column and, at a different location, a high energy stone column.

23. The dynamic probing profiles in untreated ash fill and in ash fill at a point one stone column diameter from the centre of the column are given in Figure 3(a). The two profiles are remarkably similar with a small improvement shown in the top metre of the ash fill.

24. Figure 3(b) shows the dynamic probing profiles before and after treatment (at one stone column diameter from the centre of the column) by a high energy stone column. The high energy stone column was constructed using twice the poker energy and about twice the volume of stone was used in its construction. The dynamic probing profile shows the q_d to have more than doubled in the ash and clinker but had little effect in the clays below 2.5m depth. The standard column did not appear to affect the surrounding fill whereas at one column diameter the higher energy column clearly densified the ash fill.

24. The second ash fill site consisted of 3.5m of loose ash and clinker overlying a stiff gravelly clay fill. Three treatment techniques, DC, Vibro and the RIC, were used on different parts of the site. A magnet settlement gauge was installed prior to treatment by the RIC and the settlements recorded are given in Figure 4(a) along with the ground profile (from Watts and Charles 1993). Significant settlements, and

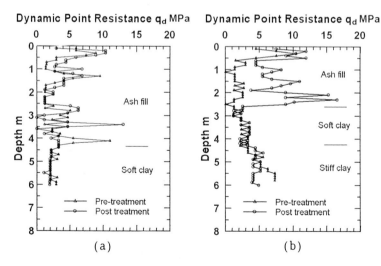

Figure 3: Ash Fill: a) Dynamic probing profiles before and after 'standard' vibro treatment
b) Dynamic probing profiles before and after 'high energy' vibro treatment

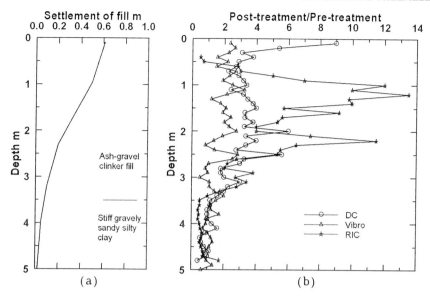

Figure 4: Ash fill: a) Magnet settlement results after
treatment by RIC
b) Comparison of the effect of treatment
technique using the ratio of post to pre
treatment dynamic probing results

hence improvements in fill density, were achieved by the RIC
down to the full depth of the ash fill.

25.   Figure 4(b) shows profiles of the ratio of post
treatment to pre treatment dynamic probing values against depth
for the three treatment techniques. This figure clearly shows
the effect on the dynamic probing profiles of the improvement
in the ash fill caused by the RIC which is also shown by the
measured settlements in Figure 4(a). The RIC did however leave
a loosened layer near the surface at the location of the
dynamic probing where the RIC did not carry out a second lower
energy treatment as it did near the magnet settlement gauge.
The DC achieved a consistent increase with depth (the DC used a
second lower energy treatment to re-compact the near surface
layers) with the vibro also showing improvement though less
than with the other treatment methods. None of these tests
detected much improvement below about 3.2m depth ie the bottom
of the ash fill.

Miscellaneous fill
26.   Two miscellaneous fill sites were tested, one treated by
vibro and the other by the RIC.
27.   The first miscellaneous fill site was a gravel pit,
between 3 and 4m deep, that was backfilled about 20 to 30 years
ago. Borehole logs revealed a variable clay fill with building
debris and some organic waste. Two pad load tests were carried
out, one on untreated fill and one on fill treated by the

Dynamic Point Resistance q$_d$ MPa

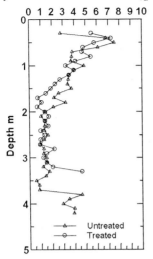

Figure 5: Miscellaneous fill:
Dynamic probing profiles
on untreated and treated
by vibro test pad
locations

installation of four 'standard' stone columns. The test pads
were 2m square and loaded by sand filled skips to impart a
bearing pressure of 49kPa. The settlement time behaviour of the
pads showed the pad on the untreated fill had settled less than
the pad on the treated fill!

28. Dynamic probings were made at the centre of the pads,
prior to loading, and showed a very similar profile (see Figure
5) with a crust within the top 1-1.5 metres of fill followed by
low dynamic point resistance down past 3m depth. The only clue
to the settlement behaviour of the pad loading tests is the
lower dynamic point resistance beneath the treated pad at

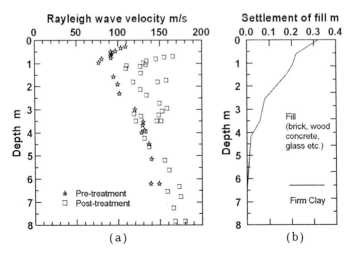

Figure 6: Miscellaneous fill: a) Rayleigh wave velocity profile
b) Magnet settlement results
after treatment by RIC

536

between 1.2 and 2m depth. These dynamic probing profiles fall within the envelope of profiles taken at other locations on the site both close to and at up to 5m from stone columns.

29.   The second miscellaneous fill site was a 6.5m deep former gravel pit where controlled tipping of builders debris had recently taken place over a few months. The fill was treated using the RIC and the compressions in the fill were measured by a magnet extensometer (Watts and Charles 1993). The dynamic probing was halted by the large inclusions in the fill. The pre and post treatment Rayleigh wave velocity profiles in Figure 6(a) show an increase in Rayleigh wave velocity down to about 3.5m which coincided with the significant settlements in the fill, shown in Figure 6(b), measured by the extensometer .

Brick rubble fill
30.   The brick rubble fill site comprised of 3 to 4m of brick and concrete rubble fill all of which passed through a 125mm sieve. Underlying the fill was 0.3m of sandy clay, and 0.8m of gravelly clay which rested on dense sand/cobbles (Figure 7(a)). The fill was placed by end tipping and dozing into place except for a small area at the edge of the site where the fill was placed to a Department of Transport specification. Otherwise the end tipped and dozed fill was either treated by DC or the RIC.

31.   Figure 7(a) shows the depth profile with the measured settlements from a magnet settlement gauge at the location of the treatment by the RIC. Rayleigh wave measurements before and after the RIC are given in Figure 7(b), before and after DC in 7(c) and in the DoT specification fill in 7(d). It is noticeable that the Rayleigh wave velocities in the underlying strata are similar at each location but the treatment methods appear to have had quite different effects on the brick rubble fill.

32.   The fill treated by the RIC had a surface layer with lower Rayleigh wave velocity than the DoT specification fill, but by 1m depth the velocities exceeded those in the DoT specification fill and then reduced to a lower level just above the sandy clay. The increase in the Rayleigh wave velocities due to the treatment by the RIC coincide with the most significant settlements measured with the magnet settlement gauge showing the geophysics to have detected the change in the density of the fill.

33.   The effect of the DC was to give a uniform increase in velocity with depth with the shallower depths having a lower velocity than the DoT specification fill which would indicate a loosening of the surface layers.

34.   Although the compressions in the ground due to treatment were only measured in the fill treated by the RIC, the geophysics did show improvements that were at the same levels as the measured compressions and the velocity profile in the fill compacted to the DoT specification was fairly constant around 210m/s as would be expected in a material placed and compacted in thin layers.

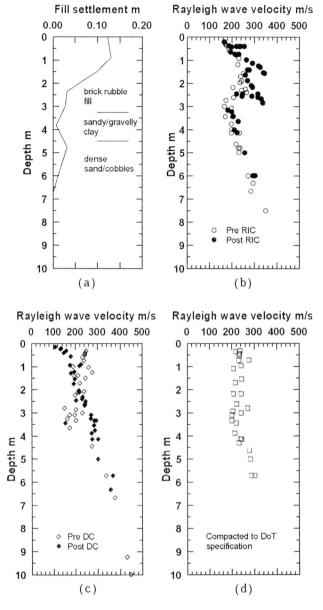

Figure 7: Brick rubble fill:
    a) Magnet settlement results after treatment by RIC
    b) Rayleigh wave velocity profiles before and after treatment by RIC
    c) Rayleigh wave velocity profiles before and after treatment by DC
    d) Rayleigh wave velocity profiles of fill compacted to DoT specification

Engineered fills

Corrigendum

The following text should appear after page 539.

surrounding soil.

41. The work on the ash fill sites has demonstrated the ability of dynamic probing to detect the different effects of each ground treatment method and that some techniques, notably the RIC and DC loosen the near surface layers on the initial treatment and need a second lower energy treatment to re-compact the shallowest fill.

42. Where the fill type, such as the brick rubble and miscellaneous fill, precluded the use of a penetration device, the Rayleigh wave geophysics were relatively successful. Rayleigh waves tend to pass through large volumes of soil so avoiding scale effects. Their penetration, however, can be restricted by surface or near surface layers with densities which contrast sharply with the surrounding material, such as the peat layer in the loose sand site. In other cases it is possible that the density of a fill can be increased by treatment without increasing the small strain stiffness of the fill. Since the Rayleigh wave velocity is related to small strain stiffness the Rayleigh wave velocity will not increase.

43. The Rayleigh wave geophysics gave the best results in miscellaneous fills, brick rubble fill and infilled swallow holes and had moderate success in clay fill, in some ash fill and in loose sand.

44. The Rayleigh wave geophysics showed the different effects of DC and the RIC on the brick rubble fill. The DC gave a looser surface layer increasing in wave velocity with depth whereas the RIC gave the same loosened shallow fill but with a higher velocity in the deeper fill than at the same depth in fill treated by DC.

CONCLUSIONS

45. A quantitative assessment of the ability of different in situ tests to detect qualitatively the effects of ground treatment has been made.

46. A range of in situ tests have been used to detect the effects of different ground treatment methods on six different fill types demonstrating the ability of in situ testing as a site control of ground treatment.

47. Penetration type in situ tests gave good results, successfully assessing the effect of the ground treatment on granular fills and loose sands. The dynamic probing test was found to be particularly suitable for assessing fills being simple to operate, economical and robust.

48. In miscellaneous and clay fills the penetration type devices were less successful but Rayleigh wave geophysics measurements detected the effect of the treatment down to about 5m depth.

49. In the end tipped clay fill the small scale in situ tests only measured the properties of the clay lumps and did not detect the reduction in air voids.

Engineered fills

Note

The order of authors on Papers 8, 12, 15 and 41 should be as follows.

Paper 8:   I. G. N. Smith, M. G. Winter, J. Oliphant, S. G. Wallis and J. M. Crowther

Paper 12:   W. F. Anderson, I. C. Pyrah, A. K. Goodwin, A. McKinlay and T. H. Salmon

Paper 15:   R. J. Jardine, H. Grace, O. Oyewumni, E. Armah and D. W. Hight

Paper 41:   D. B. Blythe, B. G. Clarke and D. B. Hughes

Other sites

35.   The ability of the in situ tests to detect the effects
of ground treatment has also been investigated on two loose
sand sites, treated by vibro and the RIC, and one site where
swallow holes have been naturally infilled which was treated
with vibro. On the vibro loose sand site the DMT, CPT and
nuclear densimeter all detected improvement up to 0.8m from a
stone column but no improvement at 1m. On the Ric treated loose
sand site the Dynamic probing, DMT and CPT all detected
improvement including the compression of a peat layer at 2.5m
depth and loose sand up to 6m depth (Watts & Charles 1993). The
Rayleigh wave measurements were not able to penetrate the peat
layer.

36.   The vibro treated naturally infilled swallow holes were
tested using CPT, dynamic probing and Rayleigh waves. The CPT
did not detect any improvement but the dynamic probing showed
improvement at near surface down to 2m and at between 8m and
12m depth in the deepest part of the swallow hole whereas the
Rayleigh waves detected improvement down to about 5m.

DISCUSSION

37.   The DMT, CPT, and nuclear densimeter had problems
penetrating some of the fills because of the relatively large
sizes of inclusions in the fills. The sizes of the inclusions
also affected the tests even when penetration was achieved.
This group of in situ tests was, however, successful at
detecting changes due to ground treatment in loose sands and in
fine grained fills with only limited success in the clay fill.

38.   Small scale in situ tests may not pick up changes in
clay fills resulting from a reduction in air voids unless the
compression in the fill exceeds 3%. The penetrometer type in
situ tests measured the properties of the clay, not detecting
the change in air voids. The penetration devices are relatively
small when compared to the sizes of the particles in some of
the fills and so can suffer from the effects of scale.

39.   The PM was used in clay fill and a miscellaneous fill
but only in the clay fill did the PM limit pressure show any
change due to surcharging. The PM limit pressure is the
estimated pressure at infinite expansion of the PM. At large
strains the effect of the reduction in air voids is most likely
to be noticed.

40.   The dynamic probing was effective at detecting changes
in ash fill, loose sand, in naturally infilled swallow holes
and moderately successful in miscellaneous fills and clay
fills. The miscellaneous fill and clay fill both contained
large air voids which were compressed by the treatment. The
clay fill was treated by surcharging but the miscellaneous fill
was treated by vibro which may not have improved the fill
sufficiently to be measurable by the probing. The dynamic
probing in ash fill showed that a vibro stone column installed
with twice the normal energy had a significantly greater
influence on the surrounding fill than a standard column which,
judging by the dynamic probing data, had little effect on the

REFERENCES
1. Abbiss, C.P. (1981). Shear wave measurements of the elasticity of the ground. Geotechnique 31, No 1, pp 91-104.
2. BS 1377:1990. Methods of test for soils for civil engineering purposes, British Standards Institution, London.
3. Burford, D. (1991). Surcharging a deep open cast backfill for housing development. Ground Engineering, September 1991.
4. DIN 4094 (1974) Dynamic and static penetrometers: dimensions of apparatus and method of operation. DIN Deutches Institut fur Normung e.V., Berlin.
5. ISSMFE (1989). Report of technical committee on penetration testing of soils - TC16 with reference test procedures. Included in Penetration testing in the UK, Birmingham, Institution of Civil Engineers, London.
6. Mair, R.J. & Wood, D.M. (1987). Pressuremeter testing: methods and interpretation. CIRIA Ground Engineering report, Butterworths, London.
7. Marchetti,S. & Crapps, D.K. (1981). Flat Dilatometer manual. Schmertmann & Crapps,Inc, Gainsville, Fla. USA.
8. Meigh, A.C. (1987). Cone penetrometer testing: methods and interpretation. CIRIA Ground Engineering report, Butterworths, London.
9. Nazarian, S. & Stokoe, K.H. II (1984). In situ shear wave velocity from spectral analysis of surface waves. Proc. of the World Conf. on Earthquake Engineering, 3, 31-38.
10. Powell, J.J.M. & Uglow, I.M. (1985). A comparison of Menard, self boring and push in pressuremeter tests in a stiff clay till. Proc. Int. Conf. on Offshore site investigations '85, SUT, London, UK. pp 201-217.
11. Sills G. (1989) Private communication.
12. Watts, K.S. & Charles, J.A. (1993). Initial assessment of a new rapid impact ground compactor. Proceedings of the Int. Conf on Engineered Fills, Newcastle upon Tyne, UK.

Embankment construction and monitoring on soft alluvium at five sites near Dartford

I. R. PAYNE, Kent County Council, UK

INTRODUCTION

1. The problems of constructing and maintaining embankments on soft alluvial soils are well known. These include possible failures during and after construction and excessive total and differential settlements during the lifetime of the embankment. There are three groups of techniques used to overcome these problems. The first is ground improvement by mechanical or chemical means. This includes stone columns, piles and grouting. The second is ground improvement by stage construction or preloading thus improving the properties of the soil by consolidation. The third is by selecting suitable fill or changing the geometry of the embankment thus reducing the average contact stress.

2. Numerous embankments have been constructed near Dartford, Kent on soft alluvial clays and peats. The opportunity has been taken to monitor five of these which involve different solutions to the problems. The first, used for comparison, is the conventional construction of a 4 m high embankment. Fig. 1 shows the location of the sites.

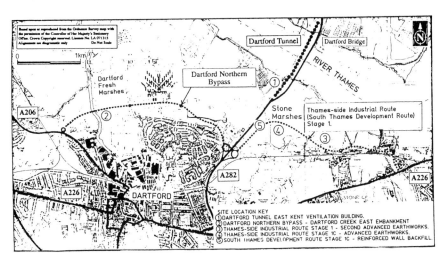

Fig. 1. Location of five case histories

Regional Geology

3. The geology underlying these sites is shown on the British Geological Survey's 1:50 000 scale geological map (ref. 1) and is discussed in the associated BGS geological memoir (ref. 2) with the following stratigraphic sequence of deposits:-

Stratum	Geological Period
Alluvium	Recent
Flood Plain Gravel	Recent and Pleistocene
Upper Chalk	Upper Cretaceous

A surface layer of Made Ground may also be expected as a result of industrial development in the area.

4. The memoir describes the Alluvium as blue grey marsh clay interstratified with beds of amorphous and fibrous peat and underlain by sands and gravels. The alluvial deposits show variation logitudinally and transversely in combination with a more familiar variation with depth.

Performance Monitoring

5. The performance of an embankment and the possibility of failure is best determined by monitoring instruments installed within the underlying soils and embankment fill. The following instruments have been used on the schemes described here.

(1) Hydrostatic profile gauges (vertical settlement)
(2) Inclinometers (horizontal movement)
(3) Magnetic extensometers (vertical settlement)
(4) Hydraulic piezometers (pore water pressures)
(5) Standpipe piezometers (groundwater levels)
(6) Surface observation markers/deep datums (settlement)
(7) Methane monitoring well (methane)

6. The data from these instruments together with the results of the site investigations have been used to produce and modify stability charts.

DARTFORD TUNNEL EAST KENT VENT BUILDING

7. Conventional embankment construction was used with no preliminary ground improvement measures at Site 1, that is the East Kent Ventilation Building which is located above the Second Dartford Tunnel 200 metres south of the River Thames (Grid Reference TQ760566). The building itself is founded upon the two lines of rigid diaphragm walls of the tunnel. The surrounding paved areas are constructed upon 4 metres of fill placed upon up to 12.5 metres of interstratified soft clays and peats of the adjacent Stone Marshes at the end of 1975.

8. A grid of surveying pins in the paved areas surrounding the Building has been monitored since 1979 with the recorded settlements of up to 570 mm between 1979 and 1993 producing

severe depressions and cracking of the bound pavement layers
as shown in Plate 1. The pattern of the surface settlements
is shown by settlement contours as illustrated in Fig. 2.

Plate 1. Dartford Tunnel East Kent ventilation building
car parking area. Typical cracking and kerb deformation.

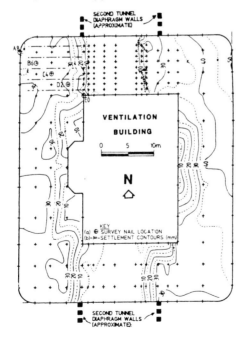

Fig. 2. Plan of Dartford Tunnel East ventilation building
showing car park settlement contours 13.04.84 to 25.07.85

9.   The data given in Fig. 3 indicates a linear relation between settlement and log time which is typical of secondary compression. High values of the coefficient of secondary compression C_O in the range 0.06 to 0.09 can be deduced from such settlement/log time graphs.

10.   The most economic solution at present to the continuous settlement is cyclic maintenance.

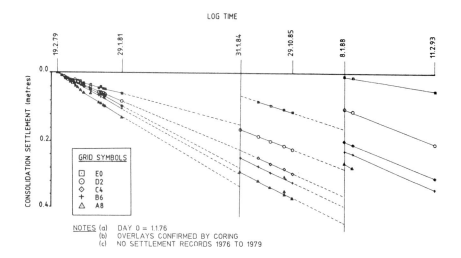

Fig. 3.   East Kent ventilation building car park settlement vs log time 1979 to 1993

DARTFORD NORTHERN BYPASS

11.   A 5m high composite embankment formed of a pfa core with granular fill haunches was constructed on up to 5m of alluvium at Site 2, the eastern approach embankment to the Dartford Creek bridge (TQ537755). Pfa was used because of the deficit of granular material. Water had to be added during construction to enable it to be compacted to the required density of about 1.4 Mg/m³. A Terram 1000 geotextile filter membrane was used between the granular fill and pfa to prevent piping.

12.   The embankment was placed on a drainage blanket composed of the same granular fill reinforced with a single layer of geogrid as shown in Fig. 4. A biaxial grid was chosen to ensure both transverse and longitudinal bridging effects across any areas of more compressible alluvial peat within the less compressible soft alluvial clay. This requires reliable jointing/overlapping of successive rolls of geogrid in both directions. A polypropylene grid was chosen to provide good interlock with the drainage blanket material.

13.   The granular haunches were stable and helped reduce the risk of plastic flow occurring beneath the embankment as the central core was infilled. Acceleration of the

544

earthworks construction was required so band drains were designed at a spacing of 1.5m on a triangular grid to accelerate pore water pressure dissipation by a factor of between 3 and 4 (ref. 3) with associated consolidation strength increases in the soft Alluvium.

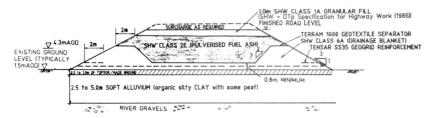

Fig. 4. Dartford Northern Bypass - Dartford Creek East approach embankment cross-section with PFA core

14. The granular haunches also affected the settlement profile as shown in Fig. 5, since the settlement across the embankment was more uniform. This is especially important when using pfa since this is brittle material which has a tendency to crack when subject to differential settlement (ref. 4).

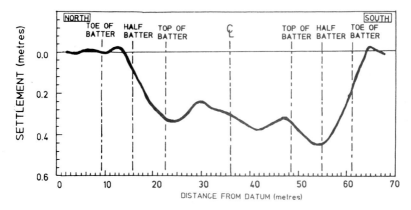

Fig. 5. Dartford Northern Bypass - Hydrostatic profile gauge chainage 1100 net settlement up to 24.08.92

THAMES SIDE INDUSTRIAL ROUTE STAGE 1

15. A 350m long 4.3m high embankment was placed as part of the advanced earthworks contract at Site 3, the advanced earthworks contract for the Thames-side Industrial Route Stage 1 (TQ572751). The embankment was constructed on a 0.3m thick construction platform formed of 125mm max size granular fill. Band drains were installed in the underlying alluvial clays and peats.

16. The heavily reinforced basal mattress (Geocell) in Fig. 6 was constructed from high strength Tensar geogrids

545

with the 1m³ cells formed into a diamond pattern. This mattress forms a stiff platform to maximise the potential bearing pressure mobilised in the underlying soft soils. It produces a more even distribution of load on the underlying foundation materials which results in a reduction in the magnitude of long term differential settlements across the width of the embankment.

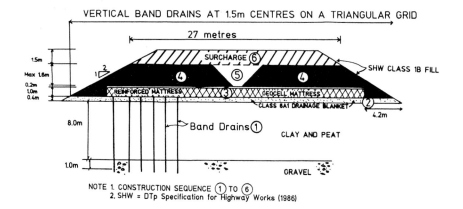

VERTICAL BAND DRAINS AT 1.5m CENTRES ON A TRIANGULAR GRID

Fig. 6. Thameside Industrial Route Stage 1 - 2nd Advanced Earthworks Contract, proposed construction sequence incorporating Tensar Geocell Mattress

17. The Geocell has the effect of forcing potential slip circle planes to pass vertically through the mattress and produces deeper critical slip planes extending into greater strength materials below the soft Alluvium (ref. 5). The low initial undrained shear strength c_u of 10kN/m² typical of the Alluvium and the use of the basal mattress suggests that instability will be by plastic failure within the soft materials (a "toothpaste squeezing" effect) rather than by a conventional rotational slip surface causing shear of the greater strength mattress materials which is considered in the Geocell design.

18. The benefits of basal reinforcement can be illustrated by a simple experiment using a jelly to model the soft Alluvium. The unreinforced model embankment in the left of Plate 2 produces bulging in the jelly beneath the fill and heave at the toes similar to records of embankment performance on soft Alluvium at Great Yarmouth (ref. 6). The addition of basal reinforcement to the model on the right-hand side of Plate 2 produces a more uniform settlement profile in the underlying jelly.

19. The end of construction HPG readings on 20.11.92. in Fig. 7 show increased settlements beneath the haunches illustrating the delayed infilling of the central section of the embankment. The current performance of the Geocell is also shown by the net HPG settlement profiles on 18.6.93. in

546

Fig. 7. The rafting effect of the Geocell has avoided the bulging beneath the embankment. This effect is demonstrated in Plate 2 in which the soil is modelled by jelly.

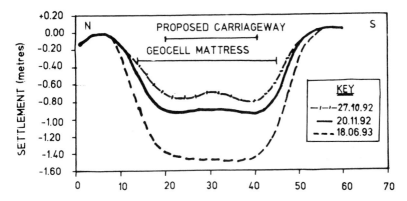

Fig. 7. Thames-side Industrial Route Stage 1 - 2nd Advanced Earthworks Contract, hydrostatic profile gauge chainage 1460, net settlement up to 18.06.93

Plate 2. Settlements of reinforced and unreinforced fills on jelly models

THAMES SIDE INDUSTRIAL ROUTE STAGE 1C

20. The techniques discussed above could not be used on this site, a link road, because of an existing adjacent carriageway and services. Piled rafts and lightweight fill were excluded because of cost and future development of the adjacent site. It was decided that a surcharge could be used

because the existing embankment would have, to some extent,
preloaded the ground.

21. Fig. 8 shows a cross-section of the site in which 3.5m
of fill were placed on 9m of alluvial clay and peats. The
top metre was removed after one year.

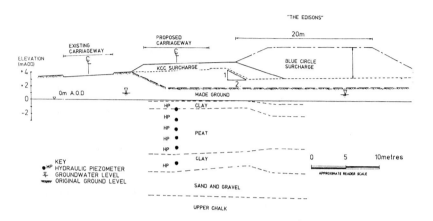

Fig. 8. Typical cross-section through S.T.D.R Stage 1C
and blue circle surcharge (Ch.850 November 1992)

22. After this surcharge was placed, part of the land
furthest from the existing carriageway was surcharged and
another part of the land adjacent to the carriageway was
excavated to form a lake. Fig. 9 shows the development of
pore pressure beneath the carriageway by the two surcharges
and Plate 3 shows the typical distress caused to the
carriageway.

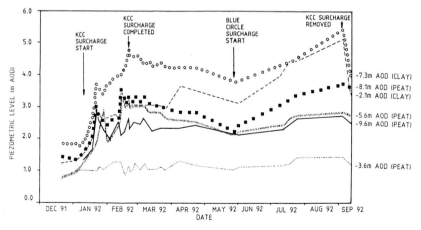

Fig. 9. Thames-side Industrial Route Stage 1C Advanced
Earthworks Contract, chainage 1000 centreine hydraulic
piezometer response

23. Thus, the technical reasons for not using polystyrene or piled rafts was justified since these would have been affected by the developments taking place on adjacent sites.

Plate 3. Thames-side Industrial Route Stage 1C Advanced Earthworks Contract, surcharges and carriageway cracking at chainage 1000

SOUTH THAMES DEVELOPMENT ROUTE STAGE 1C

24. An example of building an embankment in a restricted area is shown at Site 5, a second carriageway adjacent to an existing carriageway (TQ563754). A two metre high reinforced wall formed of polymeric grids wrapped around sandbags filled with dry mix concrete was constructed on 5m of soft alluvium as shown in Fig. 10 and Plate 4. A secondary connection was provided in the form of 0.2m long steel bars keying the grid in to adjacent bags in the event of damage to the exposed wrapped sections.

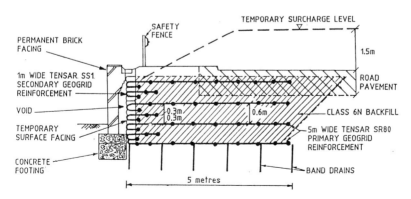

Fig. 10. S.T.D.R. Stage 1C - geogrid reinforced wall backfill

Plate 4. South Thames Development Route Stage 1C,
temporary sandbag wall during construction

25. A permanent brick wall will be formed separately from
the reinforced wall for aesthetic reasons. A gap was left
between the two walls to allow for further movement of the
reinforced wall during the six month surcharging period.
26. This fill was designed to accommodate movement so that
its final outward appearance, that is the brick wall, will
remain unaffected by the compressible nature of the soft
Alluvium.

CONCLUSIONS
27. These five case studies demonstrate the importance of
the design of the embankment in reducing potential failure
and differential settlement. The techniques will generally

550

not reduce the order of magnitude of total settlements but will keep long term differential settlements within tolerable limits.

28. It is possible to stiffen an embankment using, for example, a geocell mattress. In that case the choice of general earthworks fill is not critical.

29. If a brittle fill is to be used, for example pfa, then the use of a mattress to provide tensile resistance or haunches to limit lateral movements within the ground may be expedient.

ACKNOWLEDGEMENTS

30. In the preparation of this paper I am indebted to Kent County Council for permission to publish this data, friends and colleagues at KCC for their help and support, Ian Moore (Travers Morgan One) and John Dixon and Chris Jenner (Netlon Ltd) for productive design discussions, Dartford River Crossing for permission to continue monitoring at the Vent building. The views expressed in this paper are not necessarily those of Kent County Council.

REFERENCES
1. British Geological Survey 1:50 000 Series Geological Map Sheet 271 "Dartford". British Geological Survey, HMSO, 1977.
2. DEWEY H., BROMEHEAD C.E.N., CHATWIN C.P. and DINES H.G. The Geology of the Country around Dartford. British Geological Survey, HMSO, 1924.
3. Geotechnics Holland BV. Technical Report on vertical drainage - Mebra Drain, 1979.
4. ARBER N.R. Cracking of a PFA embankment over soft alluvium. Proc. ICE Symp. Failures in Earthworks, ICE, London, 1985.
5. JENNER C.G., BUSH D.I. and BASSETT R.H. The use of slip line fields to assess the improvements in bearing capacity of soft ground given by a cellular foundation mattress installed at the base of an embankment. Proc. Symp. Theory and Practice of Earth Reinforcement, Fukuoka, Japan, 1988.
6. BOYCE J.C. Great Yarmouth Bypass - Earthworks and Pavement Design. Jrnl. Inst. Highways and Transportation, Vol. 33, No. 1, pp 3-8, 1986.

Subject index

Author index